MODERN ELECTRIC/ ELECTRONIC MOTORS

MARTIN CLIFFORD

Prentice Hall, Englewood Cliffs, N.J. 07632

Library of Congress Cataloging-in-Publication Data

Clifford, Martin,
Modern electric/electronic motors / Martin Clifford.
p. cm.
Includes index.
ISBN 0-13-593336-6
1. Electric motors. I. Title.
TK2514.C57 1990
621.46'2--dc19 89-3708
CIP

Editorial/production supervision
and interior design: Karen Bernhaut
Cover design: 20/20 Services, Inc.
Manufacturing buyer: Robert Anderson

The publisher offers discounts on this book when ordered in bulk quantities. For more information, write:

Special Sales/College Marketing
Prentice-Hall, Inc.
College Technical and Reference Division
Englewood Cliffs, NJ 07632

Printed in the United States of America

10 9 8 7 6 5 4 3 2 1

ISBN 0-13-593336-6

Prentice-Hall International (UK) Limited, *London*
Prentice-Hall of Australia Pty. Limited, *Sydney*
Prentice-Hall Canada Inc., *Toronto*
Prentice-Hall Hispanoamericana, S.A., *Mexico*
Prentice-Hall of India Private Limited, *New Delhi*
Prentice-Hall of Japan, Inc., *Tokyo*
Simon & Schuster Asia Pte. Ltd., *Singapore*
Editora Prentice-Hall do Brasil, Ltda., *Rio de Janeiro*

To my charming and delightful

Aunt Bobby Indig

with love from Martin

CONTENTS

PREFACE

All motors are basically simple devices and depend on the attraction and repulsion of magnets for their operation. This simplicity is deceiving, for motors have become highly sophisticated. There are several reasons for their growing complexity.

For many years following its early development the motor remained a relatively elementary electromechanical device. In its first phase it found its way into industry where it was welcomed as a labor saving device. Subsequently, in its second phase, it was refined, made smaller and lighter, and became useful in home appliances. Motors are now into their third phase and are becoming combined with electronics, with the electronics external to or housed within the motor frame.

With electronic control has come shaft rotation governed automatically for fractional, precise amounts, forward or reverse. Electronics has also made it possible for mechanical motion to be self-correcting. The deviation of a laser beam having a thickness a fraction of the diameter of a human hair can be corrected immediately. And this is for motor shaft movements in terms of millionths of a meter, distances which cannot be registered by the human eye.

The arrangement of the chapters in this book follow the history and the growth of motors. The opening chapter describes the development of the motor, followed by a general analysis of its physical and electrical characteristics.

Chapter 3 has an in-depth analysis of DC motors. In its early days the DC motor was used in industry, but such motors, considerably slimmed, now have their place in the home using either batteries as their power source or electronic power supplies operating from the AC power line but having DC output.

The chapter that follows explains the operation, uses, and characteristics of the AC motor. AC motors are now available in a variety that cannot only be considered astonishing, but often confusing as well, for sometimes the difference between one AC motor type and another is simply a modification of some electrical characteristic. These motors are mostly labor saving devices, responsible for an industrial revolution in the kitchen much like the one their DC counterparts achieved earlier in factories.

Some motors are so highly specialized that they require special chapters of their own. Synchros are covered in Chapter 5, brushless DC motors in Chapter 7, and stepper motors in Chapter 8. One of the most detailed chapters in this book is the one on electronics for motors, Chapter 6. Electronics has brought with it circuitry supplying unusual motor control. This includes forward and reverse shaft rotation, automatic starting and stopping, and automatic speed control. In some of these motors the operation of the motor has become completely electric/electronic with the elimination of mechanical components such as brushes and commutators.

The final chapter covers gears, pulleys, and belts. These can be used for changing the direction of rotation of a motor shaft, for supplying a mechanical advantage by multiplying the horsepower output of a motor, and for operating two or more loads from the same motor.

Because of the inclusion of electronics, it is not enough to simply become familiar with motors. One must now have at least a basic knowledge of electronics, including solid-state devices and logic circuits.

While this book does supply an overview of motors and also gives suggestions on motor problems and how to cope with them, it is not a repair manual. It is essential for an understanding of how motors work so that repairs are not done following a cookbook approach. This understanding is a necessary prerequisite for those who want to become motor repair technicians. There are many motor problems that can be easily and quickly handled, and information on how to do so is included, not only in terms of repair but in motor maintenance.

ACKNOWLEDGMENTS

Data for this book was obtained from specification and data sheets, catalogs and advertising literature supplied, for the most part, by motor and motor parts manufacturers. Their support was considerable, and I would like to acknowledge, gratefully, their contribution. They include

Cramer Company

Eastern Air Devices, Inc.

Electric Stature, Inc.

Fairchild, A Schlumberger Company

Fasco Industries, Inc., Motor Division

Fisher Corporation
Gast Manufacturing Corporation
General Electric
Mitsubishi Electric Sales America, Inc.
Motronics Corporation
Penton/PC, Inc.
Portescap
Sony Corp.

Interestingly, some manufacturers whose components included one or more motors could supply little or no information about them. Apparently the motors were either made on special assignment or were stock items. In both instances the motors used were simply listed as such in the manufacturers parts lists with no identification other than "motor."

By contrast there were some organizations that supplied considerable useful and instructional data. One of these was RCA, and my good friend Judy Fleming at that organization made sure that I received the material I requested. Another was the Small Motors Manufacturers Association who kindly gave me permission to take advantage of the small motors' information they so promptly supplied. I would like to extend my appreciation to *Hands-On Electronics Magazine,* a Gernsback publication, for permission to use material from their article entitled "Circuit Circus," written by Charles D. Rakes which appeared in their June 1988 issue.

To all of these my thanks. I feel it is through their efforts that this book became possible.

Martin Clifford

North Lauderdale, FL.

1 DEVELOPMENT OF THE MOTOR

There are two fundamental but interesting facts about motors. The first is that all motors are based on elementary magnetic principles and the second is that such a variety of motors could have been developed from these principles.

There are three types of motors: spring loaded, electrically operated, and combined electrical/electronic motors. The spring motor depends on the gradual release of mechanical energy from a tightly wound spring. For this reason such motors have very limited applications and are used mainly in toys and clocks, but even there they are gradually being replaced by electric/electronic types.

In the spring motor there is no change in energy. The input is mechanical energy and so is the output. The electric/electronic motor, however, is a transducer. Its input consists of electrical energy; its output is mechanical energy, supplied by a rotating shaft. All electric/electronic motors, then, are energy conversion devices.

MOTOR ACTION

The behavior of a motor depends on the force of attraction and repulsion between pairs of permanent magnets (Figure 1-1) and electromagnets. This concept is as applicable to a tiny motor, a number of which can be held in the palm of a hand, as it is to motors weighing hundreds of pounds. Because the behavior of magnets is fundamental to all motors an understanding of magnets is essential to any study of motors.

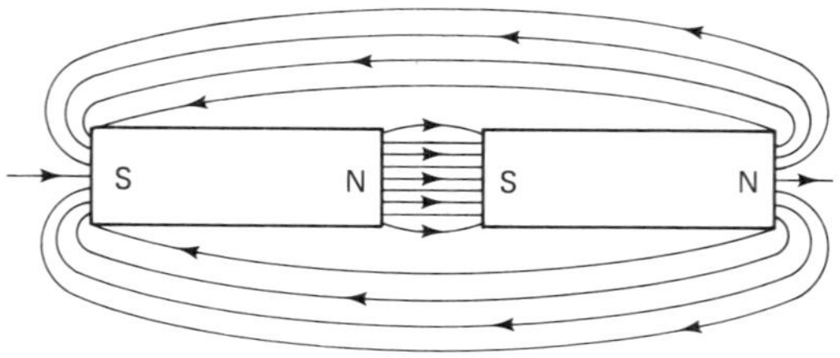

(A)

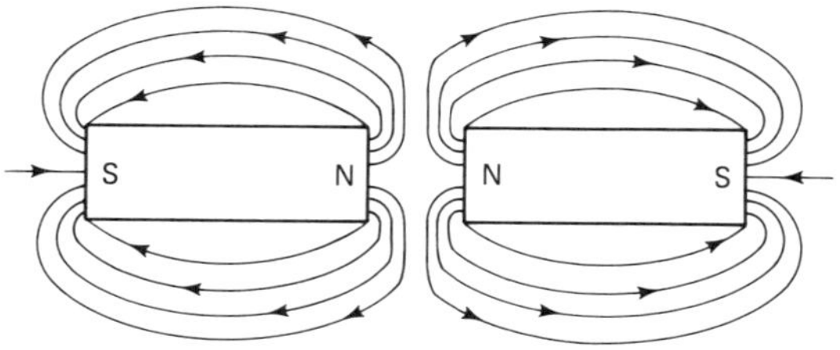

(B)

Figure 1-1 Unlike magnetic poles attract (A); similar poles repel. The arrows do not represent motion, simply the assumed direction of the magnetic lines.

MOTOR OVERVIEW

Motors are always associated with a load, using a rotating shaft to perform some action on that load. This can consist of starting the load or getting it into motion, but can also include actions such as increasing or decreasing speed, operating the load at some constant speed or at some variable speed, braking, or having the motor shaft turn a specific number of degrees. Motors can be switched on directly or remotely and are unexcelled as labor saving devices. Some motors can have their direction of shaft rotation reversed.

The input to a motor can be DC or AC. The DC can be smooth and unvarying or can consist of a series of pulses. The AC is sinusoidal and its frequency can be the usual 60 Hz, supplied by a power utility, or can be up to 400 Hz, supplied by an AC generator or electronic alternator. Of the two motor types, DC and AC, those using AC are more common because of the easy availability of AC line power, especially in the home. DC motors, however, are also widely used and are noted for their inexpensive speed control methods and their ability to supply high levels of torque. DC motors are also used extensively, especially in miniature sizes that rely on power input from rechargeable batteries like nickel-cadmium or lead-acid types. All vehicles, including autos, trucks, and vans, use DC motors.

There is no competition between AC and DC motors. Both are essential and each type has applications for which it is more suitable.

TYPES OF MAGNETS

There are two types of magnets: natural and artificial. The natural form is magnetite, a mineral consisting mainly of a magnetic oxide of iron. Artificial magnets are those which are manufactured.

The natural magnet (*lapis Magnes*) was discovered about 2,000 years ago and, suspended by a string, became the first compass. The magnet was referred to as a lodestone or leading stone because it could point toward the north geographic pole. Subsequently it was learned that a needle stroked on the loadstone could also be used as a compass. Because of its smaller mass the needle was more sensitive. By the end of the 15th century, the compass was extensively used for shipboard navigation, and Columbus depended on one for his voyages.

The magnetized needle was not only important because of its value in navigation but was the world's first motor, although at the time it was not recognized as such, nor was there any understanding of why the magnetized needle worked.

The basic principle of the motor action of the needle consists of the reaction of two magnetic fields. One of these is the magnetic field surrounding the earth, for the earth is a huge magnet. The other is the magnetic field (sometimes simply called the field) surrounding the needle. All motors we have today use this concept. The difference is that we are no longer solely dependent on the earth for supplying a magnetic field.

INVENTION OF THE MOTOR

The discovery of magnetite and its use as a compass is lost in history, but it was many centuries before investigators learned how to develop a transducer that would convert electrical to mechanical energy. Various efforts were made to take advantage of the attraction and repulsion of permanent magnets, but there was no known method of changing the polarity of a magnet, and so these attempts failed. A pair of magnets, arranged so that the opposite poles of the magnets faced each other did result in a motion of the magnets toward each other, but once that happened there was no further reaction. This problem was overcome by Hans Christian Oersted, a Danish physicist, who in 1820 revealed the connection between an electric current and the magnetic field surrounding that current. Controlling the current through a wire meant the ability to control the magnetic field around that wire. Reversing the flow of current meant changing the polarity of that field.

One of those who recognized the importance of Oersted's discovery was a British scientist, Michael Faraday, who is credited with the invention of the motor in 1821. His device consisted of an electric conductor suspended and free to rotate around the north pole (N) of a permanent magnet. This magnet was inserted through a cork at the bottom of a glass tube. The tube contained mercury so that a current could flow through it to the suspended conductor. The rotation of the suspended wire could be made clockwise or counterclockwise, depending on the direction of current movement through the wire, or by transposing the permanent magnet.

A short time prior to Faraday's invention a chemist, Dr. William H. Wollaston, was quite possibly the first to conceive of the idea of the rotation of an electromagnet. Faraday acknowledged the fact that Wollaston might have been the first who entertained the idea of electromagnetic rotation.

Once the concept was accepted that mechanical rotation could be achieved by using a permanent magnet and electromagnet there were a number of devices constructed. One of those who became intrigued by the concept was the American inventor, Joseph Henry, better known for his invention of the telegraph. His motor consisted of a coil of wire wrapped around an iron rod. This inductor was pivoted at its center by a wooden support. A pair of rectangular iron bars were positioned beneath the ends of the iron-core inductor.

The terminals of the coil were connected to voltage sources at each end. Actually, the coil consisted of two parts. When a current flowed through one half it became an electromagnet, and so became attracted to the iron bar beneath it. The current was then interrupted and allowed to flow through the second half of the coil. By timing the current stoppages, the inductor could be made to rock up and down, pivoted at its center. One of the significant differences between Henry's motor and that of Faraday was that it was a reciprocal type while Faraday's was rotational. Both techniques are used today, with reciprocal action found in solenoids, rotational in motors.

While both Faraday's motor and the one developed by Henry did demonstrate future possibilities, neither was practical. A neighbor of Joseph Henry, a blacksmith named Thomas Davenport, invented a motor in 1837 that had a rotating armature, was the world's first practical motor, and as such enabled him to get a patent. Davenport's motor was a DC type, requiring battery power input. AC power was still decades away in the future. At that time the batteries required for a motor were large, cumbersome, and discouragingly expensive. Still, the Davenport motor was successfully used for running a printing press and for operating a circular railway at Springfield, Massachusetts. Davenport built a number of different motor types, one of which used a series of fixed field electromagnets and movable electromagnets for the armature. This motor, operating with a light load, had a capability of 600 rpm.

Oddly, a German physicist, Moritz von Jacobi, invented a motor independently of Davenport, and a year after Davenport obtained his patent, von Jacobi, in 1838, was able to make a motor strong enough to power a boat.

Gradually motor construction improved but a big forward step was taken in 1873 when a Belgian inventor, Zenobe Theophile Gramme, was able to operate a DC motor from a generator. AC motors had to wait for alternating voltage to become practical. The first induction AC motor was made by the famous American physicist and inventor Nikola Tesla. Subsequently, Tesla's motor was modified by the German engineer, Michael Dolivo-Dobrowolski who is credited with the invention of the squirrel cage rotor, a motor that uses metal bars instead of coils.

PERMANENT MAGNETS

The three magnets in Figure 1-2 are permanent types and the only difference is one of shape. One way of making a magnet is to insert an iron core into a form on which a coil is wound carrying a direct current, as in drawing A in Figure 1-3. Another method, as in Figure 1-3B, is to rub a steel bar with a magnet. The bar must always be rubbed in one direction only. This is not the method used in commercial production and simply illustrates how a magnetic field can be induced in a ferrous material.

A magnet, whether a permanent type or an electromagnet, always has two poles, a north and a south pole. If the magnet is cut apart, either once or repeatedly, a new pair of magnets is always formed (Figure 1-4).

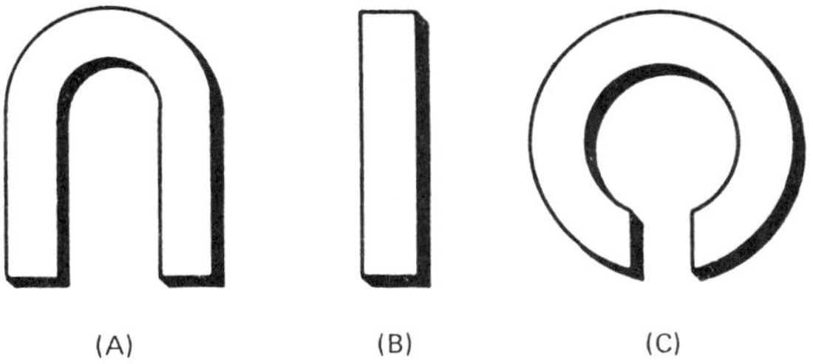

Figure 1-2 Some types of permanent magnets: (A) Horseshoe, (B) Bar, (C) Ring.

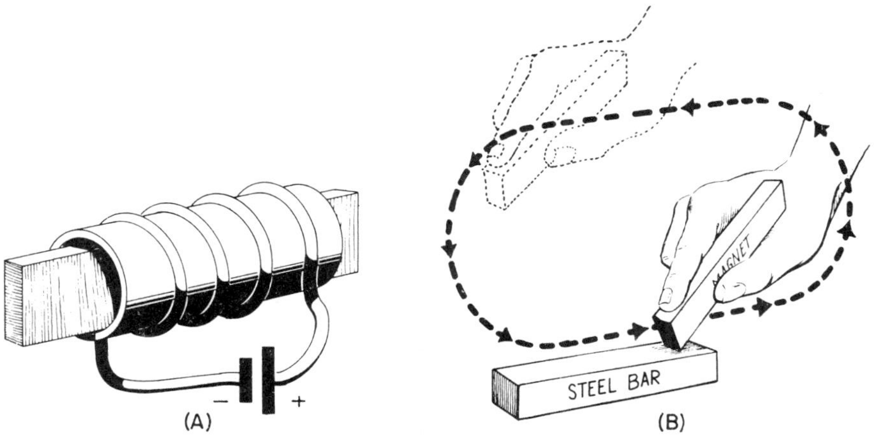

Figure 1-3 Methods for making a permanent magnet. (A) Steel bar is used as the core of a coil carrying a direct current; (B) Steel bar is stroked with a permanent magnet.

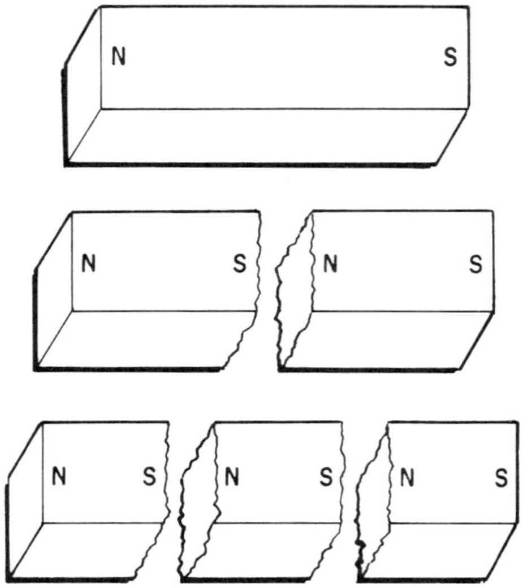

Figure 1-4 Repeated cutting of a magnet always results in pairs of new magnets.

If a pair of bar magnets are suspended, as in Figure 1-5A, and are permitted to rotate freely, they will turn until the north pole of one of the magnets faces the south pole of the other. If the magnets are physically forced into a position so that a pair of like poles face each other, they will rotate until the north pole of one magnet faces the south pole of the other. If they are not permitted to turn, and if the magnets are strong enough, one can be made to float in space above the other and can remain suspended indefinitely (Figure 1-5B). The two drawings in Figure 1-5 are the basis for all motor action, regardless of motor size, type, and whether the motor operates from AC or DC.

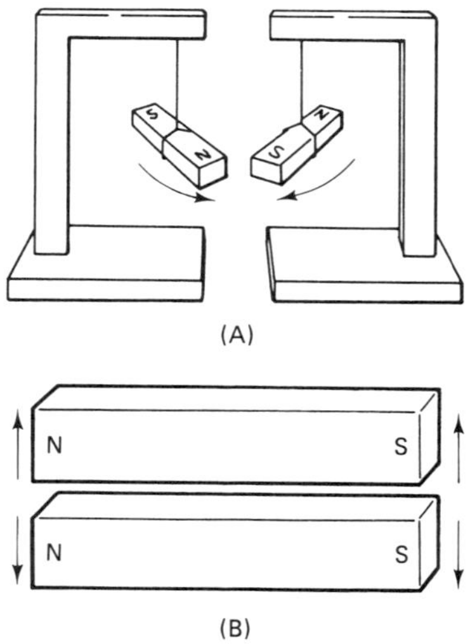

Figure 1-5 If two bar magnets are suspended (A) they will rotate until the north pole (N) of one magnet faces the south (S) pole of the other magnet. If the similar poles of bar magnets are made to face each other, one of the magnets will be suspended in space (B).

LINES OF FLUX

A magnetic field, as in Figure 1-6, consists of magnetic lines of force, also referred to as lines of flux. It is a convenient fiction to consider the lines as starting at the north pole end of a magnet. While they reach toward the south pole they do not stop there. They continue through the magnet from the south to the north pole. These magnetic lines are unbroken and there are no broken or partial lines. Like the rim of a wheel, they have neither a beginning nor an ending. Furthermore, magnetic lines never cross each other, and so each line is individual.

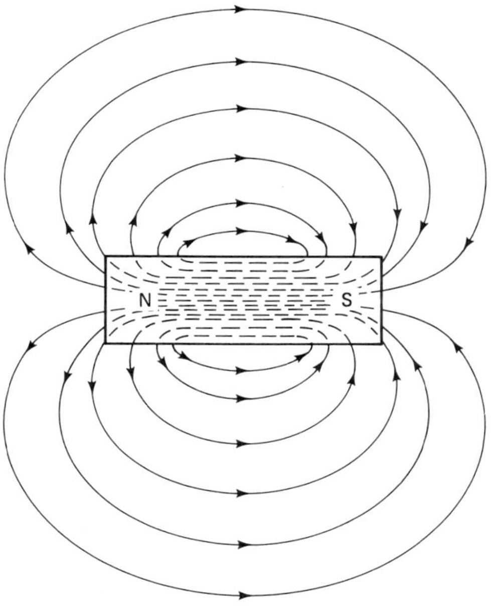

Figure 1-6 Magnetic field surrounding a bar magnet.

Depicting a Magnetic Field

A magnetic field is generally illustrated by a number of lines surrounding a magnet. Sometimes these lines are supplied with arrows, an unfortunate arrangement since it implies the lines are moving. The quantity of lines can be increased or decreased, but they are not in motion. When the lines are increased they occupy more space in and around the magnet, extending the magnetic field.

Detecting a Magnetic Field

Magnetic fields are invisible, but there are several easy ways of detecting them. One technique is to place a sheet of stiff paper or glass covered with iron filings above the magnet. The filings will arrange themselves in a pattern as in-

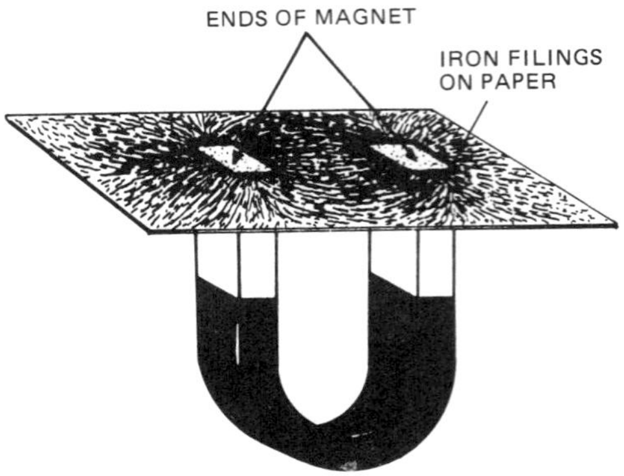

Figure 1-7 Pattern of a magnetic field shown by iron filings scattered on a stiff sheet of paper or glass.

dicated in Figure 1-7. The same method can be used with a variety of magnets.

Still another method is to bring a ferrous metal like the blade of a screwdriver near the magnet. If the magnet is strong enough the screwdriver will be attracted. A more sensitive method is to use a compass (Figure 1-8). The compass can also be used to identify the north and south poles of the magnet for the north-seeking pole (actually a south pole) of the compass will point to the north pole of the magnet.

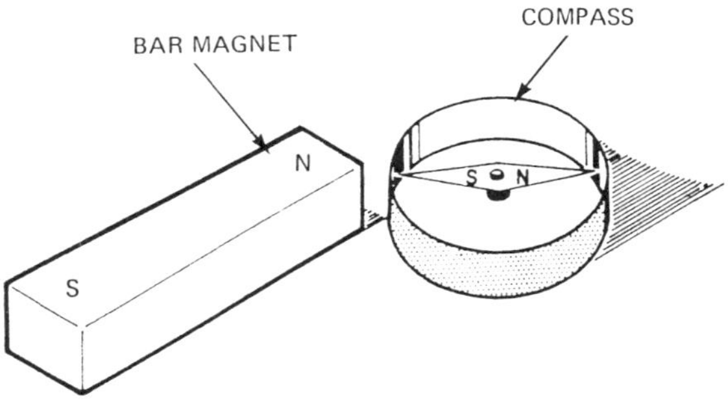

Figure 1-8 Detection of a magnetic field using a compass.

Temporary Magnets vs. Permanent Magnets

A permanent magnet is one that maintains a fairly uniform flux field for a long period of time, and if it does lose some of its strength it will do so at a low rate. At the time of its manufacture it will be artificially aged, a process that consists of making the magnet lose some of its originally induced magnetism. Permanent magnets are generally made of materials like carbon steel, chromium steel, tungsten steel, cobalt chromium steel, cobalt samarium, and alloys of aluminum, nickel, cobalt, copper, and iron. Ceramic materials are also used. It is also possible to produce magnets from alloys of elements which are individually nonmagnetic.

A temporary magnet is a substance that can be magnetized but, in the absence of a continuing magnetizing force, will either lose its magnetism completely or will retain it at a very low level only. Temporary magnets are made of ingot iron and low percentage silicon steel. A coil of wire carrying a current is a magnet but only as long as that current flow continues. Both permanent and temporary magnets are used in motors.

Retentivity

Retentivity is the ability of a magnet to keep its magnetic strength following removal of the magnetizing force used initially to set up its magnetic field. A permanent magnet has high retentivity; a temporary magnet has low retentivity. Retentivity is a dimensionless term and is used only in a general way.

Magnet Size

There is apparently no limitation on the size of a magnet. Some are so small that they are individually invisible, such as the magnetic particles used on audio and video recording and playback tapes. At the other extreme, the earth is a huge magnet.

MAGNETIC CHARACTERISTICS

A magnet can be characterized by its physical details, its size, weight, shape or the elements of which it is made, or possibly by a trade name. It can also be described by its magnetic properties.

The Maxwell

A maxwell is a magnetic line of flux. It has no specific beginning or ending and is a continuous loop. Since it does not exist in partial form, the total number of maxwells is always a whole number, never a proper or improper fraction.

In terms of a single magnet, its lines of flux form a magnetic field with the individual lines repelling each other. Since partial magnetic lines do not exist, the number of lines appearing in the space around the magnet equals the number of lines existing in the body of that magnet.

CGS and MKS Systems

Magnetic units are usually measured in one of two systems, the CGS (centimeter, gram, second) or the MKS (meter, kilogram, second). Both CGS and MKS are metric, the only difference between them being one of size.

CGS is used when the measurements are small, as indicated by its name. MKS is preferable when the quantities are large. Whichever one of the two is used is simply a matter of convenience.

SI Units

The MKS system is a subset of a more encompassing group known as SI, or International System of Units. SI units include the meter, second, and ampere, among others. Still another system is the MKSA (also known as the Georgi system). It is similar to the MKS.

Examples of SI units would include the ampere or SI unit of electrical current, the ampere-turn, the SI unit of magnetomotive force. In the CGS system the unit of magnetomotive force is the gilbert. Thus, there are two ways of describing this force depending on which system of units is used.

Practical Units

Another system of magnetic measurements is the so-called Practical or English system. Unlike the CGS or MKS it uses English units such as the foot and the square inch. A magnetizing force (H) is one ampere-turn per inch

CONVERSION FACTORS FOR MAGNETIC UNITS

The CGS system includes units such as the gauss, oersted, and gilbert, with the centimeter as the unit of length. The MKS system uses units such as the weber or webers per square meter. Hence both systems are metric. The Practical system employs units such as maxwells per square inch. Table 1-1 can be used for conversions from one of these systems to another.

It might seem odd to have a number of different measuring systems, but the fact is that no system is suitable for solving all problems in magnetics.

Flux Density

Flux density is the number of magnetic lines of force per unit area. In the CGS system that area is the square centimeter. The area is considered as being perpendicular to the lines of magnetic flux. A line of flux per square centimeter is known as a gauss and is represented by the Greek capital letter beta (β).

If a magnet has a strength of 100 gauss, the concept of area is included for 100 gauss is 100 lines per square cm or 100 lines/cm^2. A magnet used for holding papers to a refrigerator door has a strength of about 100 gauss.

The total amount of flux supplied by a magnet is equal to the the product

TABLE 1-1 Conversion Factors for Magnetic Units

Multiply	By	To Obtain
F in ampere-turns	$0.4\pi = 1.257$	F in gilberts
F in gilberts	$1/0.4\pi = 0.796$	F in ampere-turns
H in ampere-turns/in	$0.4\pi/2.54 = 0.495$	H in oersteds
H in oersteds	$2.54/0.4\pi = 2.02$	H in ampere-turns/in
B in maxwells/sq in	$1/6.45 = 0.155$	B in gauss
B in gauss or B in maxwells/sq cm	6.45	B in maxwells/sq in
B in webers/sq meter	10^4	B in gauss
B in gauss	10^{-4}	B in webers/sq meter
B in maxwells/sq in	$10^{-4}/6.45 = 0.155 \times 10^{-4}$	B in webers/sq meter
B in webers/sq meter	6.45×10^4	B in maxwells/sq in
ϕ in maxwells or ϕ in lines of flux	10^{-8}	ϕ in webers
ϕ in webers	10^8	ϕ in maxwells

of the area, A, in square centimeters at right angles to the flux, multiplied by the number of gauss, β.

$$\phi = \beta \times A \quad \text{or} \quad \beta = \frac{\phi}{A}$$

The use of area in connection with flux density does present a problem. The lines of flux around the magnet are spread out over a wider area than those inside the magnet itself. Thus, the area around the magnet is greater. But since the number of magnetic lines have remained the same, the flux density is greater inside the magnet. For this reason, we have two ways of measuring magnetic field intensity. One of these is the gauss, and it is not only used for measuring flux lines per unit area inside a magnet but in any ferromagnetic substance. Thus, a soft iron bar placed against a magnet would have the number of lines in it measured in terms of gauss.

The measurement of flux density, usually in the air surrounding the magnet, is the oersted. Thus, a field intensity of 25 oersteds is the same as 25 lines of magnetic flux per square centimeter. Both units of field intensity, the gauss and the oersted, are used, and both are important. The oersted is more frequently associated with electromagnets.

Reluctance

Lines of magnetic flux do not pass as easily through nonmagnetic materials as they do through metals such as iron or steel. Comparable to resistance in an electric circuit, reluctance is the opposition to the existence of magnetic lines in a substance. The symbol for reluctance is the script letter R ($\mathcal{R}$). The unit of reluctance is the rel.

For magnetic materials such as iron or cobalt the amount of reluctance depends on the total number of flux lines and so is a variable. For nonmagnetic substances reluctance remains constant and is independent of flux density.

Permeability

It is possible to describe the relationship of lines of flux to the substance in which it exists in two ways. One of these is reluctance or opposition to magnetic lines. It is also possible to describe reluctance in terms of its inverse, or permeability. Permeability and reluctance are opposites. Thus, permeability is the ease with which magnetic lines exist in a material. In making a comparison between air and a specific type of iron, air can be said to have a reluctance that is approximately 2,500 times that of iron, or conversely, the iron has a permeability 2,500 times that of air. In this example a magnetizing force that could produce 2,500 lines of flux in iron can develop just a single line when this bar of iron is removed and air takes its place.

The permeability of a substance can be expressed in terms of the formula

$$\mu = \frac{\beta}{H}$$

where μ is the permeability, β is the flux density in gauss, and H is the field intensity in oersteds.

An illustration of the magnetic lines surrounding a bar magnet (Figure 1-9) shows a concentration of lines inside the magnet with these lines spread externally. There also seems to be a concentration of lines at the ends of the bar and a much smaller concentration around the sides. This is due to the fact that although the total number of lines has not changed, there is a greater distribution area. The end crowding is due to the fact that the areas at these ends are much smaller than the area of the external space.

The spacing of the magnetic lines external to a magnet can be controlled simply through the use of a block of a ferromagnetic material. As shown in

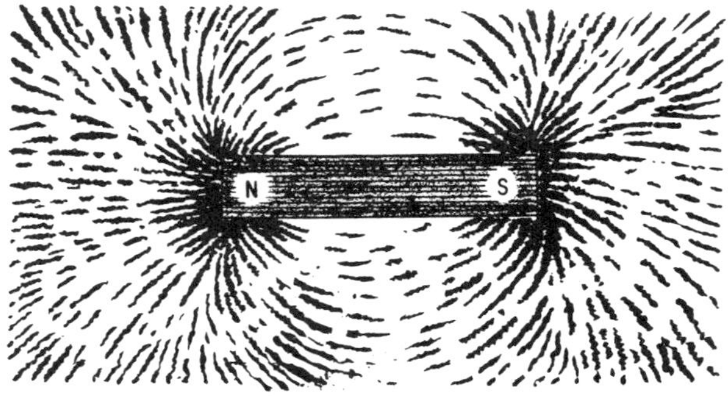

Figure 1-9 Magnetic lines concentrate at the poles.

Figure 1-10, although the soft iron bar is positioned away from the magnet and has a smaller cross-sectional area than the surrounding space, the lines of flux crowd through it in preference to the air space around it due to the much greater permeability of the iron bar than that of the air. If the bar is positioned parallel to the magnet, it is possible to get a more uniform distribution of the lines, a technique that finds application in motors.

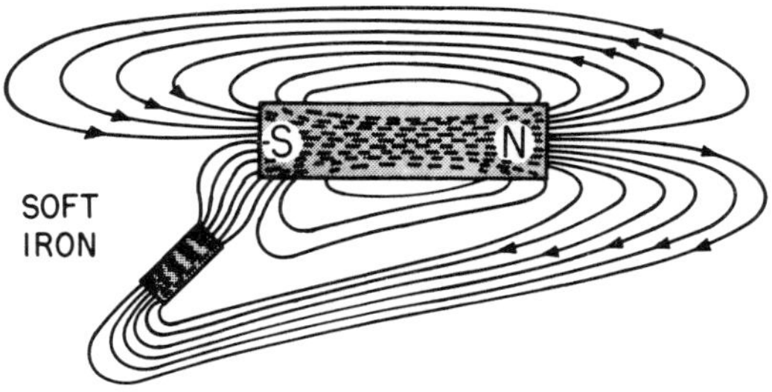

Figure 1-10 Magnetic field control.

ELECTROMAGNETS

The amount of control that can be extended to permanent magnets has its limitations. The magnet can be made to have a number of shapes, the direction of its lines of flux external to the magnet can be altered, and it is possible to have weak or strong magnets. However, while the polarity of the magnet can be changed through the application of an external magnetizing force, this is generally practical only for extremely small magnets, such as the micromagnets embedded in a binder on audio or video tape. Further, even with the use of an external force, the magnet will retain some of its original magnetization.

Copper is not a ferrous material and so cannot be magnetized. However, every electrical current, whether direct or alternating, whether steady, increasing, decreasing, intermittent or pulsing, is accompanied by a magnetic field. The wire can be straight, curved, or in the form of a helix or coil. The magnetic field supplied by the current makes the copper into a temporary magnet, temporary since the field is directly dependent on the current. If the current is stopped, the magnetic field disappears. Further, the copper unlike a soft iron bar, does not retain any portion of the magnetic field following discontinuance of current flow (Figure 1-11).

The magnetic field around a wire carrying a current can be detected in the manner described previously, using a compass (Figure 1-12). When the compass is moved around the wire, its pointer always aligns itself in the same circular direction. If the voltage source, a battery in this case, has its leads transposed, the current will flow in the opposite direction and the pointer of the compass will simply point the other way. If the voltage source is disconnected, the compass will no longer be affected by the magnetic field around

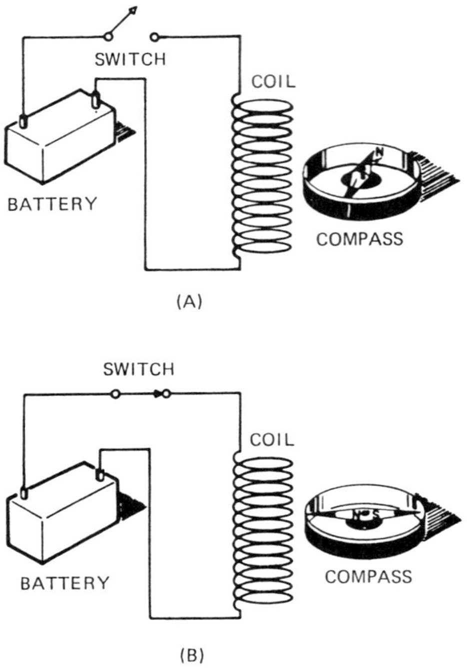

Figure 1-11 In the absence of a current (A), the coil has no magnetic field surrounding it. With current flow, the coil becomes an electromagnet (B).

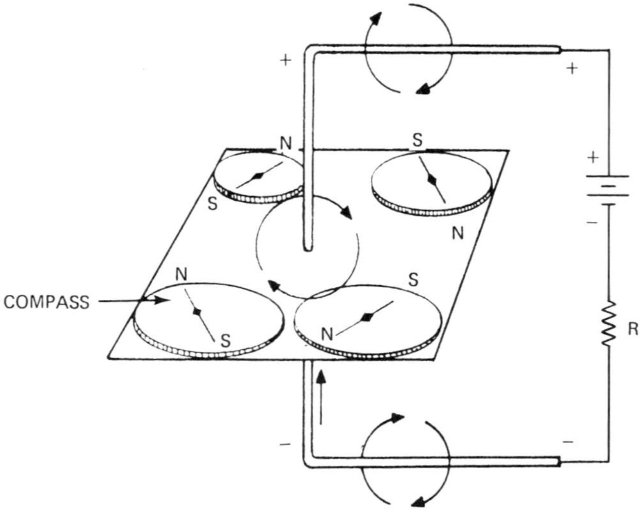

Figure 1-12 Detection of the magnetic field around a current carrying conductor.

the wire, for it will have ceased to exist, and the needle of the compass will then point to the earth's magnetic north. The resistor, R, shown in the circuit is used to limit the amount of current flow; otherwise the wire would be a direct short across the battery.

The Left-Hand Rule

The direction of the flow of current is from the negative terminal of the battery, through the connecting wire, through the resistor R, and then on to the plus terminal of the voltage source. Inside the battery it flows through the electrolyte to the minus terminal and then repeats, continuing through the wire.

To determine the direction of the magnetic field, hold the wire with the left hand as indicated in Figure 1-13. The extended thumb will then point in the direction of the current flow and the closed fingers will show the direction of the magnetic field, as in Figure 1-13A. If the leads to the battery are transposed (Figure 1-13B), the current will still flow from minus to plus, but the magnetic field will be in the opposite direction. This determination of magnetic field direction is known as the left-hand rule. The fingers represent the

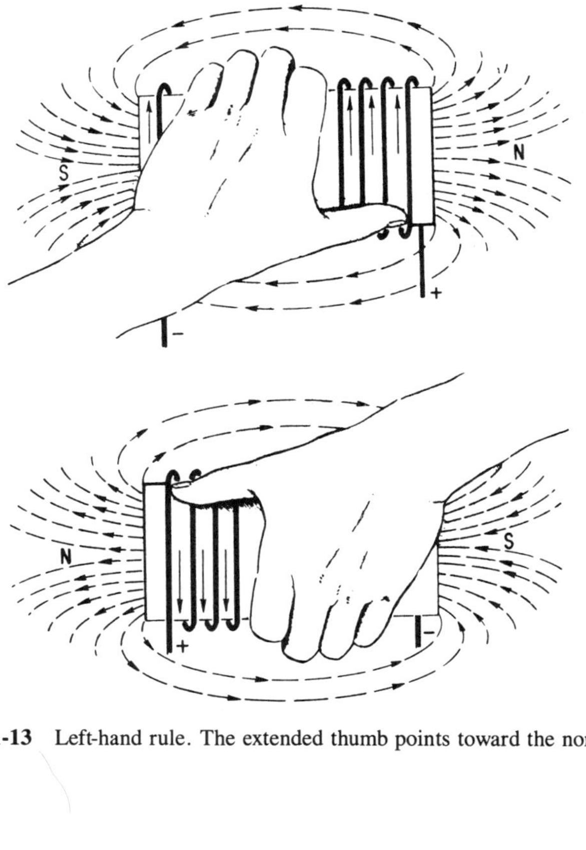

Figure 1-13 Left-hand rule. The extended thumb points toward the north pole.

lines of magnetic flux. The use of this method helps one visualize the coil's magnetic field and the position of its poles. These characteristics of the electromagnet can be further explored with the help of a compass.

It might be thought that the magnetic lines would always form circles around the coil, but the shape of the lines will depend on the reluctance of the core material (air or iron), the available space inside the core, and the reluctance of the space external to the coil. Figure 1-14 is a simplified version consisting of a one-turn loop. As a result, the magnetic intensity inside the loop has become greater. The number of magnetic lines inside the loop has not increased, but the number of lines per unit area has. Note also that the magnetic field and the current flowing through the conductor are at right angles.

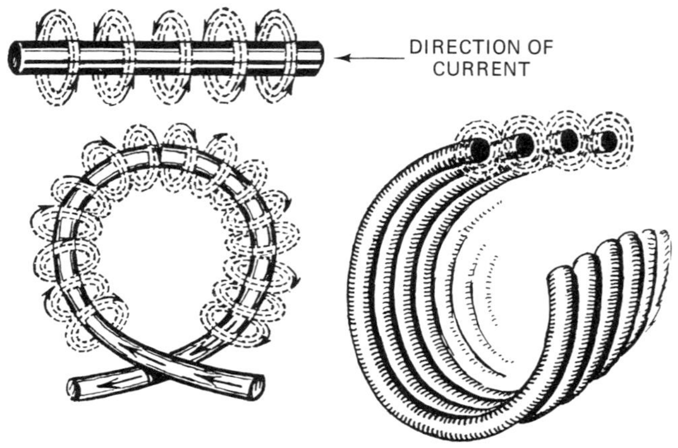

Figure 1-14 Lines of magnetic flux surrounding current-carrying conductors.

Field Intensity vs. Flux Density

The graph in Figure 1-15 is a comparison between field intensity H, that is, the magnetizing force, in oersteds, and the flux density B, measured in thousands of gauss, using a steel bar as the material being magnetized.

At the start, shown at the lower left end of the graph, a small magnetizing force results in a substantial amount of flux lines. This action continues in a somewhat linear manner until point A. At that point, the graph begins to curve and becomes somewhat more flat. This indicates that an increase in the magnetizing force does not result in a corresponding increase in lines of flux. When point B is reached, the curve is on its way to becoming flat, indicating that further increases in magnetizing produce relatively few additional flux lines. At point C any further magnetizing force does not produce more flux. This is then identified as the point of saturation.

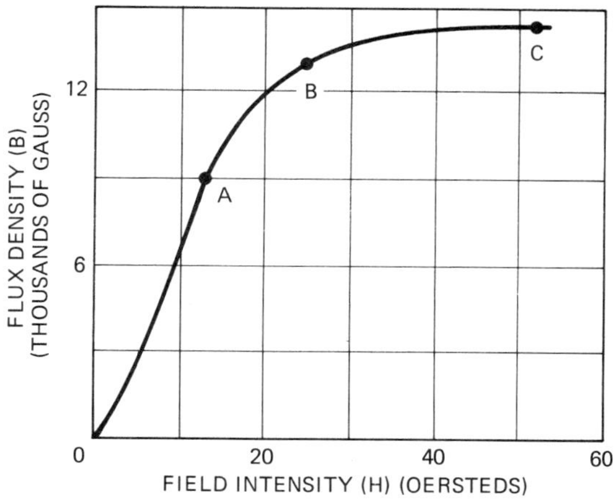

Figure 1-15 Graph of flux density versus field intensity.

Saturation

Lines of flux have no weight. An iron bar has the same weight whether it is magnetized or not. And yet there is a limit to the number of magnetic lines of force that can exist in that bar, and when that limit is reached, the bar is said to be saturated.

It may seem strange to have saturation apply to flux lines since they have no thickness. But the individual lines, extending in the same direction, repel each other, and this force of repulsion is enough to limit the space they occupy.

Theoretically, it should be possible to increase the flow of current through an iron core coil and in this way develop a magnetic field capable of lifting tons of magnetizable material. This goal is not as hypothetical as it sounds, for huge electromagnets are used in industry for this purpose.

Achievement of this objective is not as simple as it sounds for several reasons. The first is that current cannot be increased without regard to the capability of the wire of the coil for carrying that current. The second is that an iron core can only accommodate a certain number of magnetic lines of force. When this point is reached the core has reached saturation. Any further increase in current will force the extra lines of flux to take a route around the outside of the coil through the air, but since air has a much higher reluctance than the iron used for the core, the increase in flux will be quite small.

SOLENOIDS

A solenoid is a coil of wire with a length several times its width that belongs to the family of inductors. It is often equipped with a movable iron core that performs some working function. When a current flows through the turns of wire of the solenoid, its field magnetizes the core. If the core is made of soft iron, most of its magnetism will disappear when the current stops. A more permanent type of magnet can be made if the core is steel.

The illustration in Figure 1-16 shows the magnetic field surrounding a solenoid. This particular coil is not equipped with a core.

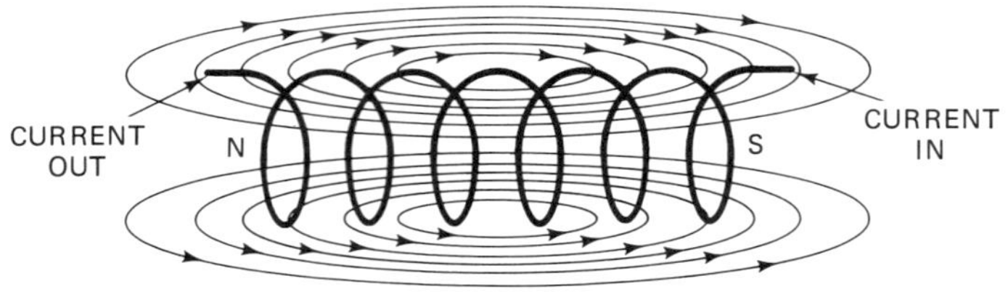

Figure 1-16 Solenoid not equipped with an iron core.

Prior to its insertion, the iron core (sometimes called a plunger) of the coil is free of magnetism. Since, in use, the core is regularly surrounded by the magnetic field of the coil, it becomes magnetized; this process is referred to as magnetic induction.

The movement of the solenoid's plunger can be used to actuate a switch. This switch can be located in a hostile environment or used to initiate some mechanical action. The current can also be controlled by a timer, and so it can be operated automatically, requiring no attention, in manufacturing processes.

Magnetic induction can be noted by stroking a bar magnet with a needle. While the needle becomes a permanent magnet, it is never as strong as the bar magnet. Similarly, the plunger of the solenoid made into a magnet by induction never achieves a magnetic strength equal to that of the electromagnet surrounding it.

Increasing Magnetic Field Strength

The strength of the magnetic field around a straight wire carrying a current can be increased in a number of ways. That field is directly dependent on the amount of current flow, and if that flow is made to increase, so will the strength of the magnetic field. Current flow can be increased by reducing the value of any resistance that may be in series with that wire or by raising the voltage applied to it. Another technique is to wind that straight wire into the form of a loop. When this is done the lines of force will be concentrated within the loop formed by the wire. The area enclosed by the loop is limited; that outside the loop is not. Thus, inside the loop there is a greater concentration of magnetic lines of flux per unit area.

If, instead of a single loop, a number of turns are used, the result is an even greater concentration of magnetic flux, assuming that the current remains the same, as in the case of the single loop. The magnetic field around the coil (Figure 1-16) then resembles that of the permanent magnet shown earlier. One end of the coil is a north pole; the other end a south pole. But unlike the permanent magnet, the field exists only as long as the current flows; hence this electromagnet is referred to as temporary.

Still another technique for increasing the strength of the magnetic field around a coil is to use an iron or steel core. The coil shown in Figure 1-16 is an air-core type, consisting of a cylindrical form (not shown in the illustration) on which the coil is wound. This is also a technique for making a permanent magnet, for if steel is used as the core it will retain its magnetism after current flow through the coil has stopped. The magnetism retained by the core is referred to as residual magnetism.

MAGNETIC POLE REVERSAL

While the electromagnet using a DC voltage source can be made to have a varying magnetic field strength, it is similar to a permanent magnet in the sense that its polarity remains constant. Its north pole is always a north pole; its south pole is always the same.

A polarity change can be obtained by using a single-pole, double-throw switch shown in Figure 1-17. The switch does nothing more than transpose the

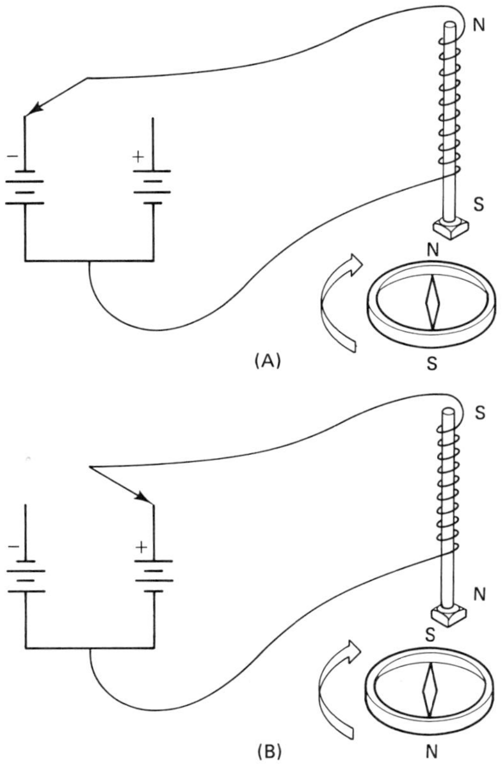

Figure 1-17 Current flows through the coil in one direction (A) but can be reversed by a switch (B).

connections to a battery, with the result that the direction of current flow through the coil reverses. Figure 1-17A shows the switch in one position; 1-17B shows the opposite position. But while this does demonstrate the possibility of polarity change, it has no value otherwise.

A better method, supplying automatic switching, is to substitute an AC source voltage for the DC input. This is done by using a sine wave voltage input (Figure 1-18).

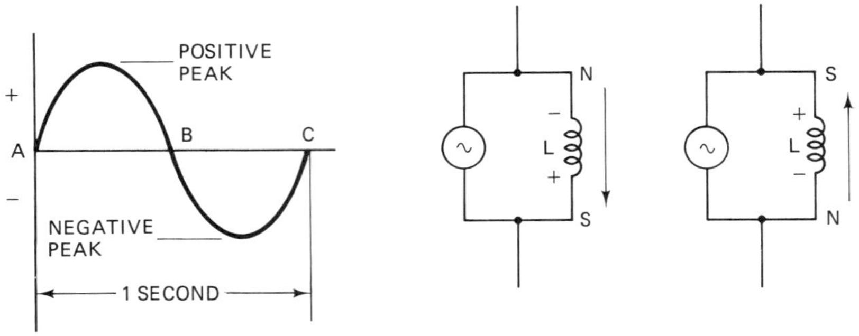

Figure 1-18 A sine wave produces automatic alternations of polarity across coil.

This wave has two alternations, one from A to B, the other from B to C. The first alternation is positive, the second negative. During the first half, the voltage starts at zero, and at this point, the magnetic field around the coil does not exist. As the voltage increases a magnetic field comes into existence and reaches a peak. The voltage then decreases and the current becomes zero at B. During the first alternation the magnetic field started at zero, reached a maximum field strength at the voltage peak, and then gradually decreased to zero

At point B, the voltage reverses and so does the current it moves through the coil. Since the current flows in the opposite direction, the polarity of the magnetic field around the coil changes. If the input AC source voltage is 60 Hz there will be 120 alternations per second, resulting in 120 polarity changes during this time.

MOTOR ACTION

If an iron bar is positioned near the AC energized coil in Figure 1-19, one effect will be the attraction of the bar by the coil. It is true that the magnetism of the coil is momentarily zero when the sine wave crosses its zero voltage point but this is for just a small fraction of a second. Regardless of the polarity of the coil, it acts like a magnet. To increase the magnetic effect, the coil uses a soft iron core. Its polarity changes in step with the changing polarity of the coil and is used in this example to increase the magnetic strength of the coil.

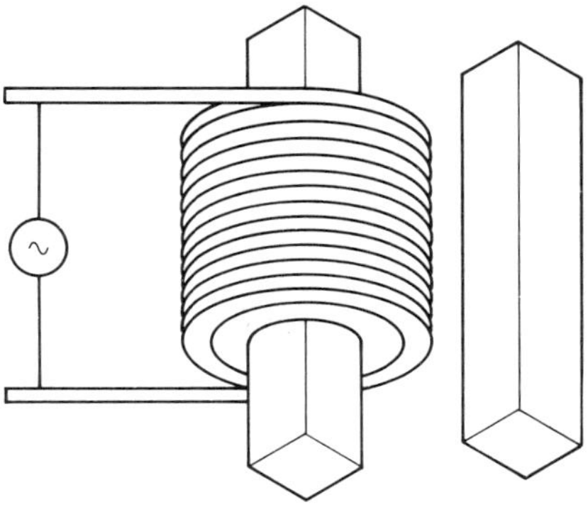

Figure 1-19 Soft icon bar is attracted by magnetic field surrounding current-carrying coil.

Figure 1-20 is almost the same as the preceding illustration, but there are a few changes. The magnetized coil, its core, and the AC voltage source remain the same. However, a permanent magnet has been substituted for the iron bar. The magnet is pivoted at its center and is mounted on a shaft supported at both ends, with the magnet free to rotate easily.

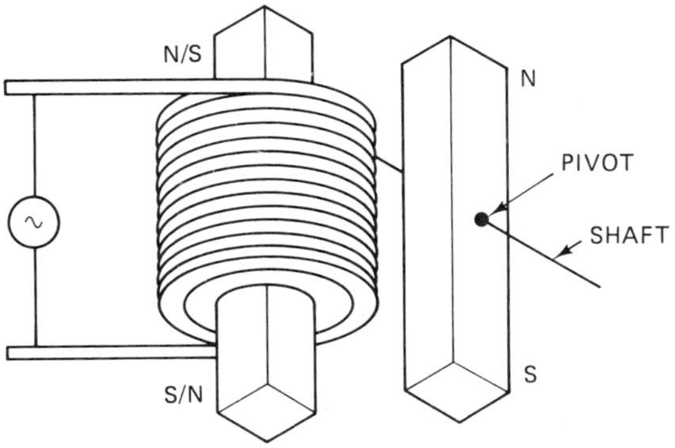

Figure 1-20 Permanent magnet is free to rotate around its pivot.

When the electromagnet develops a north pole at its top end, it repels the permanent magnet forcing it to rotate so that its south pole is at the top. But at this time, the polarity of the electromagnet changes so that it also has a south pole facing that of the permanent magnet. Once again there is a force of repulsion between the two magnets and so the permanent magnet turns away. The timing of the current flow changes in the electromagnet can be such that the permanent magnet keeps turning constantly.

The rotating permanent magnet is mounted on a shaft, and this turning shaft can be made to do useful work, possibly turning a meter pointer. Admittedly, this motor (for that is what it is) is crude and would take a number of refinements to make it work, but it does establish the possibilities for motor development.

It is the force of attraction or repulsion that supplies the operating behavior of the large number of motors now available. The effect of two independent magnets on each other varies as the square of the distance between them. If the distance between a pair of magnets is doubled, the magnetic effect of either attraction or repulsion is reduced to one-fourth of its original value.

MAGNETIC RELATIONSHIPS

The relationship between magnetomotive force and lines of flux and reluctance can be expressed in three equations:

$$\phi = \frac{H}{\mathcal{R}} \qquad \mathcal{R} = \frac{H}{\phi} \qquad H = \phi \times \mathcal{R}$$

where ϕ is lines of flux in maxwells, H is magnetomotive force in gilberts, and $\mathcal{R}$ is the reluctance in rels. From these equations it can be noted that magnetomotive force and flux lines are directly related. This indicates that the stronger the magnetomotive force the greater the number of lines of flux. At the same time, the number of lines of flux is inversely proportional to the reluctance, that is, the greater the reluctance, the smaller the number of lines.

These are the most important factors affecting magnetic strength, but there are others as well. The strength of the magnetic field is also inversely proportional to the coil's radius, applicable when the overall length of the coil is less than its diameter. This is an odd case since coil length is usually greater. However, for these special instances, H, the magnetic field intensity can be determined from

$$H = 2\pi \, \mathrm{NI}/10\mathrm{r}$$

H is the field strength in oersteds, N is the number of turns of wire forming the coil, I is the current flow through the coil in amperes, and r is the radius of the coil in centimeters. Pi (π) is a constant whose value is approximately 3.14159, but commonly used as 3.1416. The current flow must be expressed in amperes for this formula to be used. The product NI is sometimes referred to as ampere-turns.

Example

A coil of 5 turns has a diameter of 10 centimeters but a length of only 5 centimeters and carries a direct current of 8 amperes. What is its field strength?

Solution

$$H = 2\pi \, \mathrm{NI}/10\mathrm{r}$$

$$H = \frac{2 \times 3.1416 \times 5 \times 8}{10 \times 5}$$

$$= \frac{251.33}{50} = 5.03 \text{ oersteds}$$

Because this coil has a diameter greater than its length, the formula supplied above is applicable. For coils of this type, magnetic strength *H* is inversely proportional to the radius. Thus, the greater the diameter of such coils, the less the magnetic strength. Note also that the formula calls for the radius dimension, equivalent to the diameter divided by 2.

MAGNETOMOTIVE FORCE

Magnetomotive force (mmf) is the force by which a magnetic field is produced. This can be developed as explained earlier, by having an electrical current flowing through a coil or by stroking iron, steel or some other magnetizable substance with a magnet. Mmf can be expressed in two ways. In the English system of measurements it can be stated in terms of ampere turns per inch, and in the metric system, in gilberts per centimeter or in oersteds. In both instances, the letter H is used as the symbol. However, in the metric system the letter H is italicized.

The amount of mmf, represented by the italic capital letter *H*, measured in gilberts can be calculated by:

$$H = 1.257 \times \mathrm{N} \times \mathrm{I}$$

In this equation *H* is the mmf in gilberts per cm, 1.257 is a constant, N is the number of turns of wire of the coil, and I is the current in amperes. This formula is directly applicable only when the coil is wound on an air core. An air core coil, for example, would consist of turns of wire wound on a cylindrical form made of some nonconducting substance such as impregnated cardboard or plastic. The constant 1.257, sometimes represented by the letter k, represents the reluctance of air, and so a different value of k is used for iron cores. Sometimes the letter *F* is used instead of the letter *H*.

Force of Attraction or Repulsion Between Magnets

The force of attraction or repulsion between magnets is directly proportional to the product of the strengths of their poles and inversely proportional to the square of the distance between them. This is expressed by a formula

$$F = \frac{\mathrm{m}_1 \mathrm{m}_2}{\mu \mathrm{d}^2}$$

where *F* is the force in dynes between two poles separated by air; m_1 is the magnetic strength of the first pole in unit poles; m_2 is the magnetic strength of the second pole, also in unit poles; d is the distance separating the two poles in centimeters; and μ is the permeability of the substance (in this case, air) between the poles. For air (or vacuum) the permeability has a value of 1.

Unit Poles

Unit poles (Figure 1-21) are based on the measurement of the force of attraction or repulsion between magnetic poles. If a pair of poles are separated by a distance of one centimeter, and if the force of attraction or repulsion is equal to one dyne, then the strength of each of the magnets is one unit pole. This assumes that the measurement is made in air or in a vacuum. The dyne is a unit of force in the CGS system and is defined as the force that would give a free mass of one gram an acceleration of one centimeter per second per second.

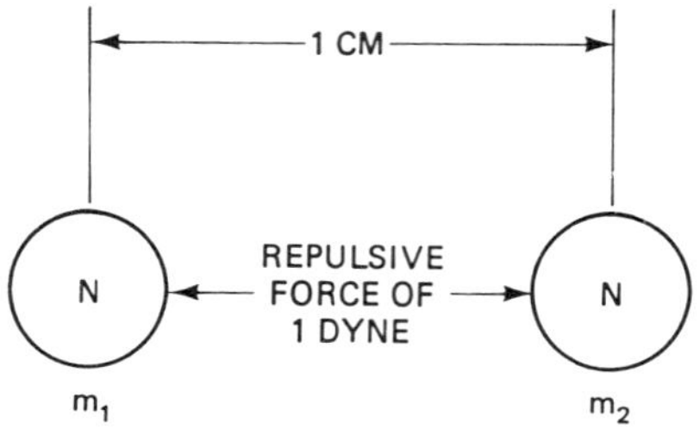

Figure 1-21 Force of repulsion between similar unit poles.

Unit poles can be regarded as positive or negative, positive for north poles, negative for south poles. If the solution to a problem involving unit poles is positive, the resultant force is one of repulsion. If it is negative, the force is considered attraction.

Example

A north pole having a strength of 60 unit poles is positioned 5 cm (in air) from a south pole with a strength of 10 unit poles. What is the amount of force existing between the two poles, expressed in dynes?

Solution

$$F = \frac{m_1 m_2}{\mu d^2}$$

$$F = \frac{(60)(-10)}{1 \times 5^2} = \frac{(-600)}{25} = -24$$

The force between these unit poles is one of attraction and is equal to 24 dynes.

The formula for magnetic force supplies the force of attraction between magnetic poles, but what if just one pole exists. This would be the situation when a magnet is used in connection with a nonmagnetized bit of a ferrous metal. The formula would still apply. Consider a magnet either one of whose poles has a strength of 20 unit poles. The bit of soft iron it attracts has zero unit poles since it is not magnetized.

$$F = \frac{m_1 m_2}{\mu d^2}$$

$$F = \frac{0 \times (m_2)}{\mu d^2}$$

$$= \frac{0}{\mu d^2}$$

$$= 0$$

Field Intensity

The intensity of a magnetic field is a measure of the force exerted by that field with the intensity specified in oersteds. The oersted is defined as the strength required to exert a force of one dyne per unit pole. In terms of the formula

$$H = \frac{F}{m}$$

H is the field intensity in oersteds, F is the force acting on a magnetic pole in dynes, and m is the strength of a magnetic pole in unit poles.

Example

What is the intensity of a magnetic field that has a force of 120 dynes on a magnet having a strength rating of 40 unit poles?

Solution

$$H = \frac{F}{m} = \frac{120}{40}$$

$$H = 3 \text{ oersteds}$$

Magnetic Induction

Magnetic induction is applicable to both permanent and temporary magnets and is the process by which one magnet produces another. If a permanent magnet is used to attract a small bit of soft iron, it is able to do so because the iron is capable of being magnetized. The magnetic field of the permanent magnet induces a small magnetic field in and around the bit of iron. This magnetic field has a polarity that is the opposite of the permanent magnet. Consequently, the permanent magnet attracts the iron for they now have opposite poles facing each other. However, since the magnetic field around the iron is very weak, the magnetic force between them is quite small.

In the case of a coil of wire carrying a current and using an iron core, the magnetic field surrounding that coil induces lines of magnetic force in the core. When removed, the core can remain a rather strong magnet and the force between it and the coil can be fairly large. However, this condition exists only as long as the current is allowed to flow through the coil.

Once iron has been magnetized, it has a unit pole, and so the force between it and the electromagnet, or permanent magnet if that was used, is no longer zero.

The force between unit poles can exist between a pair of permanent magnets, between a permanent magnet and an electromagnet, or between a pair of electromagnets.

MAGNETIC MATERIAL CLASSIFICATION

At one time all materials were classified as either magnetic or nonmagnetic. A magnetic substance was one that could be magnetized; nonmagnetic could not be. Subsequently, this classification was changed to diamagnetic, paramagnetic and ferromagnetic. This is based on the understanding that all substances react to the presence of a magnetic field in different ways. The determination as to which of these classifications is used by a substance depends on its permeability.

Diamagnetic materials. Diamagnetic materials are characterized by having a permeability of less than 1, that is, less than the permeability of air which is used as the reference and whose permeability is unity. This means that a diamagnetic material offers a greater reluctance and has a more resistive path for the existence of magnetic lines than that supplied by air. However, diamagnetic materials can be slightly magnetized when in the presence of a strong magnetic field and under these circumstances behave somewhat like ferromagnetic materials, that is, their magnetic fields have polarities opposite that of the magnetizing forces. An unusual feature of a diamagnetic material is that when it is made in the shape of a rod it will be repelled by a magnetic field and will try to align itself at right angles to that field.

Diamagnetic materials include copper, silver, gold, mercury, bismuth, phosphorus, antimony, and flint glass. Of these, bismuth has the lowest permeability known (0.998).

Paramagnetic materials. Paramagnetic materials are those that can be only slightly magnetized and have permeabilities a little more than unity. Such materials include aluminum, chromium, and platinum. Platinum, for example, has a permeability of 1.00002. Although air is used as a unity permeability reference, it can also be slightly magnetized, and so could be characterized as paramagnetic. If a paramagnetic substance is made into the shape of a rod, it can be attracted to a sufficiently strong magnet and will then align itself parallel to its field.

Ferromagnetic materials. Substances that have permeabilities much greater than unity are characterized as ferromagnetic. At one time this list comprised only iron, steel, nickel, and cobalt, but it now includes alloys such as those of aluminum, nickel and cobalt, or iron, nickel, cobalt, and rare earth metals like samarium. Cobalt has a permeability of 170, nickel 1,000, and steel 1,500, while diamagnetic substances are slightly below 1 and paramagnetic, slightly above 1.

MODIFICATION OF THE MAGNETIC FIELD

Ordinarily a magnetic field surrounds a bar magnet extending into the space around it in a uniform manner. The stronger the magnet, the greater the space its field occupies. That magnetic field can be modified (a technique used in the manufacture of motors) in a number of ways. If a nonmagnetized iron bar is positioned in that field, some of the lines of flux closest to that bar will pass through it. This is due to the fact that the iron bar has a much lower reluctance than the air, with a ratio of 1,000 to 1 or more.

In the case of a permanent horseshoe magnet, the field between the pole pieces can be increased by putting a section of soft iron between them (Figure 1-22). Since the reluctance has been decreased, the strength of the magnetic field has been increased. If the iron bar is made large enough so that its ends rest directly on the pole pieces, there will be few, if any, external flux lines, and we will then have the equivalent of a closed magnetic circuit. Sections of iron bars used in this way are known as keepers (Figure 1-23) and are used to extend the life of the magnet or to keep it from magnetizing nearby materials. It is able to do this since its magnetic lines of flux (or nearly all of them) will be confined to the body of the magnet and its keeper.

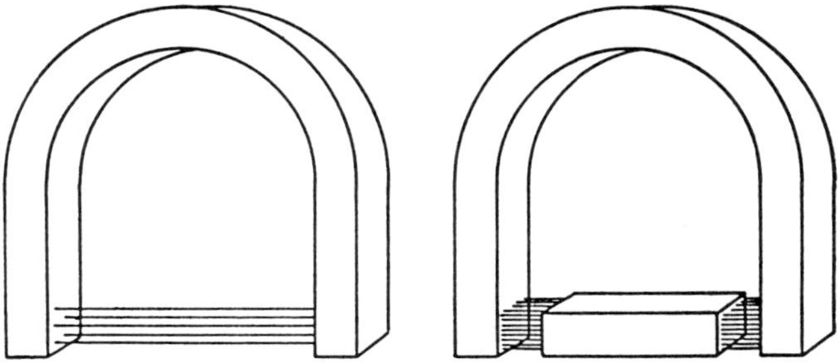

Figure 1-22 The total number of magnetic lines is increased when an iron bar is placed between the poles of a magnet, as in the illustration at the right. The effect of the bar is to decrease the reluctance of the total magnetic path. The bar makes physical contact with the ends of the magnet. An open space is shown here to illustrate the increased number of flux lines.

The iron or steel frame surrounding a motor can be regarded as a keeper, a magnetic shunt or bypass. Since the frame offers a path of much lower reluctance than the surrounding air, it helps confine magnetic lines of flux to the immediate vicinity of the motor. The number of flux lines that do "escape,"

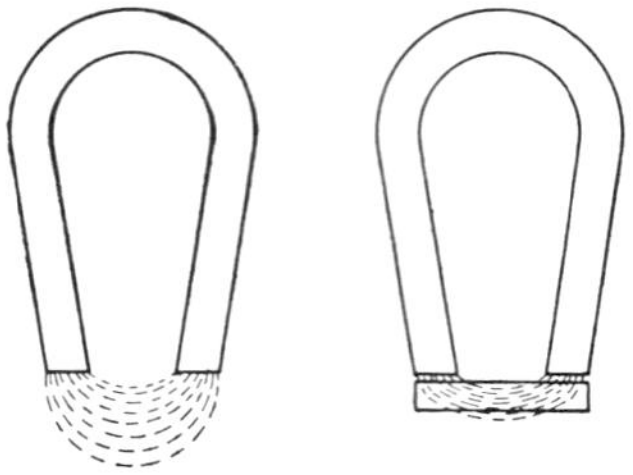

Figure 1-23 When using a ferromagnetic material as a keeper, if the keeper is allowed to make contact with the magnet there will be very few external lines of magnetic force. The horseshoe magnet makes the keeper into a temporary magnet.

depends on the strength of the magnets used in the motor and on the number and size of the openings in the motor frame.

It is possible to magnetize a ring of iron or steel so that the lines of flux are confined, as in Figure 1-24. A magnet of this kind would not have a specific, accessible north or south pole. The frame of a motor is an approximation of the ring magnet.

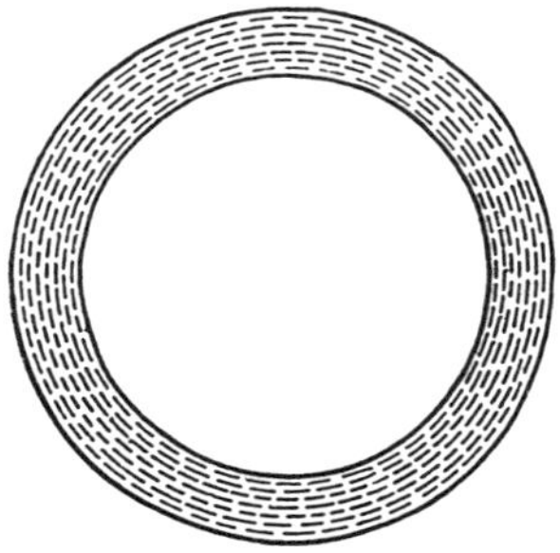

Figure 1-24 In the ring magnet, the lines of flux are completely contained and there is no accessible north or south pole.

In the case of bar magnets, the magnetic field can be confined by putting a soft iron bar in close contact with the permanent magnet, as in Figure 1-25. If the upper bar in the drawing is the permanent magnet it will, through magnetic induction, cause the lower keeper bar to become magnetized. There will be little or no external magnetic field.

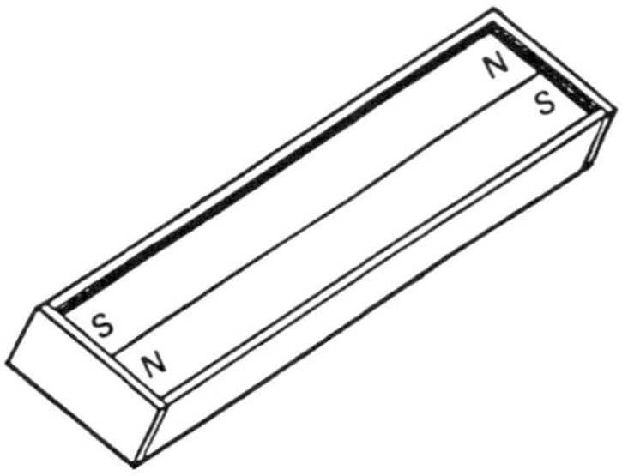

Figure 1-25 A permanent bar magnet should be stored with its keeper bar, as shown here.

Magnetic fields can also be modified by putting the magnetic poles of magnets adjacent to each other, as in Figure 1-26A. In this case a pair of poles of opposite polarity face each other. The external magnetic field is augmented, and if the two poles are allowed to be in contact, the effect will be approximately that of a single magnet with twice the magnetic field strength of one of

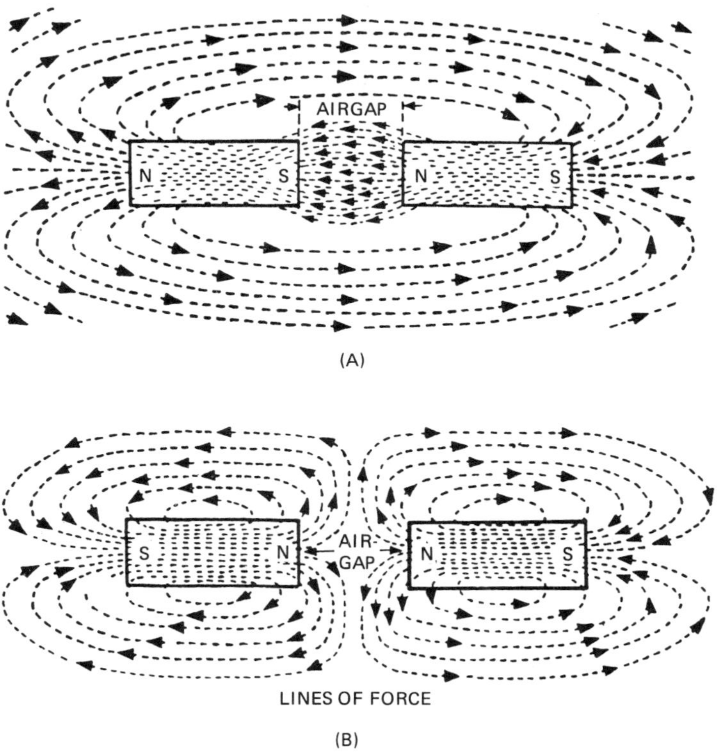

Figure 1-26 Unlike magnetic poles attract (A); similar poles repel (B). The arrows do not represent motion but are simply the assumed direction of the magnetic lines.

the magnets, assuming both magnets were originally of equal strength. Figure 1-26B shows the opposite effect, with similar poles facing each other.

The strength of a magnet can be increased by altering its shape. If a bar magnet is bent into the form of a horseshoe its strength will be increased since the path of the lines through air will be decreased. As this is done the reluctance of that path is reduced. The increased magnetic strength will be evidenced by the larger number of flux lines between the poles of the magnet.

LOSS OF FLUX

A permanent magnet can become weaker because of several factors. This can happen because of a strong increase in ambient temperature and can also be due to vibration. Both of these factors can be present because of the way the motor is mounted and can also be due to environmental operating conditions. To maintain a more equable temperature, some motors are equipped with fans mounted on the same shaft as the rotating armature or with a separate blower.

Motors can be equipped with feet or pedestals to keep vibration to a minimum. Vibration can also result from a shaft that is out of round, worn bearings, or poorly or incorrectly lubricated motor bearings. Motors, generally those that have high horsepower ratings, are usually fixed position devices, but some of those used in the home are not, including some for power operated tools such as drills or rotary saws or vacuum cleaners.

Electromagnets can be less subject to flux loss since their magnetic fields are based on current. However, they are wound on laminated iron cores and these have an influence on the strength of the electromagnet. If the permeability of these cores is affected by heat or vibration, magnetic field strength can decrease. Dropping a motor can result in flux loss.

Residual Magnetism

As long as a current flows through a coil, whether AC or DC, a magnetic field will surround it. If the coil has an air-core the magnetic lines will disappear when the current stops and the coil will not exhibit any magnetic characteristics. However if the coil contains an iron core, some of the magnetic lines will be retained. The combination of coil and core will behave like a weak magnet, but this will be due entirely to the core and not the coil. If the core is

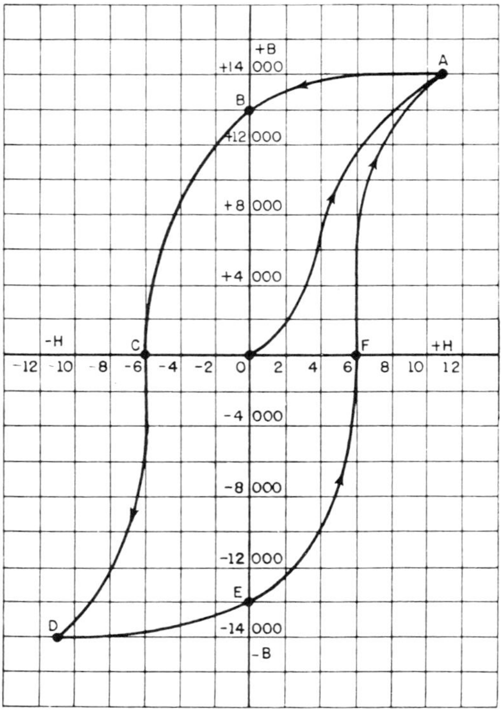

Figure 1-27 *B-H* curve.

removed, it will have a north and south pole and will be surrounded by magnetic lines. The coil, on the other hand, will not show any magnetic properties. The magnetism retained by the core is referred to as residual magnetism.

The *B-H* Curve

A magnetizing force, *H*, will produce a certain number of flux lines, but increasing that force linearly does not mean the same number of lines will always be developed. Actually, as indicated earlier, as saturation is approached, a growing magnetization force produces fewer and fewer lines of flux. This can be demonstrated visually by the *B-H* curve in Figure 1-27. The *B-H* curve extends from point A to point B in the graph and is drawn for an iron-core coil with a direct current flowing through it.

HYSTERESIS

When an iron-core coil (inductor) is energized by an alternating current its polarity will change at a rate determined by the frequency of the applied AC. What we actually have are two magnets. The coil is a magnet because of the current flowing through it. The iron core is a magnet and becomes so because of the magnetizing effect of the current.

The current will become zero twice during each full cycle of the input waveform, and at those moments the magnetic field will also become zero around the coil. This is not true of the iron core, for it will retain some magnetism even when the magnetic field around the coil disappears.

Following its zero current condition, current will again flow through the coil, but because the direction of current has changed, the polarity of the magnetic field around the coil will also change. This field will try to magnetize the iron core, but before it can do so, it must first overcome the residual magnetism. This condition, known as hysteresis, takes place every time the current reverses direction.

Hysteresis represents an energy loss and shows itself in the form of heat. Its amount depends on the frequency of the AC input and the permeability of the iron core.

The Hysteresis Loop

Using the same arrangement as that for producing the *B-H* curve, suppose the magnetizing force is reduced by gradually diminishing the amount of current flowing in the coil, until that current is zero. It is the current that produced the initial magnetizing force, and since the current is zero, the magnetizing force is also zero. But, as is shown in the graph in Figure 1-27, the flux density of the iron core is not zero and is due to the presence of residual magnetism. The amount of residual magnetism is shown as point C on the graph; this could be equivalent to about 7,500 gauss.

The amount of residual magnetism can be brought to zero by permitting current to flow through the coil once again, but this time in a reverse direction. Since the core is still magnetized, the magnetic field around the reversed current must first reduce the residual magnetism to zero. If the effect of this is plotted on the graph, it will represent the curve from C to D.

At this time, the bar is no longer a magnet; to get it back to its original condition, electrical energy has to be expended, and this produces no useful work, although work was performed in completely demagnetizing the core. Although we speak of energy loss, to be precise we should say energy waste. This waste is referred to as coercive force.

If the reverse current is permitted to flow through the coil, it can be graphed and will be the line from D to E. The core is once again magnetized but its polarity is now exactly opposite the polarity it had when first magnetized.

If the current is now reversed again, the magnetic field around the current will gradually reduce the magnetism in the core until point F is reached. At this time the current flow is zero but the core is still magnetized, once again due to residual magnetism. To bring the core back to the point where it has no magnetism requires a continuation of current flow to point G on the graph. Once again, we have had a loss of power. Thus, with an alternating current flowing through an iron-core coil, there will be two times during the AC cycle when there will be a power loss due to hysteresis.

LIMITED MOTOR ACTION

While we usually consider a motor as a device having ongoing rotation, motor action can be continuous or limited. Those with limited action are the compass, solenoids, relays, meters, dynamic loudspeakers, microphones, and other devices whose movement is deliberately restricted. But whether continuous or limited motors all work on the concept of attraction or repulsion between magnetic poles.

The Loudspeaker

A loudspeaker consists of two magnets, a permanent and a temporary (Figure 1-28). The permanent magnet is shaped so that its ends surround the coil, known as the voice coil, that forms the temporary magnet. The coil receives its current from the output audio power amplifier stage of the receiver; this current varies both in amplitude and frequency. Since the current flows in one direction only, it is DC. Because of the changing amplitude of the current, the coil becomes a variable strength temporary magnet. The reaction of the

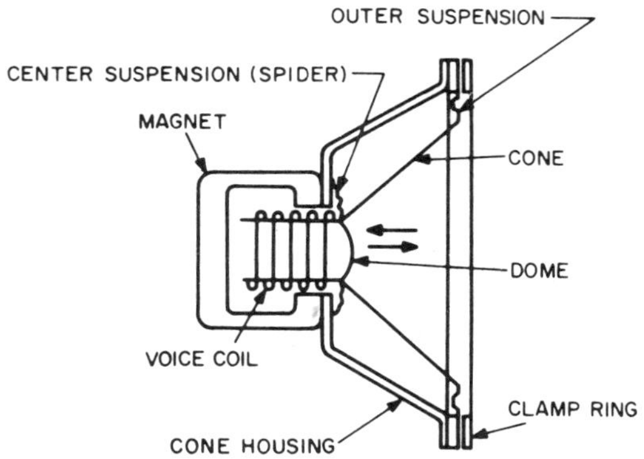

Figure 1-28 A speaker is a limited action motor.

temporary and permanent fields is such that the coil moves back and forth, but with a limited amount of motion. The coil is attached to a cone, and it is the movement of the cone that results in sound pressure which we recognize as sound.

The Meter

Like a loudspeaker, a meter depends for its functioning on the reaction between a permanent and a temporary magnet. The arrangement of a meter shows the basic concept. The permanent magnet supplies a steady magnetic field which surrounds the coil placed between its poles. The coil is pivoted at its center (Figure 1-29) and is so mounted that it is free to rotate. It will do so when a direct current is supplied to the coil with the amount of movement dependent on the amount of current supplied. When the current is discontinued, the coil is returned to its starting point by the spring.

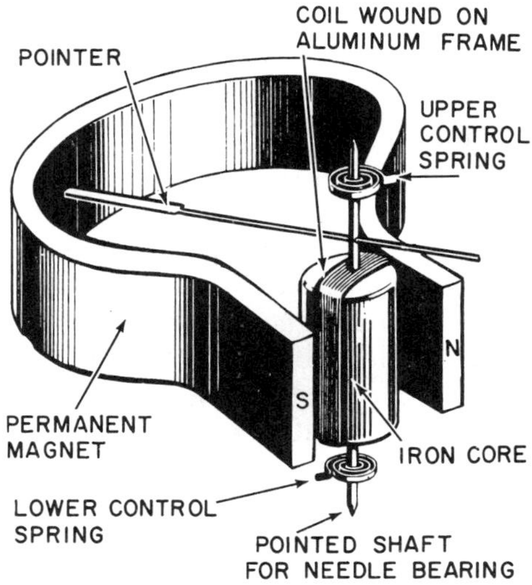

Figure 1-29 The meter is a motor with limited rotation.

THE RELAY

Relays work on the same basic principle as solenoids, but with two differences: design and function. The relay has a built in switch (Figure 1-30) and this switch can be any of the types that are also used as separate components, including the single-pole single-throw (spst), single-pole double-throw (spdt), double-pole single-throw (dpst), double-pole double-throw (dpdt), and so on

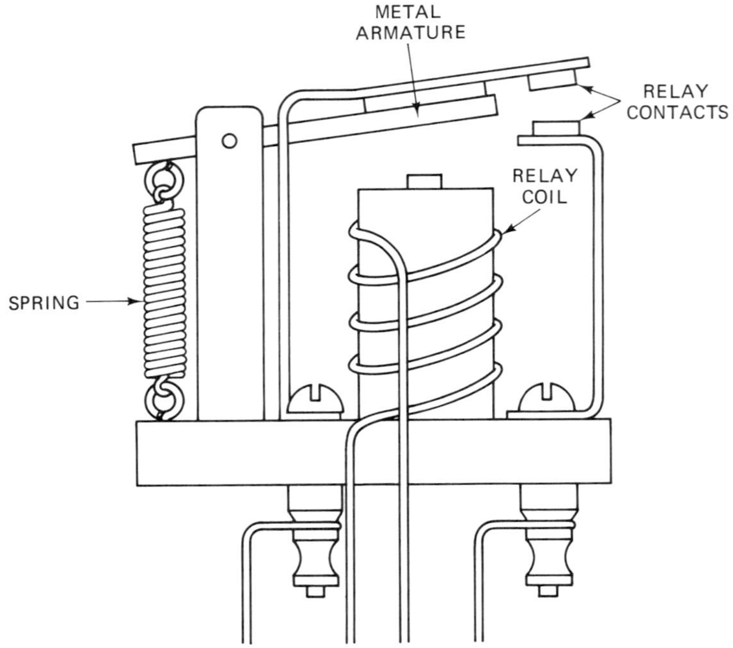

Figure 1-30 Relay structure.

As in the case of the solenoid, the relay consists of a coil made of a large number of turns of wire. The coil is wound around a soft iron core whose function is to increase the strength of the magnetic field surrounding the coil when a current flows through it.

When the coil is energized, it attracts a metallic armature pivoted at one end. The end of the armature opposite the pivot is designed to work as a switch. This switch can be normally open (NO) or normally closed (NC) depending on the design of the relay. The movement of the armature is motor action but is limited to the motion of the armature as it moves up or down.

There are numerous types of relays, just as there are numerous motors of various kinds, but all of them use the same basic principles. Relays are sometimes used in conjunction with motors, to turn them on or off immediately or after some precalculated time.

2 PHYSICAL AND ELECTRICAL CHARACTERISTICS

There are so many different kinds of motors that they could easily form a mental jumble unless they were categorized in some sort of order. As a start it is helpful to recognize that the word *motors* is also applied to engines of all kinds and so it is necessary to eliminate motors that are engines from any listing of motors that are electrical or electrical/electronic.

Motors are often arranged by function, such as phono motors, sewing machine motors, washing machine motors, cassette deck motors, and so on. A listing of this kind is not of much help to the motor technician or engineer. Another listing could be in terms of power type, and since there are only two, DC and AC, this could be a good first step toward the organization of motors into two broad categories.

Under the heading of DC motors there could be a listing of three types: shunt, series, and compound. Under the heading of AC motors the list would include synchronous and induction motors, with this latter classification including single phase and polyphase motors. (See Table 2-1 and Figure 2-1).

The family tree of motors could then be further classified in terms of power ratings, starting with fractional horsepower motors, 1/4, 1/3, 1/2, and so on, to motors rated at 1, 1-1/2, 3, 5, 7-1/2, 10, 15, 20 up to and including those having even more substantial amounts of horsepower.

TABLE 2-1 Types of Motors

Direct Current
Shunt
Series
Compound
Split field
Differential
Interpole
Universal
Tapped field universal
Brushless DC
Induction
Polyphase
squirrel-cage rotor
wound rotor
slip ring
brush shifting
shaded pole
Single phase
split-phase
repulsion-induction
capacitor
series
Electronic
brushless, slotless, coreless
Hall effect
logic circuit
electronic control types
disc stepping
power steppers
microminiature electrostatic

THE BASIC MOTOR

If two magnets are closely positioned so that their similar poles face each other, there will be a force of repulsion between them. If one of the magnets is fixed so it cannot move but the other is free to rotate readily and easily, it will do so until the force of repulsion is no longer strong enough to have any effect (Figure 2-2).

This is a demonstration of motor action, but to have a practical working unit the action must be repeated, and it must be smoothly continuous. One of the magnets can be a permanent type, the other an electromagnet, with the electromagnet capable of movement. This solves one problem but raises another in its place, and that is how to supply an electric current to a coil while it is in motion.

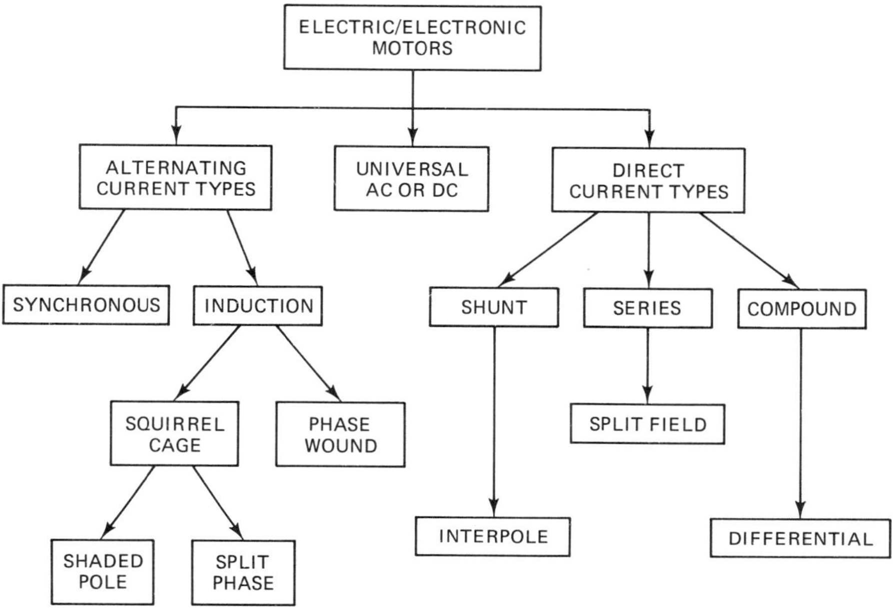

Figure 2-1 Overview of motor types.

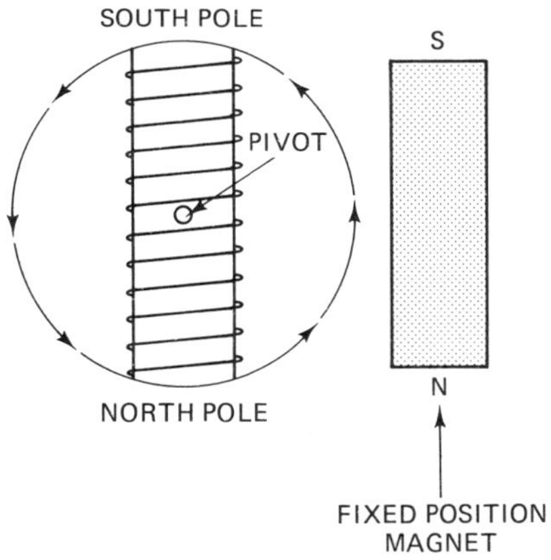

Figure 2-2 Reaction between fixed position magnets and a magnet free to rotate.

Figure 2-3 shows two views of how this can be done. One of the magnets that is used is a permanent magnet and its poles are identified by the letters N and S. For the purpose of explanation, the electromagnet is a two-turn coil. Since the permanent magnet will remain fixed in position it is known as the stator or field magnet; the two-turn coil as the rotor or armature.

Each end of the rotor is connected to what appears to be a split cylinder that is called a commutator. Made of hard copper, it is sometimes known as a split ring. The motor shown in Figure 2-3 is a simplified version, for the rotor consists of a number of coils and the commutator has more segments than just the two in the illustration.

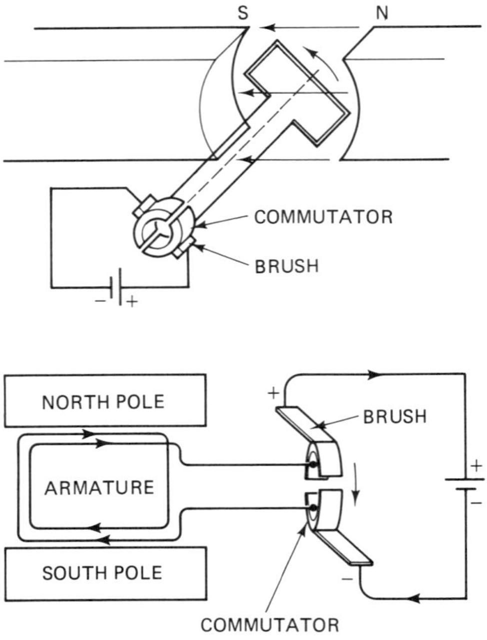

Figure 2-3 Two views of the basic structure of a DC motor.

Since the rotor coil is connected to the commutator, both will turn at the same time. Positioned on the outside section of the commutator segments are two blocks of a carbon-type material called brushes (Figure 2-4). The brushes make good electrical contact with the two commutator halves and are kept fixed in position by a brush holder. Wire conductors are connected to the brushes and these lead directly to the plus (+) and minus (−) terminals of a DC voltage supply.

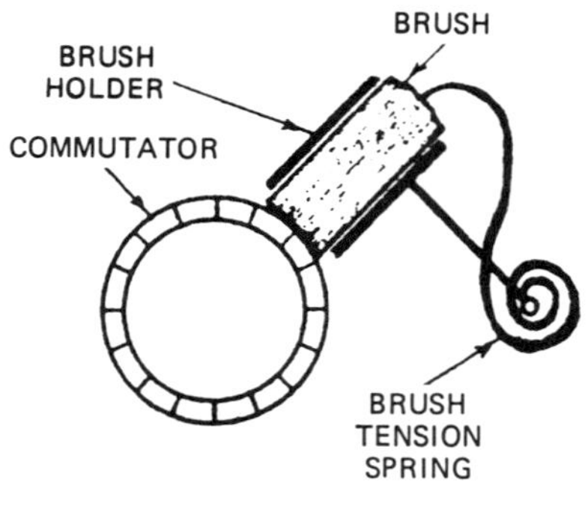

Figure 2-4 A brush, supported in a holder, makes contact with commutator segments.

In the motor in Figure 2-3, electrical current will flow from the minus terminal of the voltage source into one of the brushes and from there into one of the commutator segments. It will then continue its movement through the armature coil until it reaches the second half of the commutator, and from there it will go through the second brush, continuing on to the plus terminal of the battery.

As a result of this current movement, the armature coil will become an electromagnet, with the direction of current flow indicated by the arrows. Its polarity will be such that the upper horizontal part of the armature will act as a south pole, the lower horizontal part as a north pole. These two poles will be attracted by the opposite polarity of the poles of the permanent magnet, and the coil and the commutator segments will start to rotate in a counterclockwise direction.

As the commutator segments continue to turn, the brushes will ultimately be positioned on the open spaces between them, an action that disconnects the armature from the DC voltage supply. When brush contact is made once again, current will flow in the opposite direction through the armature, and its polarity will be completely transposed. We then have a force of repulsion between the upper horizontal section of the armature and the north pole of the permanent magnet. Similarly, the lower horizontal section of the coil will become a south pole and it will be repelled by the south pole of the permanent magnet. But by now the armature winding has rotated so that its upper half is down near the bottom end of the permanent magnet on the left side and near the top end of the magnet on the right side. The force of repulsion is such that the armature is compelled to continue turning.

This could be called a DC motor since the voltage supply is DC. However, the current flowing through the armature is AC since it flows first in one direction and then in the other. The waveform of this AC is unlike the sine wave supplied by local power utilities but is more like a series of square-wave-type positive and negative pulses.

The Field Coil

The permanent magnet can be replaced by a coil of wire using the same DC supply that furnishes current to the armature. This coil, known as a field coil, becomes a magnet when an electrical current flows through it. Its advantage is that its magnetic strength can be controlled and is determined by the iron core on which it is wound, the number of turns constituting the field coil, and the amount of current flowing through it. This is the technique followed in larger motors. Smaller motors having fractional horsepower ratings often use permanent magnets. Figure 2-5 shows the relationship of the armature to the electromagnet. For the sake of simplicity the commutator has been omitted. Initially the position of the armature is such that its top end is a north pole; its bottom end a south pole. The direction of rotation will be such that the top end of the coil will be attracted to the south pole at the right; the bottom end of the coil will be attracted to the north pole at the left, as in drawing A.

When the armature turns, its polarity will have changed by the time it reaches the position indicated in drawing B. Note that the polarity of the field coil magnets has not changed while the polarity of the armature coils has reversed. The top end of the armature is now a south pole; its bottom end a

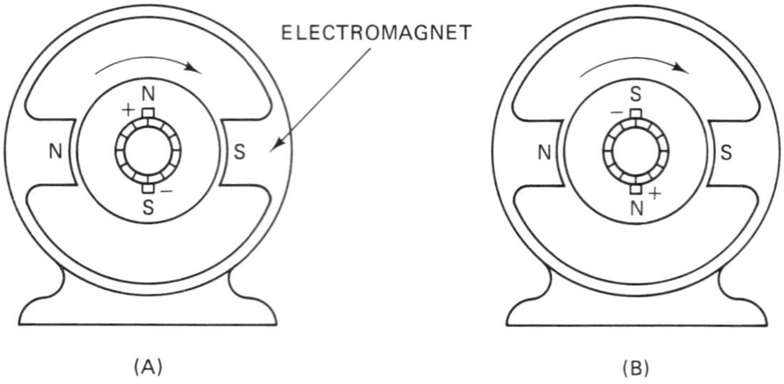

Figure 2-5 Repulsion between the poles of the field coil and armature coil compels the armature to turn (A); the polarity of the armature reverses and so the repulsive force continues (B).

north pole. The momentum of the armature is such that it continues its rotation more than an additional 90 degrees. At this time, there is a force of repulsion between the north pole of the armature and the north pole of the field coil and between the south pole of the armature and the south pole of the field coil.

While the unit illustrated in Figure 2-5 gives a general idea of motor action, it is not practical. A more realistic version appears in Figure 2-6. The number of armature magnets and field coil magnets have been doubled. As a result there is a smoother, more continuous reaction between the poles of the armature and field magnets.

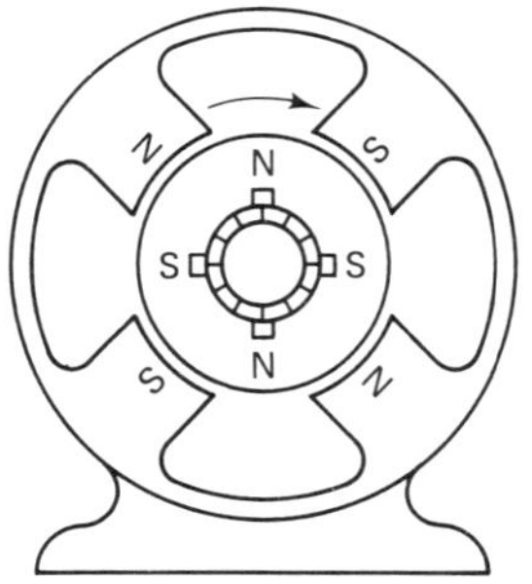

Figure 2-6 A more practical version of the motor shown in Figure 2-5.

The behavior of the commutator is similar to that of a double-pole, double-throw switch, but its action is not manual; rather it is completely automatic. The concept of the commutator as a switch is shown in Figure 2-7. Although the source is DC and in this example is supplied by a battery, the direction of current flow through the armature coil is reversed every time the switch blades are made to change their position. The reversal of coil polarity is evidenced by a compass placed near the armature. The switch and its automatic counterpart, the commutator, are both devices for changing DC to AC.

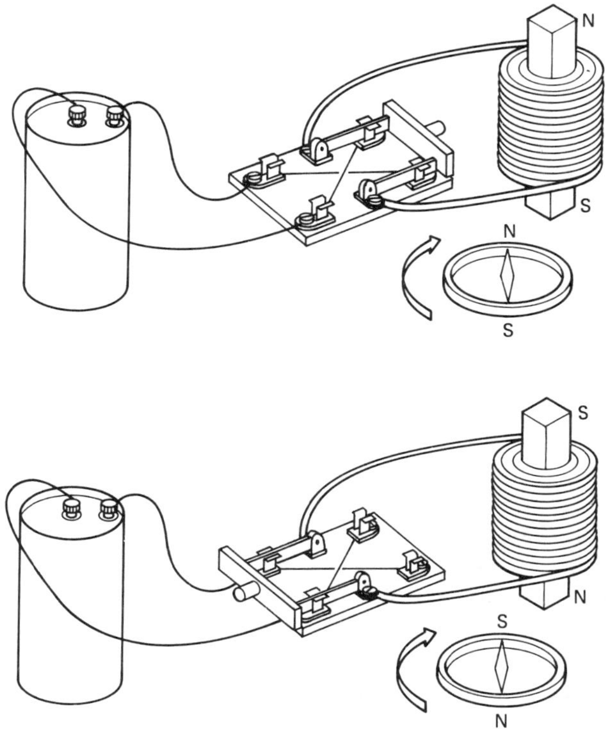

Figure 2-7 A reversing switch produces an alternating current through the coil.

CONSTRUCTION OF THE ARMATURE

The armature is the revolving part of a motor and is mounted on a shaft supported at both ends by bearings. To increase the magnetic field around the armature, it is wound on an iron core (Figure 2-8). The iron core is not solid but is made of thin sheets of laminated iron. These consist of fairly thin stampings,

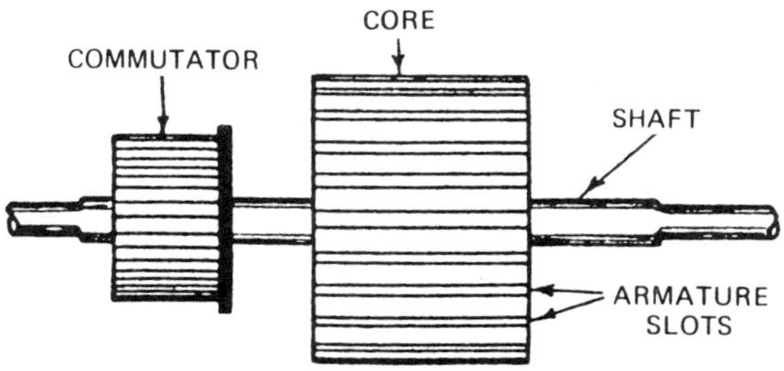

Figure 2-8 Armature core and commutator assembly prior to insertion of armature winding into core slots.

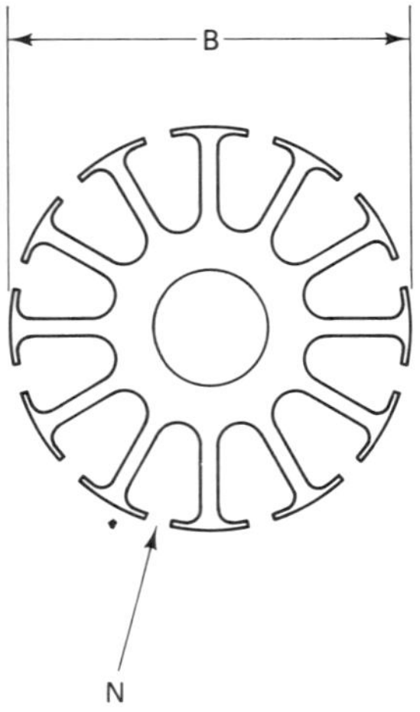

Figure 2-9 Armature lamination: Diameter (B), slot (N). (Courtesy Electric Stature, Inc.)

as shown in Figure 2-9, which may be varnished. Laminations are used to minimize the circulation of eddy currents in the armature core during its rotation. This is the same concept as that used in laminated iron-core transformers.

The diameter of the armature (B) will vary considerably and will help determine the ultimate size of the motor. The number of slots (N) is always an even number since two slots are required for each armature coil. Some armatures have as few as 12 slots, others as many as 24.

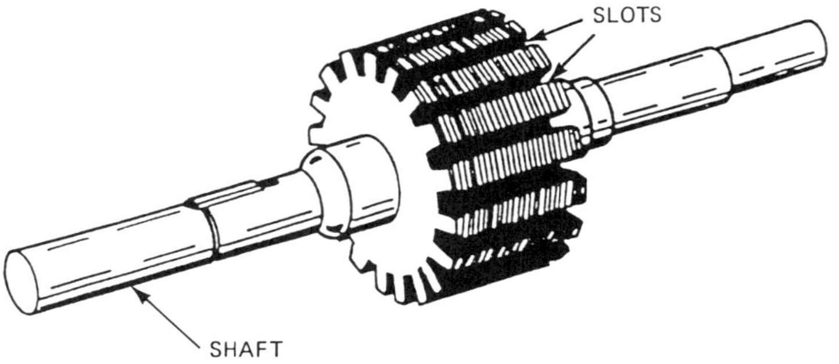

Figure 2-10 Armature laminations assembly on motor shaft.

Figure 2-10 shows the armature laminations mounted on a motor shaft. Following this assembly, the armature coils are fitted into the slots and the coil ends are connected to the commutator. Figure 2-11 shows the final assembly, with the commutator mounted on the shaft together with the armature. The ends of the shaft are supported by bearings. The construction of the armature assembly is such that it is free to rotate easily and is balanced so that rotation does not favor any armature position. The laminations are made of iron because of its high permeability and low hysteresis.

Prior to being inserted in the armature slots, the armature coils are machine formed. The ends of the coils are left bare for soldering to the commuta-

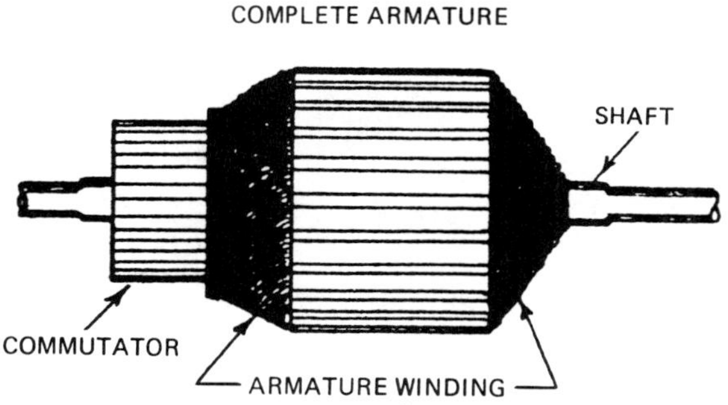

Figure 2-11 Completed armature assembly with armature coils connected to commutator segments.

tor segments. The dimensions of the coils are such that when one side of the coil is under a south (S) pole, the other side is under a north (N) pole. This refers to the horizontal length of the coil, that is, the section of the coil that is fitted into the slots. The vertical sides of the coil (the back and front ends) contribute very little to the magnetic strength of the coil.

Armature Coil Shape

It may seem strange to have the armature coils in rectangular form instead of circular, but the whole concept is to have the magnetic field resulting from the current in those coils interacting as much as possible with the magnetic field supplied by the permanent or electromagnet. By using the rectangular form, the maximum amount of length of coil wire is available. The least magnetic interaction is caused by the lengths of wire constituting the sides of the coil. Early motors used a circular type of armature; these were replaced by the rectangular winding.

Each armature coil can be considered as having two sections, one of which is horizontal, the other vertical. The horizontal section is that part of the coil that is inserted into the armature slots. The vertical part is that small section that is connected to the armature segments.

Multiple Coil Windings

One horizontal side is fitted into one slot; the opposite horizontal side into another slot. One of the sides is at the bottom of a slot; the other side at the top of a slot. In this way each slot is occupied by two different coils. These two different coils have four lead endings. Two of these endings are soldered to one pair of commutator segments; the other two to another pair. Figure 2-12 shows a pair of coils fitted into armature slots in this way. The length of armature wire, designated as a-b, is positioned on top of the horizontal wire adjacent to it. The other horizontal half of this coil is placed beneath the horizontal half of the second coil, shown adjacent. The arrows show the direction

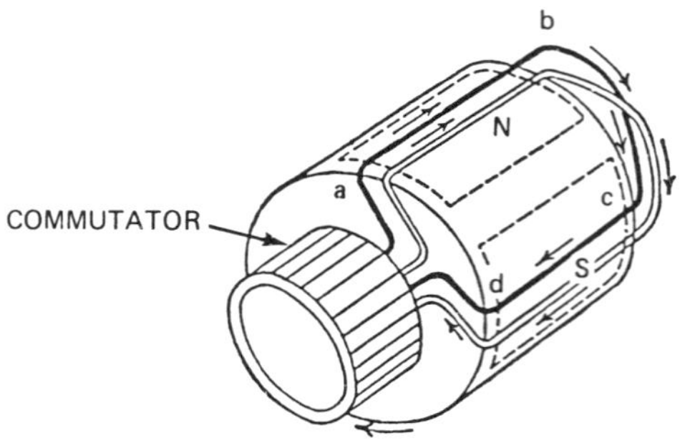

Figure 2-12 The armature coil is rectangular in shape. The most magnetically active coil sides are lengths a-b and c-d. The letters N and S indicate the north and south poles.

of current flow at a specific moment, and the direction of rotation of the armature.

An armature coil, as in the case of coils of any other type, may consist of a single turn, like the one shown in Figure 2-13A, or may have a number of turns, as in Figure 2-13B. In this illustration, the armature coil has three turns. Note, however, that this multiple-turn coil also uses a pair of commutator segments in the same manner as the single-turn coil. However, it is not necessary for the armature coil to use adjacent segments on the commutator. Depending on the motor design, an armature coil may be connected to widely separated commutator segments.

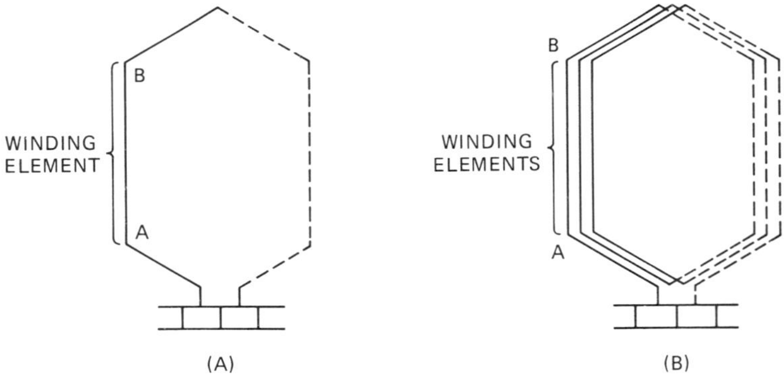

Figure 2-13 Single coil (A); multi-turn coil (B).

Armature Coil Pitch

The span of a coil, the distance from one horizontal leg of an armature coil to its other horizontal leg, is known as its pitch. The word *pitch* is also used in connection with the poles of the field magnet and is the distance separating the north pole of that magnet from the south pole. Usually, the pitch of an armature coil is equal to the pitch of the poles. If this is achieved, then when one horizontal section of an armature coil is under a north pole of the field magnet, its other horizontal section will be under the south pole of that magnet. In this way, there will always be maximum interaction between the poles of the two magnets, of the field magnet, and of the armature coils. However, there is a slight improvement in commutation if the coil span is somewhat less than the pole pitch, an arrangement known as a fractional-pitch winding. Fractional-pitch windings range between about 90 to 99 percent of full pitch.

Front and Rear Armature Designations

The front of the armature is the side adjacent to the commutator; the other end is called the back or rear. The term *pitch* is also applied to the vertical segments of the coil winding, but is somewhat different from the span of the horizontal segments. Thus, back pitch consists of the number of elements that the coil advances on the back of the armature and is actually determined by the span of the horizontal elements. At the front of the armature, the commutator end, the number of elements that are spanned is the front pitch. Front pitch may be greater or less than back pitch, but the two are never equal.

Progressive and Retrogressive Windings

If the front pitch is greater than the back pitch, the winding is designated as retrogressive. This means that the winding advances counterclockwise when viewed from the commutator end of the armature. If the reverse situation prevails, that is, if the front pitch is less than the back pitch, the winding is known as progressive and the winding advances clockwise, again when looked at from the commutator end.

This concept is illustrated in Figure 2-14. Horizontal element 1 is connected on the back of the armature to element 10. Element 1 is at the upper

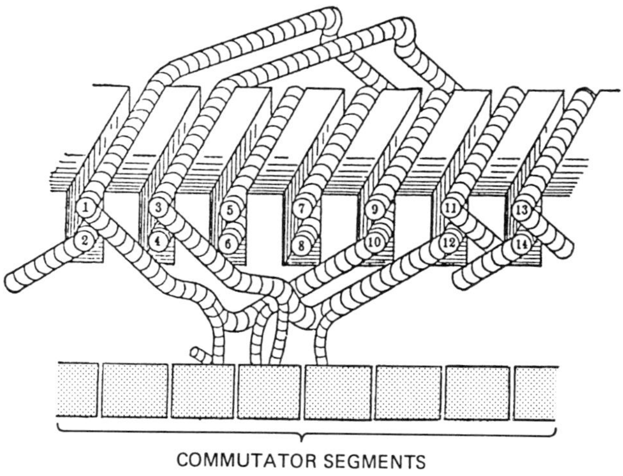

Figure 2-14 Armature with a back pitch of 9 and a front pitch of 7.

part of the slot; element 10 is at the bottom end of its slot. The next horizontal element, 3, is connected to horizontal element 12. In this example, all even-numbered conductors are positioned at the bottom of the slots; all odd-numbered conductors at the top. The back pitch is 9 since $10 - 1 = 9$, and $12 - 3 = 9$, and so on for the remainder of the windings. Note also that a horizontal element at the top of the slot is connected at the rear to a horizontal element at the bottom of a slot. Similarly, elements at the bottom are connected to elements at the top. The front pitch in this drawing is 7. Both the front and rear pitch have odd values. This is a requirement if the winding is to be correctly positioned in the armature slots. If both the front and rear pitch were even, it would be impossible for one horizontal element of a coil to be at the top of one slot and the other horizontal element at the bottom of another slot.

Symbolically, we can designate pitch by the letter P. Front pitch would then be Pf and back pitch Pb. The relationship between front and back pitch could be shown as

$$Pb = Pf + 2.$$

The plus sign in this equation indicates a progressive winding; a minus sign, a retrogressive winding.

The drawing in Figure 2-14 shows a back pitch of 9 and a front pitch of 7; hence it is a progressive winding.

Figure 2-15 supplies a comparison between a progressive and a retrogressive winding. The drawing in A is progressive; that in B is retrogressive. A characteristic of the retrogressive winding is that the finish lead crosses the start lead. If a winding is changed from progressive to retrogressive the effect will be to change the direction of rotation of the armature. The same effect will be noted in changing from retrogressive to progressive.

The horizontal copper conductors are sometimes called winding elements. If a pair or more of copper conductors are used as winding elements they may be held together with tape preparatory to insertion in the armature slots. They are then regarded as a single conductor and may be shown as such in a drawing of an armature winding. Armature slots may contain a single horizontal element, or two, with one element on top of the other.

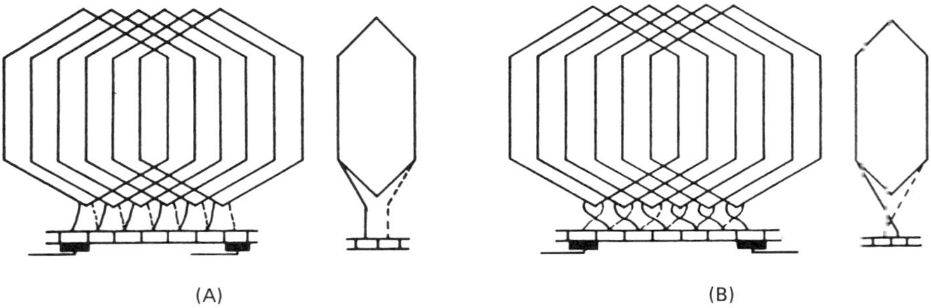

Figure 2-15 Comparison of progressive (A) and retrogressive (B) windings.

The Armature Coil Symbol

The coil symbol (Figure 2-16) is a hexagon, but while it helps simplify the drawing of a winding it can be misinterpreted. The symbol gives the impression that the coil itself is a closed loop and hence short circuited. The coil is just part of a path that starts at the minus terminal of a DC voltage source, moves to one brush, and then to the commutator segment on which it rests. From this segment there is a connection to one terminal of a coil winding, through that coil, to another commutator segment. From that segment the path goes through another brush, and from that segment to the plus terminal of the DC voltage source.

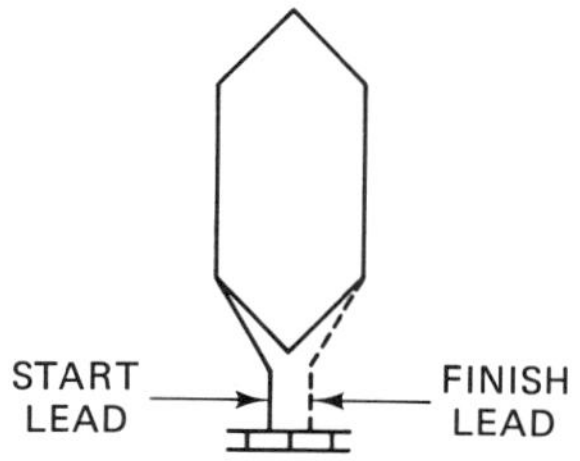

Figure 2-16 The symbol of a multi-turn armature coil does not suppy information about the number of turns.

Number of Elements

Figure 2-17 shows the relationship between the horizontal elements of a winding and the commutator segments. Figure 2-17A shows the arrangement of a single element type, so-called since just one element is put into a slot. The dark vertical sections represent the slot. Two slots are required for each coil and two commutator segments are needed. The solid line connected to the coil shows the connection from the commutator side of the coil; the dashed line shows the connection from the back. The coil symbol at the bottom apex implies that the coil is closed, but as indicated earlier, there is no connection here and so one side of the coil is connected to one armature bar; the other side of the coil to the other armature bar.

For a two turn (two element) coil (Figure 2-17B) a pair of slots can still be used, with one horizontal coil element positioned on the top of the other.

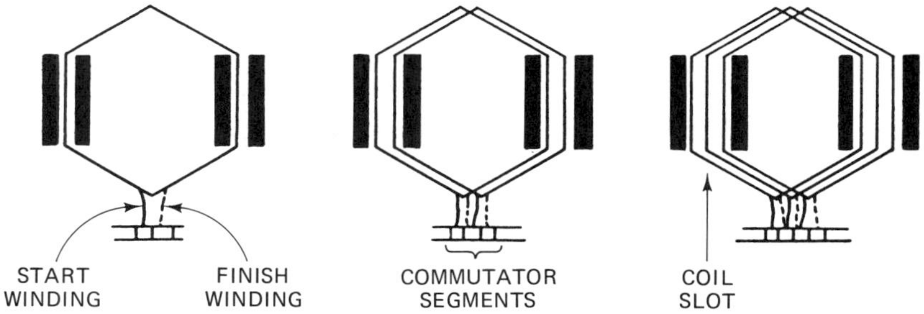

Figure 2-17 Single element (A), two element (B), and three element windings.

The same number of slots will be used as in the single element. The two coils are wired in series as indicated by the connections to the commutator bars. This means that the termination of one coil is also the start of the next. The series coil concept is continued in drawing C. Here there are three horizontal coil elements per slot, but by tracing the solid and dashed line connections to the commutator segments, it is evident that the three coils are in series.

The complete wiring diagram of a motor (Figure 2-18A) can look quite complicated but is not too difficult if each section is analyzed one at a time. The two short parallel lines in the center of the drawing are a pair of wires leading to a DC voltage source. One of these wires is the plus (or positive) lead connected to the plus terminal of a battery or other DC voltage source. The other lead is the minus (or negative) connection.

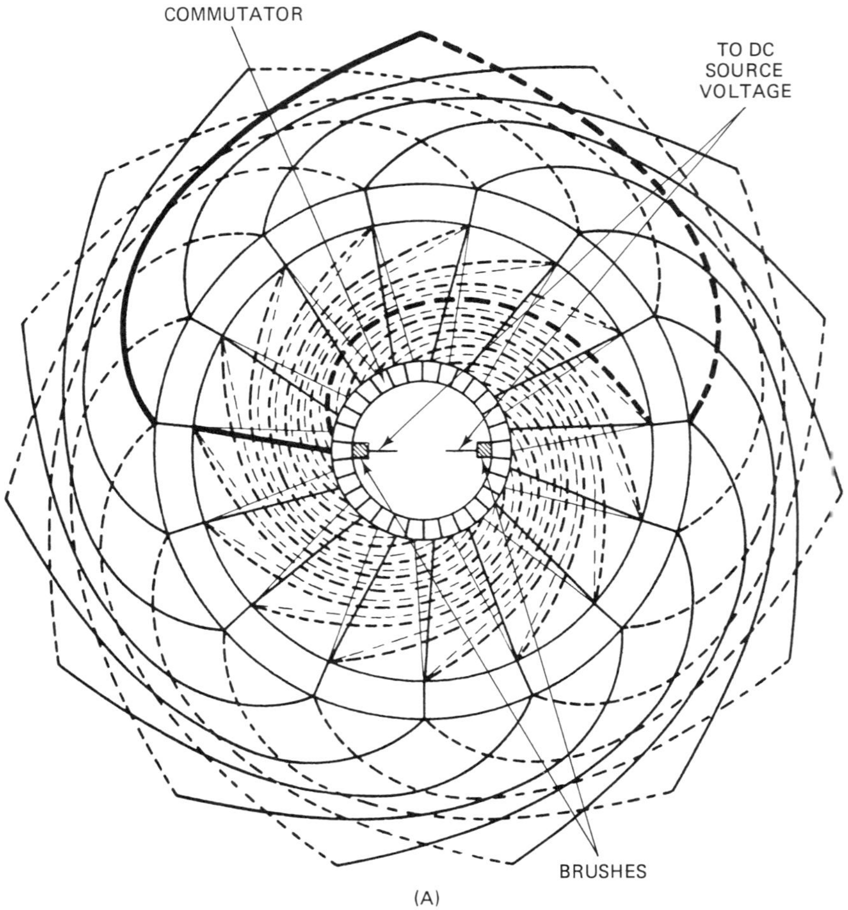

Figure 2-18A Motor wiring diagram. The solid lines represent wires at the top of a slot; the dashed lines are wires at the bottom of a slot. This is a progressive simplex lap winding using 15 slots. It is a two-pole motor with a 1-8 coil span.

The four cross-hatched blocks are the brushes. The block at the left is wired to that at the right; that at the top is wired to the block at the bottom. If the two brushes positioned horizontally are connected to the north pole end of the armature coils, the other two brushes positioned vertically will be connected to the south poles of the coils. By starting at brush number 1 and moving clockwise, there is a north pole, a south pole, a north pole, and then the south pole again. Thus, this is a four-pole machine.

Still moving in an outward direction, each brush rests on a commutator bar, and there are 24 of these; there are also 24 slots. The slots are shown by the outermost circle of rectangles. The solid lines passing through the slots represent one coil element; the dashed lines, another coil element. The coil span of this winding is 1 − 7; the winding is progressive and is a single-element type.

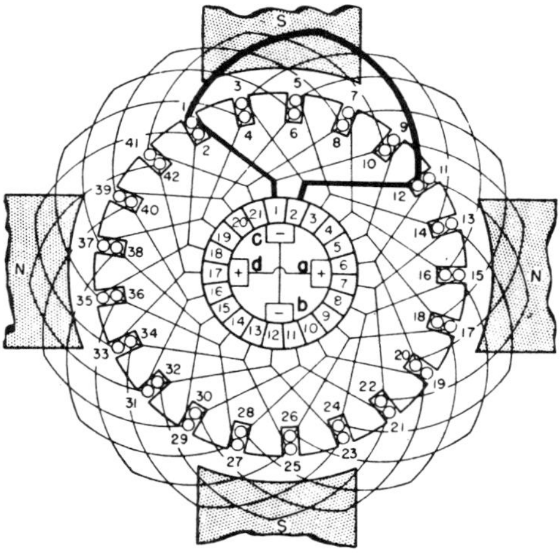

(B)

Figure 2-18B Simplified version of a wiring diagram. The wire coil starts at commutator segment 1 and passes through slot segment 1. It wraps around the armature and continues to slot segment 12. It is then connected to commutator segment 2. Each slot carries two wires. Odd numbers are for the upper position of the slot; even numbers are for the lower position.

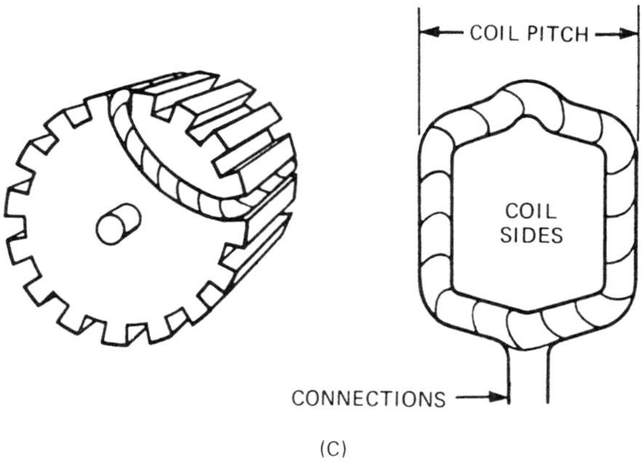

(C)

Figure 2-18C Insulated armature coils fitted into armature slots.

For all armatures, the winding arrangement must be such that if one section of a coil is opposite a north pole of the field magnet, the other side should be opposite a south pole. Figure 2-18B is a simplified wiring diagram. Figure 2-18C shows how armature coils are fitted into their slots.

LAP WINDINGS AND WAVE WINDINGS

The beginning of an armature coil is at a commutator segment, while the end of the coil can be connected to an adjacent bar or one that is relatively close. Such a winding is referred to as a lap (Figure 2-19A). If the ending of the coil is connected to a commutator segment relatively removed from the start segment, the winding is known as a wave type (Figure 2-19B). The polarity of the magnetic poles on the coils is determined by the direction of current flow.

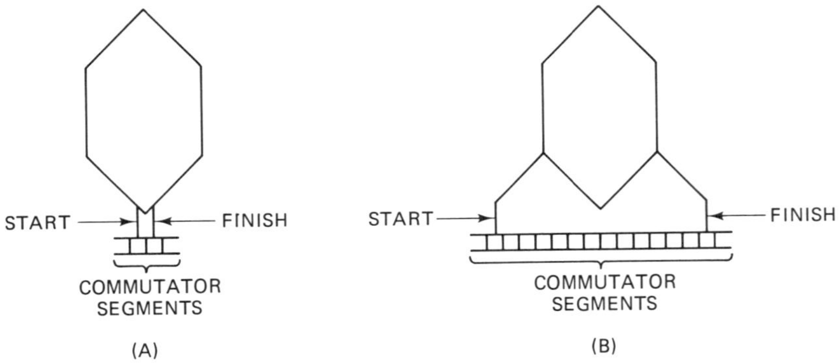

Figure 2-19 Lap winding (A); wave winding (B).

SYMMETRICAL AND NONSYMMETRICAL CONNECTIONS

The connections from an armature coil can lead directly to the commutator bars using the shortest possible length of connecting copper, as in Figure 2-20A. This arrangement is called symmetrical. Drawing B shows a nonsymmetrical connection. Here the commutator bars are offset from the coil, and so the two leads from the commutator coil are not of the same length.

The position of the brushes with respect to the armature coil is determined by whether the connections are symmetrical or nonsymmetrical. If the coil is symmetrical, the brushes must be positioned on center with the pole of the field magnet. If it is nonsymmetrical they are positioned between the field poles.

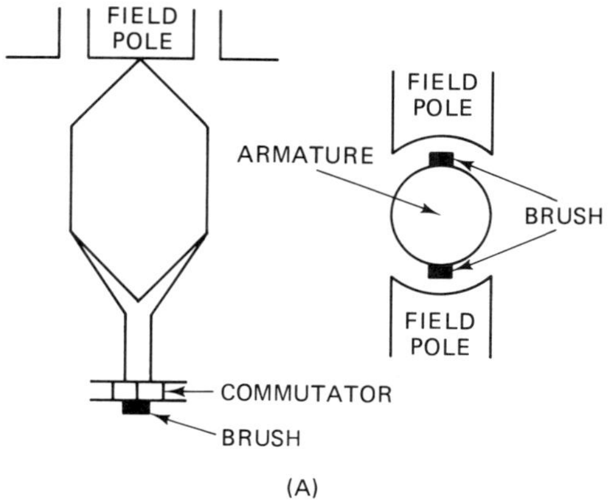

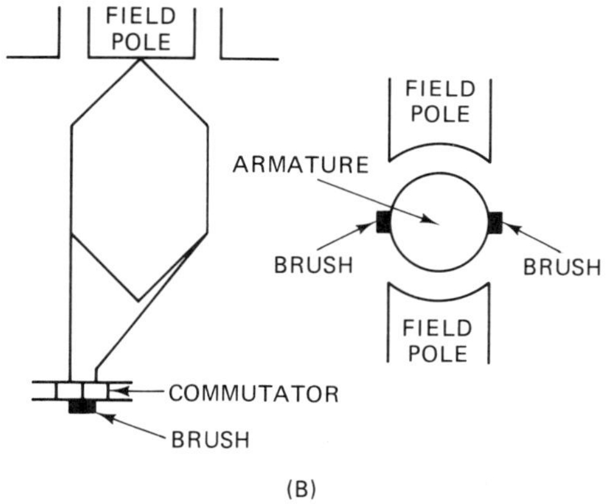

Figure 2-20 Symmetrical connections (A); nonsymmetrical connections (B).

THE ARMATURE CURRENT PATH

The armature coils are connected in a closed network consisting not only of the coils, but the conducting links to the commutator segments, the brushes, and the DC voltage source. The DC resistance of this network consists of the sum of the resistances of all the individual components. This would be the internal resistance of the voltage source, the connecting wires from that source to the brushes, the brushes and commutator segments, the resistance of the connecting wires to the armature coils, and the resistance of the coils as well. In addition there is the matter of contact resistance for the brushes have a sliding contact with the armature bars.

Despite the large number of components and contact points in this network, the total resistance is very small and, because of the large size of the conductors, is often much less than one ohm, with a value of 0.05 ohm fairly typical. Most of the resistance is due to the armature coils.

If the source has an EMF of 12 volts, the amount of current flowing in this circuit, based on Ohm's law (I = E/R) is 12/0.05 or 240 amperes. If, instead of working from a 12-volt battery, the source voltage should happen to be a 110-volt DC line, the current flow would be 110/0.05 or 2200 amperes. It does not take much of an increase in source voltage to raise current flow to values that would easily burn out the armature coils.

This armature current exists when power to the motor is turned on and takes place in the second before the armature starts to rotate. With an overall resistance of only 0.05 ohm the armature is practically a short circuit across the DC voltage source.

Counter EMF

As the armature starts to rotate a voltage is induced across the armature coils. Known as a counter electromotive force (CEMF), it is opposite in polarity to the voltage supplied by the DC source and as such opposes that voltage. Thus, if the source is 12 volts and the induced armature EMF is 10 volts, the effect is equivalent to 12 − 10 or 2 volts. Under these circumstances the armature current is reduced to 2/0.05 or 40 amperes. Whether this amount of armature current is satisfactory depends on the motor design. The armature must be capable of carrying this current.

Calculation of CEMF

CEMF can be calculated by assuming that all the resistance of the armature circuit is concentrated in the coils. If it is 0.05 ohm and if the current through the coils is 40 amperes, the voltage drop across those coils is $E = I \times R = 40 \times 0.05 = 2$ volts. Since the CEMF is equal to the source voltage minus the voltage drop across the armature coils, the CEMF is 12 − 2 or 10 volts.

To overcome the lack of CEMF when power is first applied to the DC motor, a resistor is inserted in series with the external DC voltage source

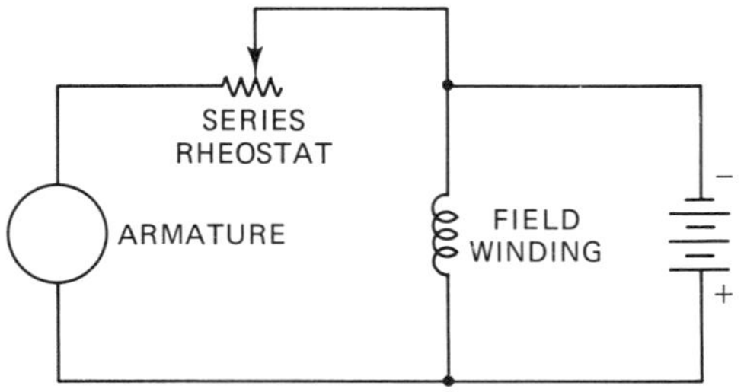

Figure 2-21 A series resistor limits armature current.

(Figure 2-21). The resistor is variable and its resistance is gradually reduced as the armature increases its speed. At full armature speed the variable resistance is reduced to zero.

ARMATURE POLES

An armature will have a number of poles depending on the number of independent armature coils. The minimum number of poles is two, a north and a south pole. The number of poles is sometimes used to identify a motor: thus, a two-pole, four-pole, six-pole motor, and so on. Armature poles, as shown in Figure 2-22A, are positioned midway between the field poles twice during each complete armature revolution. If we consider this as the 0 degree point, with clockwise rotation, there is a force of attraction between the north pole of the armature and the south pole of the fixed position field magnet. Similarly, there is a force of attraction between the north pole of the other field magnet and the south pole of the armature. This force of attraction changes to a force of repulsion when the armature reaches its 180 degree point. If either the direction of current flow through the field coils or through the armature coils is reversed (but not both), the direction of rotation will be changed and will be counterclockwise.

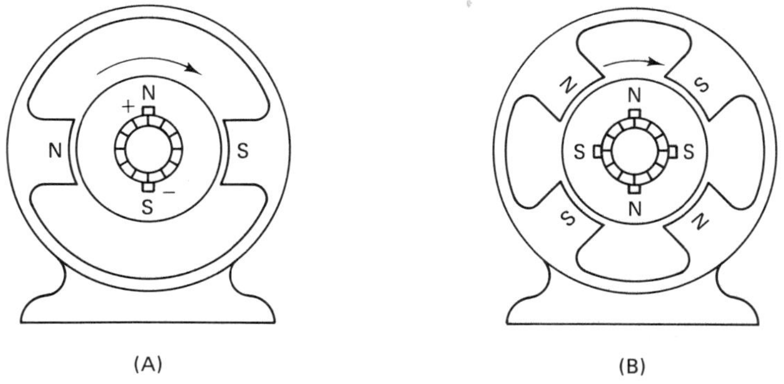

Figure 2-22 Relationship of armature and field poles: two-pole machine (A); four-pole machine (B).

If the number of armature coils is doubled it will have four poles, as indicated in Figure 2-22B. However, increasing the number of armature coils means the number of field magnets must also be doubled, since for each pair of armature poles we must have a corresponding number of field poles. A two pole armature will not work in a motor that has four field poles, and vice versa. And, as in the case of a two-pole motor, reversing the flow of current through either the field or armature coils will result in reverse movement of the rotor.

Slot Usage

For smaller motors it is not uncommon for a slot to carry one or two horizontal coil elements. For larger motors slots can accommodate as many as four, six, or eight elements. Using too few slots (only one or two) presents design problems. If fewer slots are used, slot size is increased, with a consequent reduction in coil winding costs.

Although two or more coils can be positioned in the slots, they are independent of each other but are taped together to form what would appear to be a single coil. Such associated multiple coils are often called double coils, triple coils, and so on. As far as the slots are concerned, a multiple coil is handled as though it was a single.

Commutator Bars

The commutator is made of rectangular lengths of hard-drawn or drop-forged copper in the shape of a wedge. The wedge shape is required since upon assembly the commutator segments, also known as bars, must form a tight circle. The bars must be insulated from each other and this is done by separating them with thin layers of mica. The bars are subjected to centrifugal force when the motor shaft rotates and so are kept in position by flanges which pull the bars inward in opposition to the centrifugal force. The flanges, made of metal, are prevented from short circuiting the commutator bars by insulating mica collars. Each bar has a longitudinal slit. The end of an armature coil is force fit into the slit.

AMPERE-TURNS

The ampere-turns of a coil is the amount of current, in amperes, flowing through the coil multiplied by the number of complete turns. If the current is in milliamperes or microamperes, these must be converted to amperes prior to making a determination of ampere-turns. The letter symbol for current is I, and N is for the number of turns. Ampere-turns is the product of these two.

$$\text{ampere-turns} = N \times I$$

If a current of 1/2 ampere flows through a coil having 10 turns, $N \times I = 10 \times 1/2 = 5$ ampere-turns (see Figure 2-23). If the current is increased to 1 ampere and the number of coil turns is decreased to 5, the ampere-turn rating remains the same.

$$\text{ampere-turns} = N \times I = 5 \times 1 = 5 \text{ ampere-turns}$$

The NI value of a coil is for an air core type. It is indicative of the field strength intensity, with the oersted as the unit. For the CGS system the am-

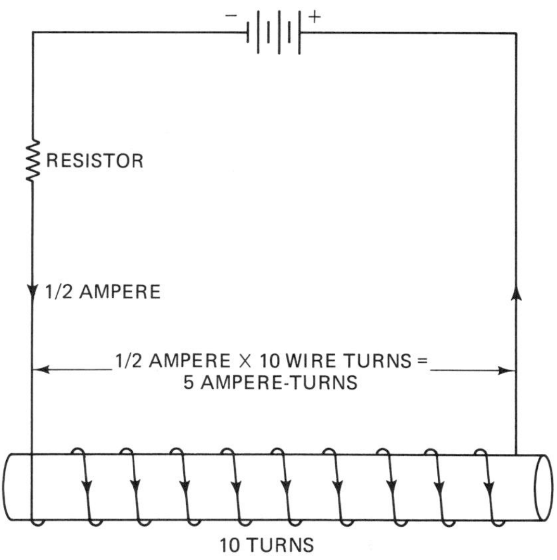

Figure 2-23 Determination of ampere-turns.

pere-turn per centimeter is 1.257 oersteds while in the practical system the ampere-turn per inch is 0.495 oersteds. If, for example, the total length of a coil is 20 inches using 500 turns, and the coil has a current of 1 ampere flowing through it, the NI value is $500 \times 1 = 500$ ampere-turns, or 25 ampere-turns per inch. The field intensity, indicated by the letter H is $25 \times 0.495 = 12.375$ oersteds, in the practical system of units. This can be converted to the CGS system by changing inches to centimeters: 1 inch = 2.54 cm, 20 inches = $20 \times 2.54 = 50.8$ cm. The ampere-turn/cm = 1.257 oersteds: $50.8 \times 1.257 = 63.86$ oersteds. Although the actual field intensity is the same for both the CGS and practical systems, note how different the answers are. It is essential, then, in solving such problems, to first determine the system to be used and to use that system throughout.

DEVELOPMENT OF AN INDUCED EMF

Electromotive force (EMF), sometimes known as potential difference (PD), is more often referred to as voltage. An induced EMF is one that is produced by the interaction of a magnetic field and a conductor, and is called induced to differentiate it from other methods of developing a voltage.

An induced EMF is always produced when there is relative motion between a conductor and a magnetic field. Figure 2-24 illustrates one method. This drawing shows a conductor positioned in a magnetic field. As long as neither the conductor nor the magnetic field move, no voltage is induced across it. If the conductor does move but does so parallel to the lines of flux, no induced

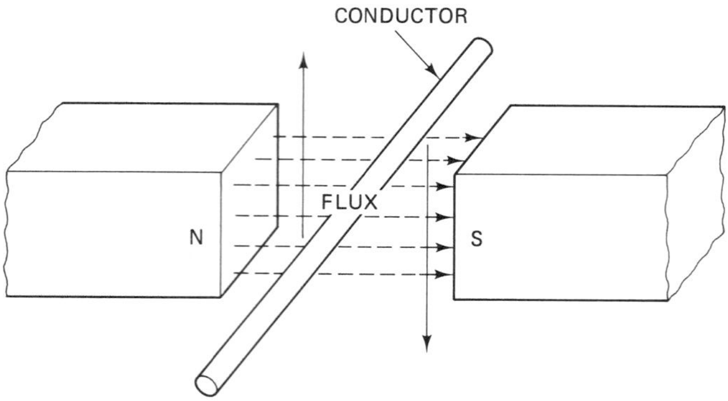

Figure 2-24 A voltage is induced across the conductor as long as the lines of magnetic flux cut across it.

voltage is produced. However, if the conductor moves down, as shown by the arrow, or moves up, a voltage is induced across the conductor.

There are other alternatives. The conductor can remain fixed in position and the magnet can be moved up or down. Another technique would be to have the magnet and the conductor move in opposite directions. In either case an induced voltage would be generated across the conductor. Thus, an induced voltage is produced every time a conductor cuts across magnetic lines of flux. Further, the greater the number of lines of flux, the greater the induced voltage. Although the conductor shown in the drawing is just a single length of copper, it could consist of a coil of wire. If the coil consisted of a single loop, the induced EMF could be called E. For a two-turn coil the voltage would be 2E. The larger the number of turns, the greater the voltage. The voltage across the turns are additive.

This is the situation that prevails in a motor. The armature consists of loops of wire that form a coil. This coil rotates in the magnetic field produced by a magnet. As a result a voltage is induced across the coil, and the greater the number of turns of the coil, the greater the induced voltage. In the case of a lap winding, the start and finish of the coil are to adjacent commutator segments. If the induced EMF is great enough, the voltage may break across the insulting mica barrier between the two copper segments. This is much less likely to happen with a wave winding. Both lap and wave windings have advantages and disadvantages. With the lap winding, voltage breakdown across commutator bars is a possibility, but against this is the advantage that a lap winding can be designed to have a higher current carrying capacity.

Voltage Induced across a Conductor

This formula can be used for calculating the electromotive force induced across a conductor as it moves through a magnetic field. The assumption is made that the conductor is at right angles to all the lines of flux.

$$e = BLv \times 10^{-8}$$

where e is the induced emf in volts, B is the flux density in lines per square centimeter, L is the length of the conductor in centimeters, and v is the relative velocity between the flux and the conductor in centimeters per second.

Force Applied to an Armature Conductor

When the flux lines of a magnetic field are at right angles to the conductor of an armature coil, the force that is exerted on the conductor is given by

$$F = \frac{BLI \times 10^{-7}}{1.13}$$

where F is the force exerted on the armature conductor in pounds, B is the flux density of the field in lines per square inch, I is the conductor current in amperes, and L is the length of the conductor in inches. Note that this equation assumes an orthogonal (mutually perpendicular) relationship between field and conductor.

The force applied to a conductor in a magnetic field can also be expressed as

$$F = BLI$$

where F is the force in newtons, B is the magnetic density in webers per square meter, and L is the length of the conductor in meters. The torque is the product of this force and the radius of the rotating armature.

Eddy Currents

The armature of a motor consists of two main conducting elements: the coils of wire that are fitted into the slots of the armature and the iron core of the armature. The coils of wire rotating in the magnetic field supplied by electromagnets or permanent magnets experience an induced voltage. But the core itself, since it is a conductor, will also have an induced voltage. This voltage will cause a current to flow in the core, referred to as an eddy current. This eddy current does no useful work but results in a loss since its electrical energy is converted to heat energy, thus contributing to the overall heat of the motor.

To reduce eddy currents to a point where energy loss is at a minimum, the armature is not made as a solid mass of metal but is in the form of laminations which are insulated from each other. Each lamination or thin sheet of metal has much less conductivity than a solid armature, or has a much higher resistance. This higher resistance limits eddy current flow to a much smaller amount. The laminations are about 0.0156 inch thick, but this is variable. To insulate between the laminations, they are often coated with a thin layer of lacquer or varnish. In some instances the laminations are deliberately allowed to rust. The formation of this layer of iron oxide supplies a high amount of resistance, practically equivalent to insulation.

THE FIELD COIL

A motor is the result of the reaction between magnetic fields. One of these magnetic fields is supplied by a stator, the other by a rotor. The magnetic field of the stator can be supplied by a permanent magnet, a group of permanent magnets, or an electromagnet. This electromagnet consists of fixed position coils with current from a DC voltage source supplied by a battery, by a DC power line, or by an electronic power supply whose input is connected to the AC power line and whose output is DC.

The field coil electromagnet may consist of one or more coils and the magnetic field it supplies, as in the case of permanent magnets, has a fixed polarity. Since it does not rotate, connections to the DC voltage source are easy.

The field coils, like the armature, are supplied with a soft iron core (Figure 2-25). Here, the effect of the iron core is to increase the inductance of the winding, producing a stronger magnetic field.

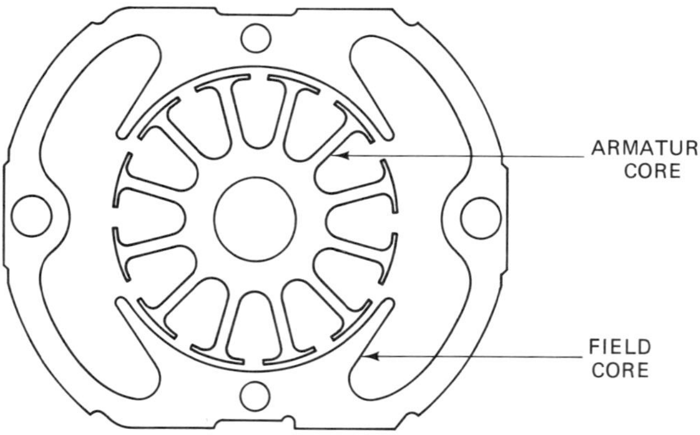

Figure 2-25 Relationship of armature and field coil cores.

The field coils are held in position against the inner periphery of a yoke or frame (Figure 2-26). The armature fits into the open space with a minimum amount of separation between it and the field coils. The yoke not only supplies mechanical support for the field coils but is part of their magnetic circuit. Without the yoke, the magnetic lines of flux would need to pass through the much higher reluctance of an air path. Because the yoke helps reduce the overall reluctance of the magnetic path of the field coils, their magnetic strength is increased.

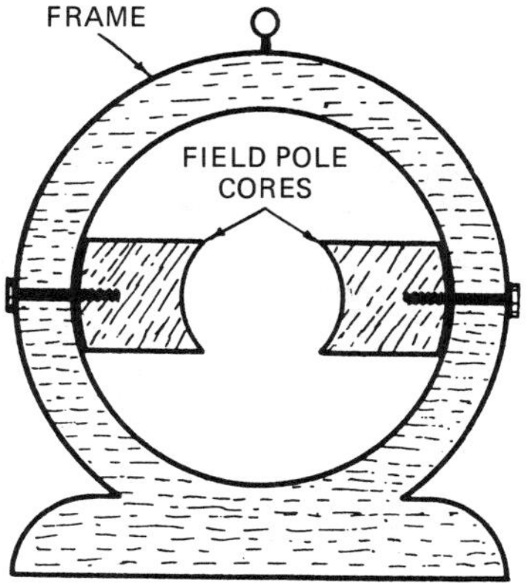

Figure 2-26 The field core is fastened to the motor frame.

Field Coil Insulation

Both the stator slots and the field coils are insulated. The insulation for the slots not only insulates the coil windings from the slots but supplies mechanical protection for the wires at the same time. Materials for slot insulation include substances such as black or yellow varnished cambric, varnished silk, and fish paper. Tapes for insulating wire splices include cloth and plastic types of friction tape and rubber tape. The wires of the coils can be insulated with cotton, linen, silk, or glass impregnated cloth. Insulating compounds include air-dry shellac, which has the advantage of not requiring baking. Varnishes that require baking include clear and black types. Field coils are often wound with cotton covered enameled wire. This is a double type of insulation supplied by both the cotton and the enamel. There are various substances, like Formvar, that resemble enamel but are extremely hard and much less subject to flaking.

Figure 2-27 shows the laminated core assembly. It is equipped with a

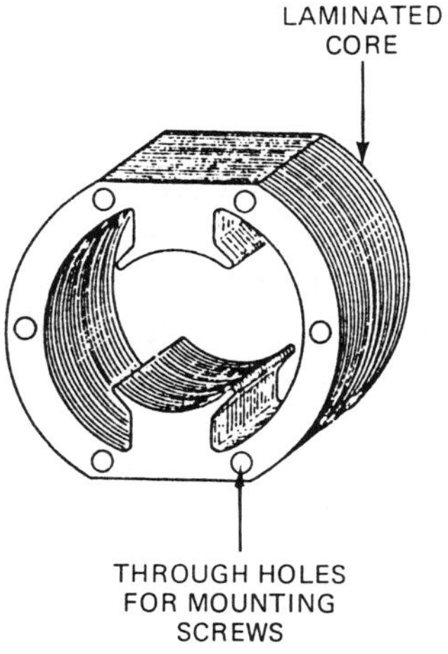

Figure 2-27 Laminated core assembly for field coils.

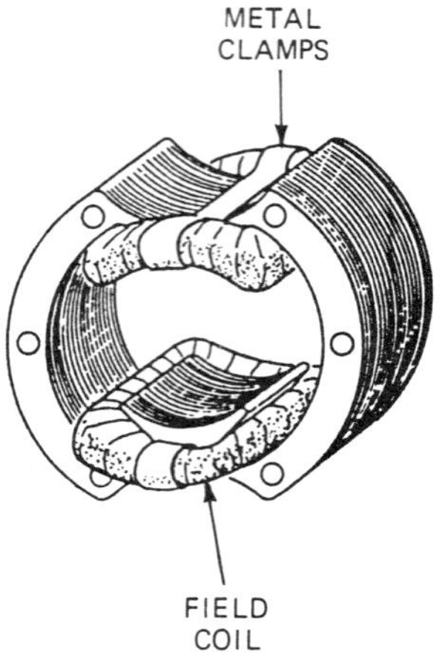

Figure 2-28 Field coils can be held in place by metal clamps.

number of through holes so that the assembly can be held together by bolts. Figure 2-28 illustrates a method of holding the field coils in place using metal clamps.

Still another method, shown in Figure 2-29, is to use a metal pin and have it pass through a hole extending through the full length of the laminated field coil core.

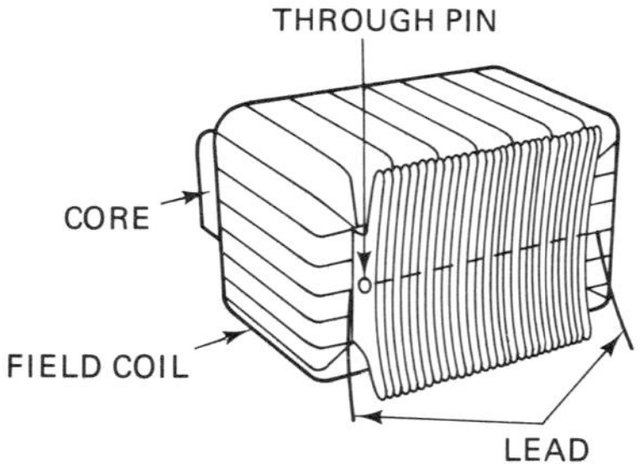

Figure 2-29 Through pin method of holding the field coil in place.

Field Coil Shoe

To bring the magnetic fields produced by the field coils as close to the rotating armature as possible, the ends of the field piece are fitted with soft iron extensions whose ends facing the armature are curved, forming a shape identical to that of the outline of the outer circumference of the armature (Figure 2-30).

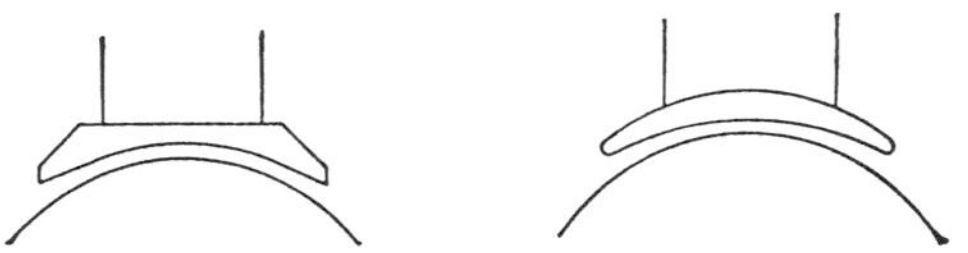

Figure 2-30 Types of shoes used as field pole pieces. The one at the left is chamfered. Both types follow the curvature of the armature closely.

The magnetic field produced by the field coil depends on its core material, the number of turns of wire, the proximity of these turns to each other, the amount of current flowing through the coils, and the permeability of the pole pieces. Field coils, depending on the design of the motor, can consist of a few turns of fairly heavy wire or many turns of fine wire.

The magnetic flux produced by the field coils can be considered as though it consisted of two parts. One part is the flux that cuts across the coils of the armature; the other does not reach the armature, that is, it does not cross the air gap between the field coils and the armature. It is known as leakage flux. The total flux produced by the field coils is the sum of the leakage flux and the armature coil cutting flux.

The basic motor relies on the reaction between a pair of magnets; the simplest is the permanent magnet (PM) motor used to supply a magnetic field. The PM motor is a DC type and is generally used where fractional horsepower is required. The field permanent magnet may be a ceramic type, desirable since it produces a strong magnetic field for its size. Because of its compactness, it is used in automobiles for operating windshield wipers, power windows, power seats, and so on.

BRUSHES

Brushes are the physical link between a voltage source and the commutator bars and are necessary to be able to deliver current to a moving element, the rotor. Brushes are supported in a holder which can be spring loaded to hold the brush firmly against the commutator segments.

Brushes are made from a number of materials including carbon, a mixture of carbon and graphite, electrographite, and metal graphite. These various kinds of brushes have different characteristics which make them suitable for particular applications.

Carbon Brushes

Carbon brushes are made of hard carbon and have relatively high abrasiveness. Their specific resistance is high as is their contact drop. Contact drop is the IR (current multiplied by resistance) drop or voltage loss at the area where the brush is positioned against the commutator. The voltage applied to the rotor is equal to the source voltage minus the voltage drop across the brushes. This voltage drop is the sum of the voltage loss across both brushes. The carbon brush has a low current carrying capability. From an advantage viewpoint they are able to tolerate substantial shock, vibration, and current reversal. They are also noted for their ability to keep the commutator clean.

Graphite Brushes

Graphite is a softer form of carbon and is the writing material used in lead pencils. Its specific resistance is low and its abrasiveness is less than that of carbon brushes. It has excellent lubricating properties. Its IR drop (voltage loss) is moderate, and it is often used for motors having a high rotational speed.

Metal Graphite Brushes

The metal graphite brush consists of a mixture of carbon and copper. It has low specific resistance, low hardness, and its IR drop is also low. Because of its copper content, it is characterized by having a high current carrying capability. This type of brush is often used with low voltage motors because these usually require large amounts of armature current.

Electrographitic Brushes

Electrographitic brushes are carbon brushes having a high content of graphite. They are rated as having a medium to high specific resistance, current carrying ability, and hardness. Because of their graphite content they have good lubricating qualities and so are suitable for high speed motors.

Brush Voltage

Brush voltage is the amount of voltage measured across brush pairs and is characterized as low, normal, or high. A source ranging from 6 to 32 volts is regarded as low; normal when the source is rated at 100 to 130 volts, and high when it is between 200 to 250 volts. These voltage ratings do not overlap since voltages between these values are not often used.

Brush Hardness vs. Abrasiveness

Hardness is not the same as abrasiveness. A soft brush can be abrasive. The higher the abrasiveness of a brush, the greater the wear on the commutator. Abrasiveness is sometimes regarded as a completely unwanted brush characteristic, yet it is necessary for two reasons: to keep the commutator segments clean and polished and to keep the surface of the brush clean.

Brush Position

The brush may be a rectangular block with one of the shorter sides of that block resting directly against the commutator bars. This side may have a slight curvature to make a snug fit across the commutators (Figure 2-31). In

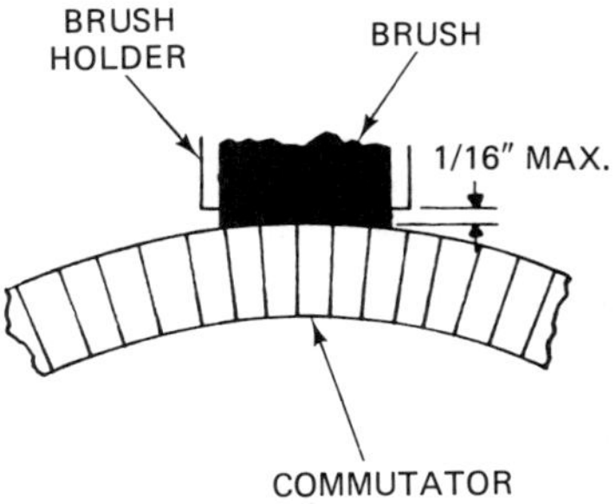

Figure 2-31 The brush extends $^1/_{16}$-inch or less from its holder.

some installations the brush faces the commutators at an angle and the brush is cut so that its face is flush against the commutator bars.

The brush holder should hold the brush almost completely, except for about 1/16 inch. This gives the brush as much support as possible. If the brush does not have this support, a corner of it may become broken.

In inexpensive motors the brush holder is simply a small rectangular metal block integrated with the brush. When the brush wears away, the entire assembly must be replaced. The brush is not movable. In most motors, though, the worn brush can be taken out of its holder and replaced.

Brushes can be set at either a trailing angle with respect to the direction of rotation of the armature or radially. Figure 2-32 shows how the brush is held in position by the holder. In turn, the brush holder is held in place by an adjustable brush stud. The brush holders simply work as brush guides and permit the brush to slide freely across any irregularities in the commu-

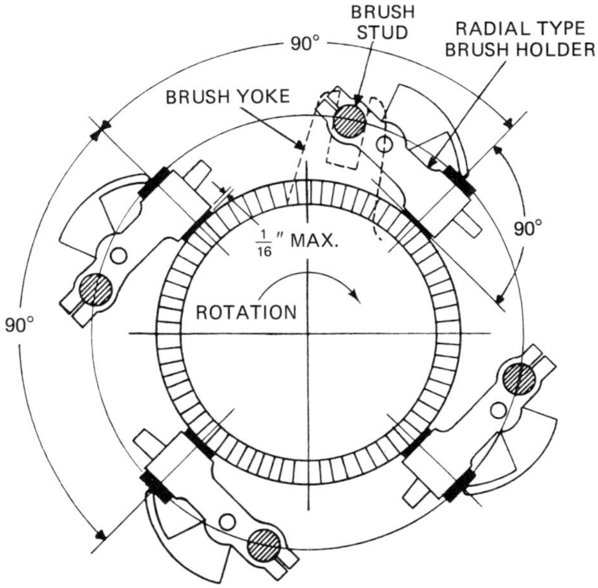

Figure 2-32 Assembly using four brushes.

tator surface. The brushes are held against the commutator by springs. These do not carry any current but help prevent brush vibration, a possible cause of sparking.

It might be thought that having a strong spring pressure against the brushes would be the most desirable arrangement. Usually the tension of the springs is adjusted until the pressure of the brushes against the commutator is 1.5 to 2 pounds per square inch of contact area. If the spring pressure is increased above this the amount of contact resistance is not lowered appreciably. However, sliding friction does go up, increasing the heating of the commutator. Since the commutator is made of copper and has a positive temperature coefficient of resistance, the I^2R loss does go up.

That part of the brush in contact with the commutator is identified by heel and toe. The toe of the brush is that part that makes the initial contact with the commutator as it rotates, that is, the rotation of the commutator must be against the toe of the brush. The heel of the brush is the last part to contact the commutator.

BEARINGS

The shaft of a motor supports the laminations of the armature, the coils positioned in the slots of that armature, and the commutator. While the brushes are held in a brush holder, they do press against the commutator and this pressure is also supported by the shaft. To avoid shaft eccentricity, all of these objects must be balanced so that the shaft has uniform pressure on it around its circumference. Thus, the brush holders must be equally spaced around the commutator.

The shaft rests on end bells which support bearings. The purpose of the bearings (Figure 2-33) is to support the shaft and to do so with minimum friction. There are various kinds of bearings: ball, roller, sleeve. It is essential for correct lubricants to be used, for these actually become part of the motor system. The difference between the outside diameter of the shaft and the inside diameter of the bearing is between 0.0002 inch and 0.001 inch and the shaft clearance ranges from 0.0001 inch to 0.0005 inch. These dimensions do not hold for all motors but give an indication of the tolerance tightness involved. The shaft, all the components mounted on it, and all components pressing against it must be supported by the film of oil in the space between the shaft and its bearings.

Bearing oil cannot be selected haphazardly and should be the type recommended by the motor manufacturer. Sleeve bearings, for example, use pure mineral oil only. One factor concerning oil not commonly recognized is that the correct oil can be more economical than lower priced substitutes since it will not require as frequent replacement. Two of the factors determining the oil to use are the operating temperature and ambient temperature. Oil viscosity is

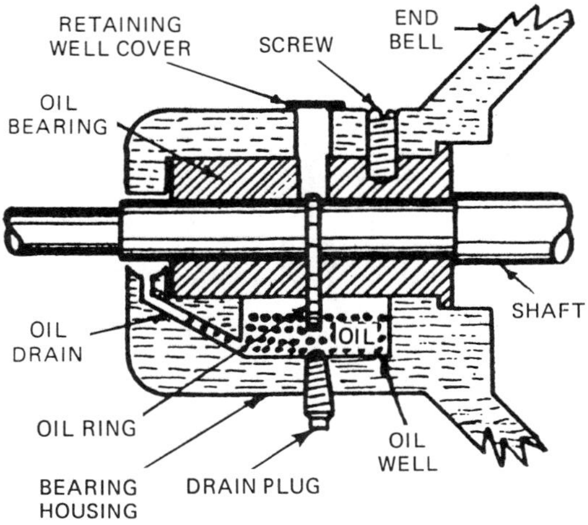

Figure 2-33 Cross-section of bearing.

another consideration. An oil that is too heavy will impose a burden on the motor and will lower its efficiency. Further, it may hinder motor starting.

The oil must not only be used to minimize friction to prevent unnecessary wear, but should prevent rust and supply cooling. This latter factor is important when the shaft and its bearings are made of different metals, such as steel for the shaft and bronze for the bearings. These have different temperature coefficients of expansion. If the bearing cools faster than the shaft it will contract more rapidly, possibly resulting in a seizing condition.

MOTOR SHAFT

The motor shaft, made of hardened steel, may be available in various lengths, diameters, and shapes. Three of the more commonly used shafts are shown in Figure 2-34. Figure 2-34A is a round type, 2-34B a flatted round, and 2-34C a knurled round.

The letters adjacent to the shafts in these drawings indicate dimensional specifications. A, for example, is the shaft diameter. B is the free shaft length; C is the width across the top of the shaft; D is the flat length; E is the length of the knurled portion for those shafts so manufactured; F is the knurl setback; G is the bearing outside diameter; and H is the bearing height.

The Source Voltage

The DC source voltage can be supplied in three ways. The first is by a battery, and this can be by a single unit, batteries in series, in parallel, or in a series-parallel arrangement. Another method is to connect the motor to a 110-volt DC power line supplied by a DC voltage generator or by an electronic

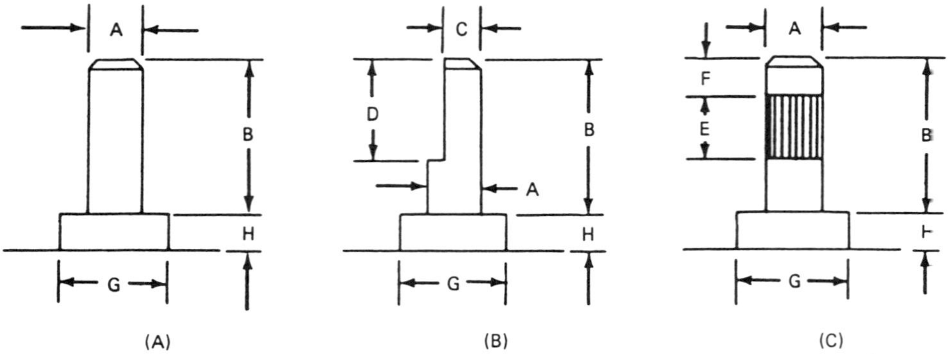

Figure 2-34 Representative motor shafts: round; (A), flatted round (B), knurled round (C). (Courtesy Cramer Company)

power supply. In all instances the DC voltage sources must be capable of meeting the current and voltage requirements of the motor.

All of these methods come equipped with problems. With a single battery, the internal resistance of the unit increases as the battery is used over a period of time. Because current flows through it, the battery develops an internal voltage drop. In effect this reduces the amount of current available across its terminals when the battery is under load. Further, the internal resistance can reduce the amount of current available for the armature coils. The problem can be eliminated by recharging the battery or by replacing it with a new unit.

The problem of internal battery resistance is also present when battery combinations are used, whether in series, in parallel, or in series-parallel. In any combination of batteries, all the other units will first try to recharge the weakest, thus reducing available voltage and current.

If the source voltage is supplied by an electronic power supply, the regulation of that supply must be considered. The regulation is a measure of how much its output voltage will drop when a load is applied, and in this case the load is the motor.

MOTOR FRAMES

A motor frame (Figure 2-35), also called an enclosure, has several functions. It is a support for all the motor parts, laminations, coil windings, bearing holders, end bells, and so on. This is its mechanical characteristic. But it has others as well. It protects the motor against dust, dirt, corrosive elements in the atmosphere, and a possibly hostile environment. Its large exterior surface works as a heat sink helping keep the interior temperature rise within working limits. It also has electrical properties, restricting the radiation of noise signals generated by the motor. At the same time, it limits the pickup of external signals. The frame contributes to the inductance of coils since it adds to the core sub-

Figure 2-35 Motor frame with field coils in position.

stance of coils. It also acts as a base for the support of the name plate, possibly an on/off switch, and in some cases, for a capacitor.

There are a number of different kinds of motor frames; each is intended for a specific purpose.

The open frame. This enclosure has large openings to permit the escape of substantial amounts of heat when the frame, working as a heat sink, is incapable of doing so. It may be associated with a blower that supplies a draft of air that can reach heat-generating rotating parts and is useful for motors that must be operated on a continuous rather than an intermittent basis. Continuous operation can result in steady heat buildup.

Totally enclosed frame. This is the opposite of the open type motor, for the frame encloses the motor completely. This does not mean that circulating air has no access to the motor but simply that the motor is being protected against a hostile environment. This can include dust, dirt, and chemical fumes. The frame may be cooled by an external blower.

Totally enclosed frames can be used by large motors as well as miniature and microminiature types. In small sizes, such as motors used in toys, the enclosure is intended as protection of fingers.

The protected frame. Some motors may be equipped with vents to allow for air circulation; these vents covered by metallic screening. The function of the screening is to help guard against accidental contact with the motor's rotor. This frame is called semienclosed.

The drip proof frame. This type of frame may be used in installations where the possibility of liquid drip exists and in which a frame ventilated motor must be used. It does not offer complete protection against liquids arriving at some angle instead of completely vertically. A drip proof frame is built so that liquid falling on the motor at any angle at or less than 15° off vertical cannot enter the motor.

The splash proof frame. A splash proof frame is one in which liquids or solid particles falling vertically or at an angle of 100 degrees or less from the vertical cannot enter the motor.

The dust tight frame. This frame is constructed so that the motor is completely enclosed, not permitting dust to enter from any angle.

Watertight frame. This frame encloses the motor completely and is constructed so that water applied in a stream from a hose cannot enter.

Explosion proof frame. This frame is a variation of the totally enclosed type. The frame is constructed so that it can withstand an internal explosion. It is intended to prevent it from igniting gas or dust in the motor's environment.

Frame variations. The motor frame may have its outside ribbed for greater heat dissipation.

Vented: Vents are on the front of the motor and is the exit portion for the motor shaft.

Internal Fan Cooled: These motors have vents around the edges of the outer covering.

Cooling Band: This is a completely shielded frame with a heat sink enclosing the motor on the outside circular part of the frame. It may consist of a ribbed structure or metal strips forming bands.

Totally Enclosed, Fan Cooled: This consists of a circular frame but with the frame separated a short distance from its circular end plates.

Open Construction: The motor is completely exposed and does not have any enclosure.

Motor Frame with Conduit Box: This frame is designed to accommodate a conduit box for electrical connections.

Totally Enclosed: While this motor is enclosed, it has provisions for side mounted electronics.

The motor frame may be equipped with a rigid base, mounting lugs, resilient mount, weld bolts, belly bands, face mounting nuts, extruded and tapped holes, and through bolts.

SUPERCONDUCTIVITY FOR MOTORS

Motors, especially those rated at 1 HP or more, require large amounts of current. An industrial, 2 HP motor will need about one and a half kilowatts of power, and with an input of 110 volts, single phase, this could mean a current demand of 13.6 amperes. Even a small amount of resistance in the armature and field coil windings results in a substantial energy loss, expressed as I^2R, manifesting itself as heat.

To eliminate this loss and produce more energy efficient motors, experiments are being made with superconductive materials in place of copper for the armature and field coil windings.

Materials lose their resistance with a decrease in temperature, but they do so at varying rates. At absolute zero resistance disappears. The absolute zero of temperature is taken as the start of the absolute or Kelvin scale, but on the Celsius scale, absolute zero is approximately −273 degrees or 0 Kelvin (K).

Certain substances that reach an extremely low value of resistance, or practically zero resistance, at some temperature above absolute zero are known as superconductors. For example, various ceramic compounds have shown superconducting characteristics at temperatures above zero Celsius (C). The ceramic involved contains oxygen, copper, barium, and the rare earth yttrium. Later a compound of oxygen, copper, barium, and lanthanum (a metallic element) was tried. A claim was made that the combination became superconducting at 35 K.

Superconductivity was discovered in 1911 by a Dutch physicist, Heike Kamerlingh Onnes. He cooled a wire made of mercury, liquid at room temperature, to 452 degrees below zero Fahrenheit, about 4 degrees warmer than absolute zero, corresponding to about −269 degrees C. He called the temperature at which superconductivity occurred the critical temperature.

If made cold enough, metals such as aluminum, zinc, tin, and uranium show superconducting properties. Alloys of the metal niobium are being used to make superconducting magnets, with the possibility that these could be used in motors. To achieve superconductivity liquid helium is used, which can be chilled to about −452 degrees F. Liquid nitrogen can be cooled to −321 degrees F, a much more practical cooling approach. In the meantime, many other elements and compounds are being tried, including calcium and lutetium.

3

DC MOTORS

THE MOTOR

While this chapter has DC motors as its subject heading, it is essential to remember that all motors are AC, that is, they depend for their operation on the reaction between permanent magnetic fields and magnetic fields having a regularly reversing polarity. The description DC, as applied to motors, simply means that the input operating voltage is DC, but through the mechanical switching action of the commutator, is changed to AC. This is also applicable in the case of brushless DC motors except that the change is accomplished electronically.

Advantages and Disadvantages of DC Motors

Some types of DC motors have excellent armature rotational speed regulation, despite the imposition of varying loads. Further, their rotational speed can be governed by electronic methods, and their torque, the strength of rotation of the shaft, is also constant, despite controlled variations in speed.

Still another advantage of the DC motor is the ease with which its speed can be controlled. The speed can be governed by current flow through the armature and the field coils, and this can be adjusted by the DC voltage input Thus, when variable rotational shaft output is needed, possibly in some manufacturing process, DC motors are often preferred. Motor speed control can also be governed electronically, lending itself well to automatic control.

When DC motors are used in subway cars or electric trains they can be used for dynamic braking. Since the DC motor can be used either as a motor or as an electric generator, trains and subways can take advantage of this characteristic for braking. When going downhill or coming into a station, the motor, working as a generator, will put power back into the DC line and in so doing will slow down the train, in effect, braking.

One disadvantage of the DC motor is not so much due to its design and construction as it is to the fact that AC is so much more readily available. Homes and industry are equipped with AC supplied by power lines. AC's advantage is that it is readily transportable and over long distances results in much lower losses than DC. AC can be stepped up or down by power transformers and so high-potential lines of 33,000 and 66,000 volts are not uncommon. The step-up action means that less current is transported, resulting in a much lower power loss in the lines.

Another disadvantage of the DC motor is its tendency to "run away," that is, to reach an uncontrollable speed if the field winding should happen to open. The speed could be such that the motor would destroy itself.

During the time it is at work, the armature builds up a substantial magnetic field. If the switch controlling input power to the motor is suddenly turned off, the collapsing magnetic field is capable of inducing an extremely high voltage across both the field and armature coil windings. This voltage can cause extreme sparking across the brushes, possibly damaging them or the commutator segments or the field coil or armature coil windings.

Finally, DC motors require more operational maintenance than most AC motors.

MOTOR CLASSIFICATION

Motors can be classified in a number of ways and one of these, classification by type such as universal, synchronous, series wound, and shunt wound, was supplied in Chapter 2. It is also possible to select a specific characteristic of motors and use that as the basis for forming a comparative analysis.

Classification by Speed

Constant speed. This is a motor whose speed varies very little with changes in load.

Varying speed. With this type of motor the speed is dependent on load conditions. As the load increase, motor speed decreases, and vice versa.

Adjustable constant speed. This motor is equipped with a speed control, either one that is smoothly variable or a click-stop type. Speed can be selected over a fairly wide range, but once a wanted speed has been selected it will remain constant despite load variations.

Adjustable variable speed. Like the adjustable constant speed type this motor is also equipped with some form of speed control. Once the desired speed is selected, it is constant only for the load at the time the choice is made. A change in the load will result in a change in motor speed.

Multispeed. A multispeed motor is one that is equipped with a control to permit selection of previously determined speeds.

There are a number of other ways of classifying motors and the resulting listing may be separate and independent of any other or may be combined with one or more.

Classification by Construction

Size or weight. This is actually a determination of horsepower output. As a general rule (but not always) the smaller the motor, the less horsepower it can supply. In terms of size it can be designated as flea, mini, micro, regular, standard, and so on, with the description selected by the manufacturer. In terms of horsepower a motor can be rated as fractional (less than 1 horsepower) or integral (1 horsepower or more).

Usage. Motors can be categorized by their function, such as sewing machine, washing machine, mixer, domestic use, commercial use.

Frame. A motor is sometimes described by the kind of frame it uses such as totally enclosed, drip proof, explosion proof, open, or dust proof. It may or may not be air cooled.

Mounting. The motor may be enclosed in the casing, as, for example, oil burner motors, vacuum cleaner motors, and sewing machine motors. For commercial use the motor may be designed for floor, wall, or ceiling use with the motor frame not enclosed.

Classification by Characteristics

Motors can also be classified by specific mechanical, electrical, or electronic characteristics. These could include starting torque, running torque, normal speed, speed regulation, input voltage, input frequency, operating temperature, type of running control, weight, cost compared to other motors, efficiency, starting current, running current, type of input, whether AC or DC or both, ease of mounting, and availability of a gear train or pulley system. A set of gears may or may not be included in the motor housing.

DC vs. AC MOTORS

Since all homes are wired for AC it is not surprising to find substantial numbers and varieties of motors using this as the power source. However, because of technical advances in battery design the use of DC motors has increased. DC motors, for example, are not only used in portable electronics equipment, but also in cars, trucks, buses, trains, planes, and so on.

Motor Power Variations

Generally, motors are AC or DC types, but there are some that are AC/DC, that is, capable of working from either type of power input. In some installations when the operating characteristics of a DC motor are wanted, a DC motor can be used with a motor power supply connected between the DC motor and the AC line. The power supply rectifies and filters the AC input, converting it to DC. And there are also some AC motors designed to work from two AC voltage inputs, 120 and 240.

THE RIGHT-HAND MOTOR RULE

The basic operation of all motors is very simple and is based on the application of the right-hand motor rule (Figure 3-1). In this drawing a section of wire carrying a direct current is positioned between the poles of a permanent magnet. The horizontal lines shown between the two poles represent the magnetic flux supplied by this magnet.

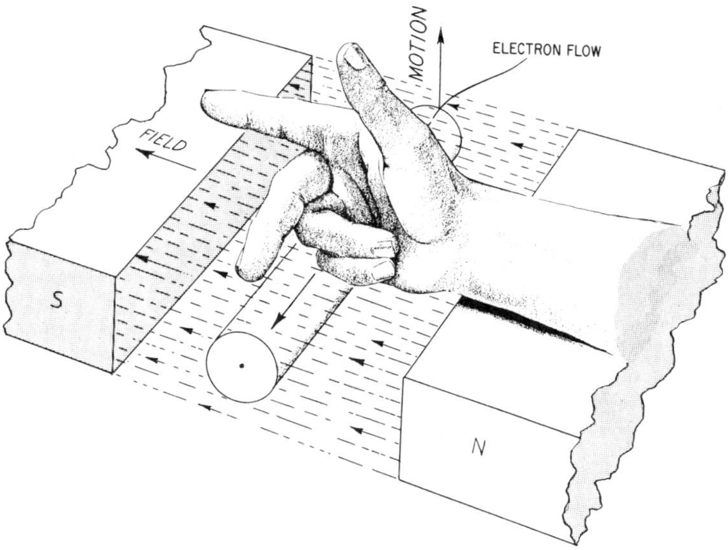

Figure 3-1 Right-hand motor rule.

As a result of this arrangement we now have two magnetic fields: one supplied by the permanent magnet and the other by the conductor. These two fields interact in such a way that a force exists that pushes the conductor out of the field of the permanent magnet.

The right hand shown near the magnet has three fingers pointing in a different direction. One finger shows the direction of current flow through the conductor; this is from the minus terminal of the battery, through the conductor, to the battery's positive terminal. The finger pointing straight up illustrates the direction of movement of the conductor. This is also shown by an arrow positioned directly above the conductor. Finally, the forefinger shows the direction of the magnetic field of the permanent magnet.

In an actual motor, the wire that is pushed upward would be immediately replaced with another wire and so the motor action would be continued indefinitely.

Figure 3-2 shows the resultant motor behavior derived from the interaction of a pair of magnetic fields. The dark circle in the center represents a cross section of a current carrying conductor. The circular lines around this conductor are the magnetic lines of flux produced by this current.

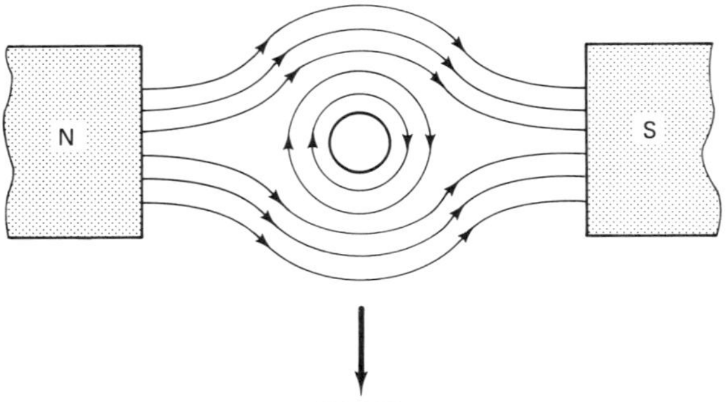

Figure 3-2 Interaction between magnetic fields results in motor action.

The permanent magnet identified by its poles, N and S, has a magnetic field which interacts with the magnetic field of the centrally located conductor. Above the conductor, all the magnetic lines reinforce each other and so the overall magnetic field becomes stronger.

An opposite situation prevails in the region below the conductor. Here the lines of magnetic flux oppose each other, indicated by the arrows pointing in opposite directions. As a result, the magnetic field below the conductor is weakened.

With a strengthened magnetic field above the conductor and a weakened magnetic field below it, there is a resultant force pushing the conductor in a downward direction.

THE SERIES-WOUND DC MOTOR

The basic series-wound motor consists of a field magnet wired in series with the armature (Figure 3-3).

Since the armature and the series field have very low resistances, most of the voltage drop appears across the variable series resistor R. As the amount of resistance is reduced, more current flows through the armature and it develops a counter EMF. Finally, the series resistance is made practically zero and the full current is available to the armature.

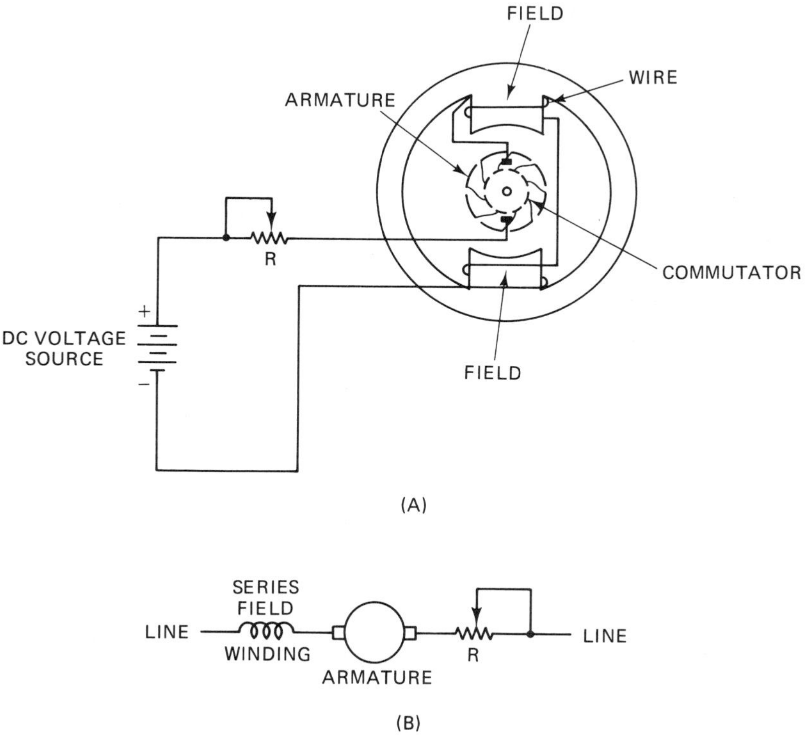

Figure 3-3 Series-wound DC motor: pictorial diagram (A) and schematic diagram (B).

Characteristics of the Series DC Motor

This motor is load sensitive. The rotational speed of the armature will be less with a heavy load and will run faster as the load is reduced. The variable resistor in series with the armature can be used to control motor speed. The greater the amount of resistance, the slower the speed. It is possible for the armature to race excessively if the load is removed.

Other characteristics include quick starting and maximum torque at low rotational speeds. The reason for high torque is that low rotational speed means a smaller counter EMF and consequently much greater armature current. As the load is decreased, the armature increases its speed. But when the load is reduced the magnetic flux is also decreased, and therefore the armature must increase its speed in order to build up its counter EMF. The motor then begins to race until shaft and bearing friction and wind resistance form an internal load that may be sufficient to slow the armature or at least to keep its speed from increasing further. If this action is not adequate, it is possible for the series motor to destroy itself.

Since the series DC motor has a high starting torque it finds application for use in elevators, cranes, hoists, and large blowers for equipment that requires heat dissipation. Such motors are also desirable for subway, trolley car, and railway operation. It finds application in aircraft and in automobiles for engine starting, and in aircraft for raising and lowering landing gear and operating cowl and wing flaps. In general, series DC motors are used when a high starting torque is required and the load is a heavy one. The shaft output of this motor is usually connected to its operating load by a gear train.

Although the load imposed on this motor can vary, requiring substantial changes in operating current, the DC source is such that its output voltage remains constant, that is, the source has excellent voltage regulation.

Effect of a Short Circuit in the Armature

If one or more coil turns of the armature winding should become shorted, there could be a variety of effects, including excessive sparking between the brushes and the commutator or an increase in the overall temperature of the motor with the motor laboring under its normal load. The most significant result would be excessive current flow in the shorted winding, resulting in a possible blowing of the line fuse or circuit breaker. There would be a substantial magnetic field surrounding the short-circuit current with this field working in opposition to the magnetic field of the armature. The result would not only be a slowdown of the armature but a reduction in torque.

Reversing the Direction of a Series DC Motor

There are two ways of reversing the direction of rotation of the armature of a series motor. The first is to reverse the polarity of the connections from the DC voltage source to the armature, thus reversing the direction of current flow through the armature coils. The effect of this will be to change the polarity of the magnetic field around those coils.

The other technique is to reverse the connections to the winding of the series field coil. This can be done, as shown in Figure 3-4, through the use of a double-pole double-throw (dpdt) switch. Since the field and armature coils

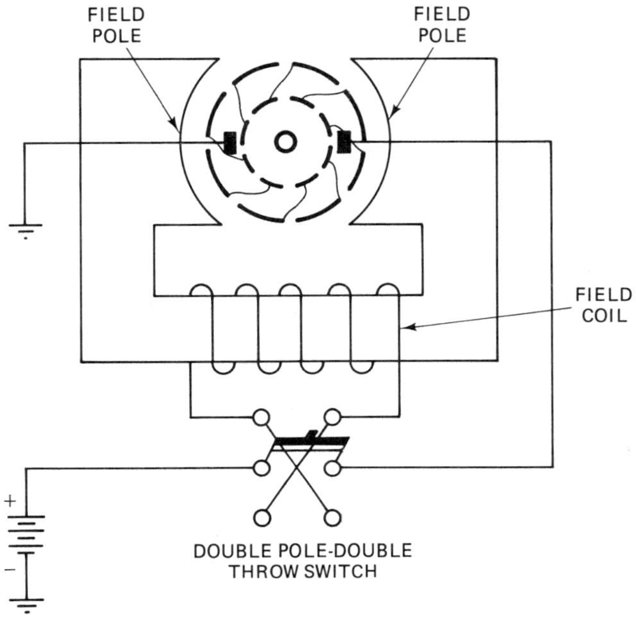

Figure 3-4 Circuit for reversing armature rotation for a series-wound motor.

are in series, current reversal means a change in the direction of current through both field and armature.

Starting the DC Motor

The circuit in Figure 3-5 shows two resistors and these are wirewound units. One of these, in series with the armature and field coils, is a rheostat, while the other is a fixed value resistor in shunt with the series field windings.

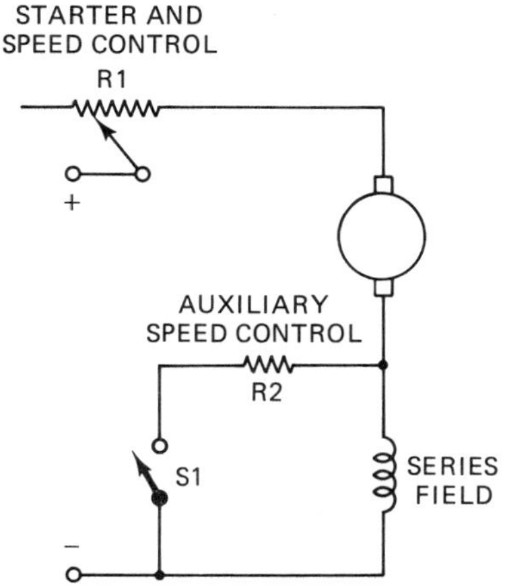

Figure 3-5 Series-wound motor equipped with main and auxiliary speed controls.

A single resistor, such as R1, could be used alone, but the combination of R1 and R2 does provide for smoother control. R1 has the entire armature and field current flowing through it, while R2 works as a bleeder. It carries a smaller current load than R1. R2, the auxiliary speed control, is connected to a switch, S1, and so can be inserted or removed from the circuit as required. While R2 is shown as a fixed resistor type, it can also be a variable unit.

A starting box (Figure 3-6) is used in some industrial applications. The switch, connected to the DC voltage source, is a double-pole double-throw (dpdt) type and is followed by a pair of fuses, one in the minus line, the other in the plus line. These fuses can be cartridge types but must be rated at least at the full current capabilities of the armature. The starting box has a pair of terminals for external connections. One of these, marked L, is connected to one side of the DC voltage source through one of the fuses and one blade of the dpst switch. The other terminal, marked A, is connected to one lead of the armature winding.

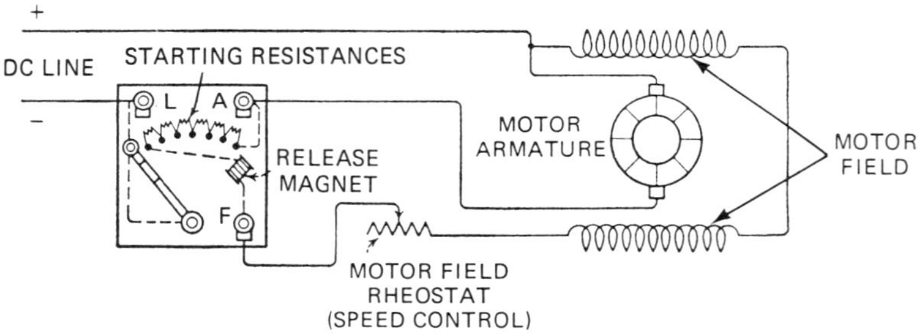

Figure 3-6 Series-wound DC motor starting box.

Behind the front panel are a number of high wattage fixed resistors. When the handle (H) is pushed toward the right it will make contact with resistors in the series chain. As the handle is pushed forward, less and less resistance is used, until the handle finally makes contact with magnet M. The field around this magnet is strong enough to hold the handle in position, and then the maximum current flows through the armature, with the current limited only by the counter EMF developed by the armature, its DC resistance, and the resistance of the brushes.

If, for any reason, current flow through the armature should decrease substantially, the electromagnet would no longer be able to hold the handle H in position and it would return to its stop position automatically. Shown in the drawing is a strong spring which would return the handle to its stop point. Similar types of starting boxes can also be used with other types of DC motors, such as the shunt and compound.

Because the starting box governs the armature current, it works as a speed control, but its basic function is that of giving the series DC motor an opportunity to develop its counter EMF. The starting box is a speed control in the sense that it controls motor rotation only until the armature reaches its operating speed. Once the handle is held in place magnetically, the speed is controlled by the load, and if the load is variable, so will be the speed.

The starting unit in Figure 3-6 may be a separate box mounted near the motor, or it can be part of a motor system. In a trolley or a train, for example, it may simply consist of a sliding arm used by the engineer with the resistor elements hidden below.

While the starting box protects the motor by limiting current until the counter EMF builds up, it offers no protection against current runaway. If anything, excessive current flow into the motor will make the holding magnet stronger, thus ensuring continued current buildup.

A way of overcoming this problem is to use an overload relay (Figure 3-7). With normal current flow, current passes from the line, through the coil of the overload relay. This current flow, however, is not strong enough to operate the relay. In the event of excessive current flow, the magnetic field of the coil will attract the shorting bar of the relay. This will put a direct short across the terminals of the holding coil, causing the magnetic field around the holding coil to drop to zero. The effect will be to release the slide arm which will drop back to its off position. This action will cut off all current input to the motor.

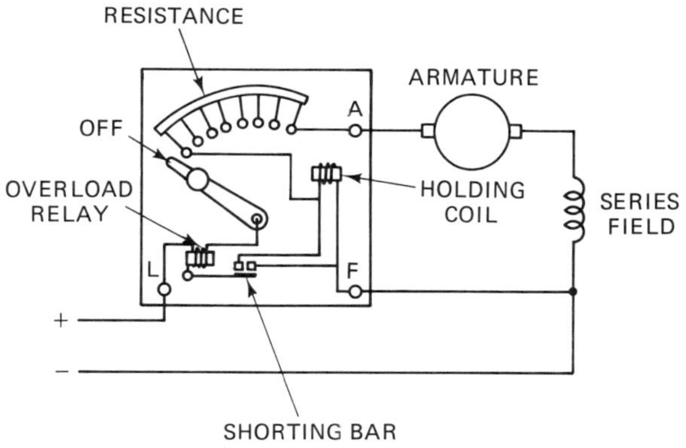

Figure 3-7 Protection against excessive motor current.

While the slide arm method of controlling a DC motor has been widely used, a number of alternative techniques are available that consist of pushbuttons for starting and stopping a motor. Depressing an "on" button removes resistors in series with the DC power line and does so progressively, that is, one resistor after another is removed from its series connection in the DC power line. This is done by magnetic switches, also called contactors. When the "off" button is pushed, the resistors are automatically reinserted into the DC power line. The basic idea of using resistors to limit armature current remains the same. All that has changed is the method of inserting or removing those resistors. There are a number of techniques for doing this.

The manual starting box has several disadvantages. The movement of the handle takes the starting resistors out of the circuit one at a time, but if done too rapidly can result in excessive starting current with possible damage to the field coils and armature. On the other hand, if it is done too slowly it can produce starting resistor burnout or excessive power consumption.

Automatic starters are better since motor current control is more accurate and is more essential for larger DC motors. The smaller DC motors can be started by direct connection to their voltage source.

There are various types of automatic starters, including the time-element starter, the circuit-element starter, and the CEMF starter. In the time-element starter the current-limiting resistors are removed from the circuit by time delay devices. The current-element starter is preferred for motors driving variable loads, where starting conditions vary and where the rate of acceleration with heavy loads is lower than that with light loads. There are various arrangements of current-element starters, but one type uses series relays for removing the series resistors from the starting circuit. Still another starter is the counter EMF. With this starter the time taken for the removal of resistors from the starting circuit is controlled by the time it takes for the counter EMF to rise to some predetermined amount. The starting time also depends on the load.

Figure 3-8 is the circuit of a time-element starter. The series starting resistance has two resistors. The first is cut out by relay switch 2R while both are removed by relay switch 3R.

This starter is activated by closing pushbutton C. When this is done the auxiliary switch, C1, is also closed since it is mechanically linked to C. This action energizes timing relay TR. The numbers 1 and 2 above TR are the contacts of this relay. These operate independently of each other. When contact 1

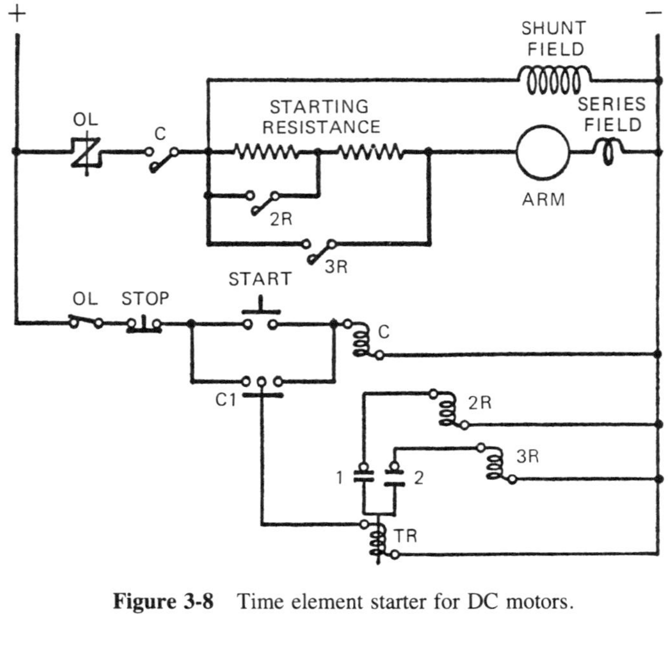

Figure 3-8 Time element starter for DC motors.

closes, coil 2R is energized, but this coil is part of a relay which closes switch 2R. As a result, the first starting resistor is cut out of the circuit and current flow to the armature and its field coils is increased. After a further time interval, contact 2 of the starting relay, TR, closes. Current flows through relay coil 3R, thus closing switch 3R. As a result, both starting resistors are shorted and the full line voltage is applied to the motor.

The pushbutton starters are sometimes called accelerating starters. In some instances the starting switches are closed by a small motor.

COMMUTATING POLES

Also known as interpoles, these are small auxiliary field poles located halfway between the field poles of a DC motor. The interpoles consist of a few turns of large diameter wire.

Their function is to reduce sparking at the commutator due to changing loads on the motor even though the brushes have been set at their neutral points. With the use of interpoles it is not necessary to move the brushes.

It is necessary for interpoles to have the correct polarity. Interpoles, as indicated in Figure 3-9, always have the same polarity as the armature pole that is adjacent. This is true regardless of the direction of rotation of the armature. Physically, commutating poles are much narrower than the main poles. Because of the presence of the commutating poles, there are three sources of magnetic fields in the air gap between the stator windings (the field coil windings) and the rotor windings (the armature windings) and the commutating poles. The effect of the commutating poles is to twist the field forward in proportion to the current demand on the DC voltage source.

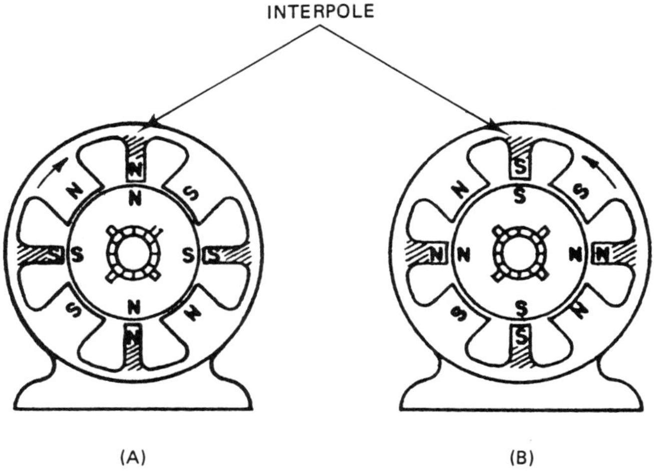

Figure 3-9 Polarity of interpoles (A) is reversed when direction of current flow is changed. Armature rotates in opposite direction (B).

The windings of the interpoles (Figure 3-9) are connected in series with the armature coils, and so the same current flows through both. Because of these two factors, the interpoles are properly regarded as part of the armature circuit. When the flow of current through the armature coils is reversed, its polarity is also reversed, but the same is true of the commutating poles. Hence, the polarity relationship of the commutating poles and the armature poles remains unchanged. The number of interpoles can be the same as the number of field poles. Alternatively, it is possible for the number of interpoles to be one-half the number of armature poles. Interpoles are not only used on DC motors, but on generators as well.

THE SPLIT-FIELD SERIES-WOUND MOTOR

The split-field series motor is a variation of the single-series field type. The split-field has a tap at the electrical center of the field coil, as shown in the diagram in Figure 3-10. Its advantage is that a change in the direction of armature rotation can be easily obtained through the use of a single-pole double-throw switch.

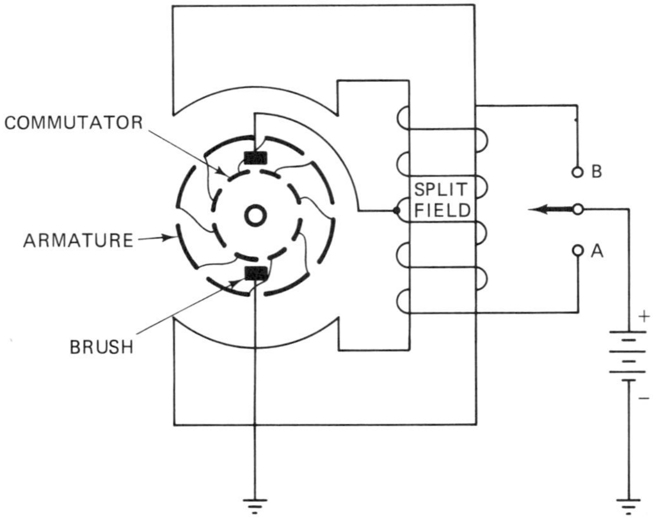

Figure 3-10 Split-field series-wound motor.

When the switch is in position A, only the lower half of the split winding is used. The current flow through this part of the field coil produces a magnetic field with a north pole at the bottom and a south pole at the top. The armature will then rotate in one direction.

When the arm of the switch is moved to position B, current will move through the upper half of the field coil. But the direction of flow of the current is the opposite of that when the switch was in position A and so the polarity of the upper coil is also opposite. In this position, current flows so that there is a magnetic field with a north pole at the top and a south pole at the bottom. This polarity change in the field coil results in a reverse rotation of the armature.

The use of a switch to reverse armature rotation was also described previously in connection with the series wound motor illustrated in Figure 3-4. When this switch is in its up position, current flows through the field coil in one direction; that direction changes when the switch is moved down. There is a difference in switches, for the one in Figure 3-8 is a spdt while that in Figure 3-4 is a dpdt.

THE SHUNT-WOUND DC MOTOR

In the shunt-wound DC motor (Figure 3-11), the field coil is connected in parallel with the armature, and so the entire voltage of the input DC source voltage is directly across it and the armature. The shunt winding, unlike that of the field coil in a series-wound motor, consists of many turns of small diameter wire. Also, unlike the series motor, the shunt-wound type has practically constant speed under widely varying load conditions, although the motor will slow slightly when the load is made quite heavy.

If the shunt motor is operated at full speed without a load, the rotating armature coils will develop a counter EMF almost equal to that of the line voltage, that is, the DC source voltage. As a result, the effective voltage, the difference between the DC source voltage and the counter EMF, will be quite small. If, at this time, a load is placed on the motor, the armature will rotate more slowly. However, as a result, the counter EMF will decrease and both the effective voltage applied to the armature and the armature current will increase. The torque supplied by the motor will also increase. The reverse is true if the load is lightened. As a result, the armature tends to run at a constant speed, following the variations in the load.

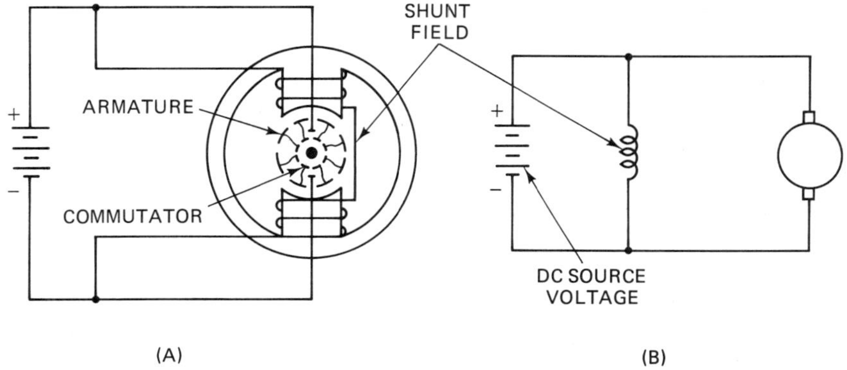

Figure 3-11 Shunt-wound motor: pictorial (A) and circuit diagram (B).

Motor Speed Control

The resistor, R, in the field circuit in Figure 3-12 can be adjusted to obtain speeds higher than normal. Speeds lower than normal can be reached by putting a resistor in series with the armature. In this arrangement, a variable resistance is put in series with the armature with the field in shunt with these two. The variable resistor must be able to carry the full current used by the armature. As the armature speed is lowered so is its counter EMF, but this can never be the same as that of the DC supply. The disadvantage of this method is that the power losses across the resistor, in the form of heat, can be substantial.

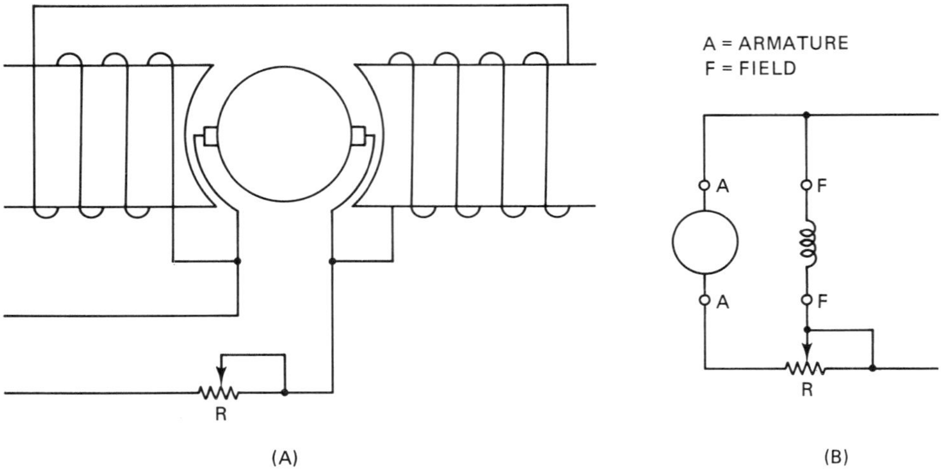

Figure 3-12 Rheostat (R) serves as speed control for shunt-wound motor: detail diagram (A) and circuit diagram (B).

The effect of a resistor in series with the shunt field is to reduce current flow through the winding, thereby lowering the flux density around the field coil. As an example, if the field strength is cut by 50 percent, the rotational speed of the armature will be doubled, but this is just an approximation. By using a pair of variable resistors (Fig. 3-13), the armature speed of the shunt-wound motor can be increased or decreased by using both armature and field coil variable resistors. While this method is simple, it can result in substantial power losses. However, the field current is small compared to that of the armature and so the two power losses are not the same. That of the resistor in series with the armature is much greater. The power loss varies as the square of the current (in amperes) multiplied by the ohmic value of the resistance ($P = I^2 R$). While the resistance of the rheostat in series with the armature is lower than that of the resistor in series with the field coil, it is the amount of current flow through the two paths that is the dominant factor in power loss.

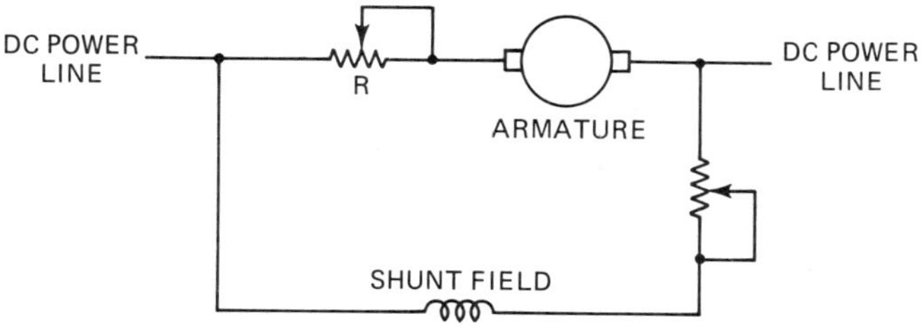

Figure 3-13 Alternate method of shunt-wound motor speed control.

Another method of controlling the speed of a DC shunt-wound motor, either increasing or decreasing the speed, is illustrated in the diagram in Figure 3-14. The change here is in the DC voltage source. Instead of using a constant voltage source, such as a battery supply, the DC voltage for the motor is supplied by a motor generator (G). A motor generator consists of a motor which can be connected to an AC voltage source. This motor drives a DC generator which, in turn, supplies the DC voltage to the shunt-wound motor. The output voltage of the DC generator portion of the motor generator can be governed by controlling its field voltage, and this can be done by using a variable resistor. While this setup does eliminate the need for external power resistors for shunt-wound motors and eliminates their power losses, there is an additional cost in the inclusion of the motor generator and the electrical energy required to operate it.

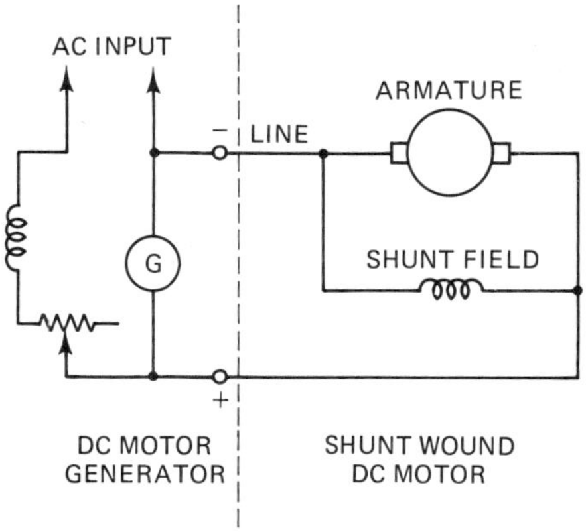

Figure 3-14 DC motor generator, a power source for a DC motor, can control the speed of a shunt-wound motor.

DC Source Voltages

The source voltage for DC motors can be supplied by a DC power line, although these are now quite rare, by batteries, by a motor generator, or by an AC power supply.

The AC Power Supply

The AC power supply (Figure 3-15) consists of an autotransformer, a rectifier for changing the input AC into pulsating DC, and a filter system for converting the pulsating DC into a much smoother DC.

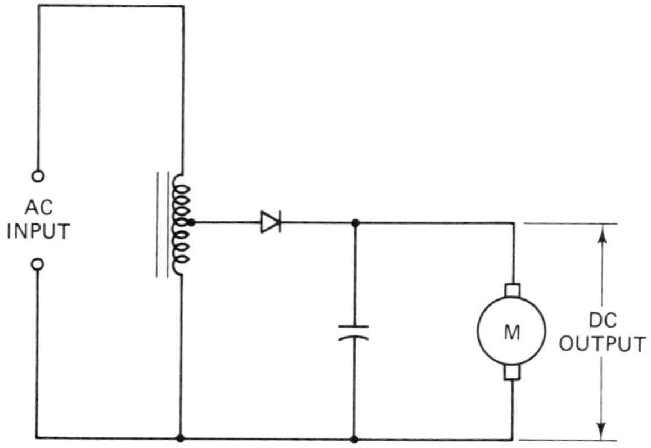

Figure 3-15 Autotransformer power supply for DC motor.

The output voltage of this power supply is controlled by the autotransformer which is connected to the AC power line. The autotransformer is a single winding type and can be a step-up or step-down type depending on the ratio of turns of the secondary to the primary winding. The transformer can use fixed taps on its turns or can be a variable unit.

Figure 3-16 shows another electronic method for supplying DC to the field and armature coils of a shunt-wound motor. The turns ratio of the power transformer determines the amount of DC voltage available at the ouput of the power supply, a full-wave rectifier type using power semiconductor diodes. The field coil is shunted across the armature with the amount of field coil current controlled by a field rheostat. The entire motor assembly is directly across the DC ouput of the power supply.

The amount of current available for the motor is determined by the wire gauge of the secondary winding of the transformer and the current handling capability of the power diodes, D1 and D2. The filter capacitor C is used to smooth the DC output of the power supply.

The field shunt coil is wound with a higher gauge wire (wire having a smaller diameter) than the comparable field coil in a series wound motor. Since it has more turns it produces a strong magnetic field even though its current requirements are more limited. This is an important factor in determining transformer size, since the flow of current in the secondary winding of the transformer is equal to the total current flow through the motor.

The armature and field coils for this motor, as well as for other motors, are all iron-core types. The electrical symbol for these coils is somewhat dif-

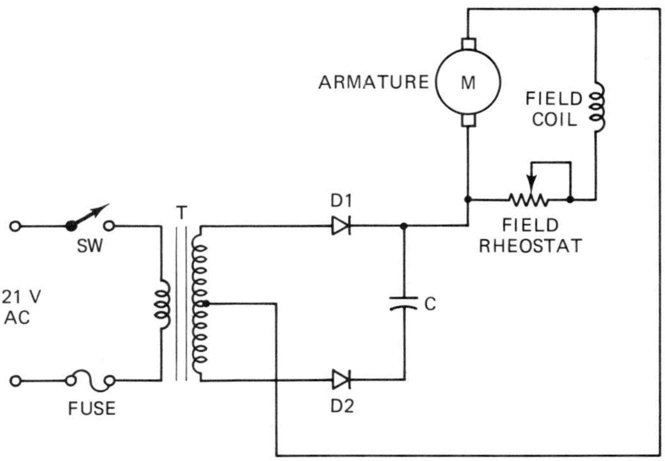

Figure 3-16 Full-wave AC power supply for a shunt-wound DC motor.

ferent than that used for coils in electronics, since the core symbol is missing. Note that the straight parallel lines for the power transformer represent an iron core, while the armature and field coils do not use such lines.

The speed of a shunt-wound motor remains fairly constant and will do so even if the load is removed, or it may increase very moderately. The starting torque of the shunt-wound motor is less than that of a series-wound type, assuming a comparison between equally sized motors. Under certain circumstances the armature of the shunt-wound motor can "run away." If, for example, the field winding should open with the motor operating without its load, the rotating armature would keep increasing its speed. Under these circumstances the line fuses or circuit breaker would open; otherwise the motor could self destruct.

With the opening of the shunt wound field coil, the counter EMF of the armature will disappear, and consequently, its speed will rise almost without limit. In normal operation the armature current is substantially limited by the counter EMF it develops because of the presence of the field coil's magnetic field.

When starting a shunt-wound motor it is important for the field winding to receive its excitation current first, for without the magnetic field around the field coil, as indicated in the preceding paragraph, the armature could quickly reach a dangerously high speed.

Although this motor is noted for its constant speed characteristic, it is sometimes used in applications where some speed variation is required, as in the case of machine tools. This is most often done by varying the current through the field coil. If, for example, resistance is added to the field coil circuit, the amount of current flow through that coil will be lessened, and the armature's speed will increase.

Problems Involving the Shunt-Wound DC Motor

Since the armature coils of a motor, when running, move through the flux lines supplied by the field coils, a voltage develops across those of the armature windings. This voltage, referred to as a counter EMF (CEMF), is directly in opposition to the voltage applied to the motor from its operating source. The actual operating voltage, then, is the applied voltage minus the counter EMF. It is this difference in voltage that is responsible for the amount of armature current.

Example

The resistance of the armature winding of a shunt-wound DC motor is measured at 0.04 ohm. The current flowing through this winding is 20 amperes and the applied line voltage is 230. What is the amount of CEMF developed across the armature winding?

$$E_a = I_a \times R_a$$

where E_a is the armature voltage; I_a is the armature current and R_a is the resistance of the armature winding.

Solution

$$E_a = 20 \times 0.04$$
$$= 0.8 \text{ volt}$$
$$\text{Counter EMF} = \text{line voltage} - \text{CEMF}$$
$$= 230 - 0.8 = 229.2 \text{ volts}$$

Note that the counter EMF is almost as large as the original applied voltage and differs from it by only 0.8 volt. In the absence of a counter EMF, the current is extremely high. This is always the case when the operating voltage is initially applied or if for some reason the armature is prevented from rotating.

Example

What is the amount of current flow through the armature of a DC shunt-wound motor at the instant it is turned on? The line voltage is 230 volts and the armature resistance is 0.04 ohm. Ohm's law can be used to solve this problem.

$$I = E/R$$

where I is the armature current, E is the applied line voltage, and R is the resistance of the armature.

Solution

$$I = 230/0.04 = 5{,}750 \text{ amperes}$$

While this current is only momentary and decreases as soon as the armature starts turning, it is still enough to burn out the armature. It is for that reason that a starting resistance is used. The value of resistance is gradually reduced and becomes zero when the armature reaches its full rotational speed.

Quite often solving a motor problem is simply a matter of selecting the right formula, plugging in the data that has been supplied, doing the arithmetic, and arriving at an answer. There are times, however, when the solution involves the use of a number of formulas. This is often the case when the problem requires a number of answers.

Example

A shunt-wound DC motor is connected to a 121 volt DC power line. The power used by the motor is 18.7 kilowatts (18,700 watts). Power losses in the armature in the form of heat are 395 watts. The resistance of the field windings totals 92.8 ohms. Determine the amount of line current, the field current, the resistance of the armature winding, and the CEMF.

Solution One of the power formulas is

$$P = E \times I$$

and the line current can be calculated from

$$I_L = P_t/E_L$$

where I_L is the line current, P_t is the amount of power taken from the voltage source, and E_L is the line voltage.

$$I_L = 18.7 \text{ kw}/121 = 18{,}700/121 = 154.55 \text{ amperes}$$

The field current can be calculated from

$$I_f = E_L/R_f$$

where I_f is the field current in amperes, E_L is the line voltage, and R_f is the resistance of the field winding.

$$I_f = 121/92.8 = 1.3 \text{ amperes}$$

The armature current is

$$I_a = I_L - I_f$$

where I_a is the armature current, I_L is the line current, and I_f is the field current, all in amperes.

$$I_a = 154.55 - 1.3 = 153.25 \text{ amperes}$$

One of the basic formulas for the calculation of power is $P = I^2R$. The armature resistance can be determined by using this formula with a transposition, that is, $R = P/I^2$. Thus the armature resistance is

$$R_a = P_a/I_a^2$$

$$R_a = 395/153.25^2 = 0.01682\ \Omega.$$

The counter EMF can be calculated from

$$\text{CEMF} = E_L - I_a R_a.$$

$$\begin{aligned}\text{CEMF} &= 121 - (153.25 \times 0.01682)\\ &= 121 - 2.577665\\ &= 118.422 \text{ volts}\end{aligned}$$

THE COMPOUND-WOUND DC MOTOR

A strong magnetic field can be obtained in two ways: by using a large current flow through a coil having relatively few turns of large diameter wire, or a smaller amount of current through a coil having a large number of turns but using wire of a smaller diameter. In both instances the coil is wound around an iron core having a high permeability. The shunt-wound motor uses a field coil having many turns while the series wound is just the opposite using a field coil of much fewer turns.

The compound-wound motor (Figure 3-17) is a combination of the series- and shunt-wound types and has good speed regulation. In the compound-wound motor there are two sets of field coils, connected so their magnetic fields aid each other.

The speed of armature rotation of the compound motor is greater than that of the series type but not quite as high as in the shunt motor. The speed can be controlled by inserting a resistance in series with the shunt field winding, obtaining greater than normal armature rotation. Speeds below normal can be obtained by putting resistance in series with the armature.

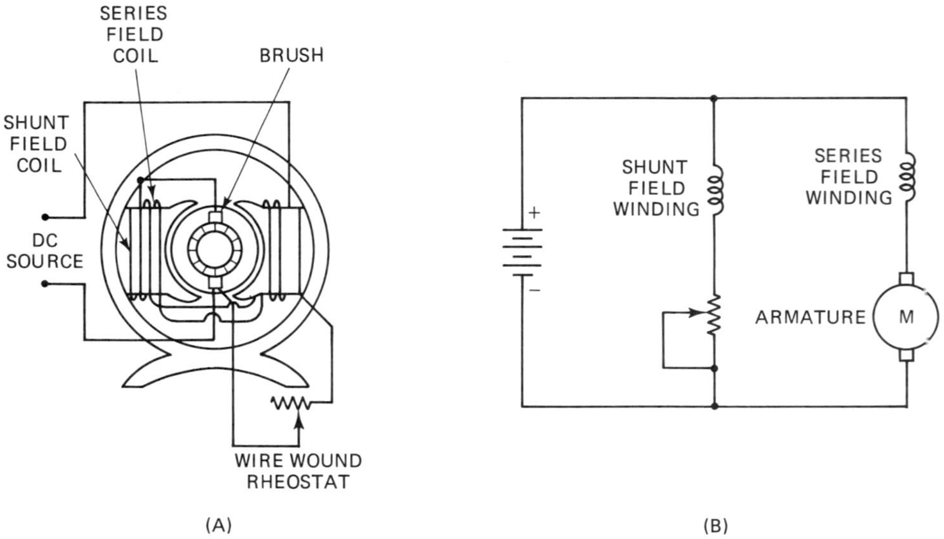

Figure 3-17 Compound wound motor: pictorial (A) and schematic (B).

The series-wound motor is much better for starting a heavy load; the compound is between the shunt and series types. Compound motors are selected for working conditions where the machine must be started under load, where speed constancy is not essential but the wide speed variation of the series motor would be objectionable.

Uses for the Compound-Wound Motor

This motor finds application in aircraft when constant motor speed combined with high starting torque are required. It is also used in elevators where comparatively rapid starts and stops are operational characteristics. These can be modified depending on the motor use. The armature coils can be connected in series with some of the field coils and in parallel with the rest. This motor is also used as part of a motor-generator combination where a varying load is imposed on the generator but constant motor speed is required.

In other applications the compound motor is put to work in escalators, "walking sidewalks," and printing presses. It is also desirable in situations where the regulation of the DC source voltage is either poor or where it is subject to variations.

In the compound motor, a slowdown of the armature means more current will be drawn through the series field which is wound in opposition to the shunt field. This could be caused by an increase in the load. Weakening of the field means the armature develops a smaller counter EMF, but this indicates more current flow through the armature, causing it to increase its speed. Consequently, the armature speed is self-correcting in the face of varying loads.

Reversing Armature Direction

There are two methods that can be used for reversing the direction of rotation of the compound motor. One is to reverse the connections to the armature only. The other is to reverse the connections to both the shunt and series field windings.

In operating conditions when speed or load variations are expected, the DC motor, regardless of its type, is usually equipped with commutating poles to reduce brush sparking. The winding of the interpoles is in series with the armature.

The Differentially Wound Motor

This motor is a member of the compound-motor family. Usually, in a compound-wound machine the magnetic field surrounding the shunt winding aids that around the series field. In the differentially wound motor (sometimes called a flat-compounded motor) the two fields, series and shunt, are wound so that their magnetic fields are in opposition. The advantage of this arrangement

(Figure 3-18) is that it supplies automatic regulation of the armature speed under varying load conditions.

Assume the load on the machine is such that the armature is rotating at its normal speed. If the load is increased, the speed of the armature will slow slightly. This will result in a smaller value of counter EMF, allowing more current to flow from the source voltage. But the increased current through the armature will also flow through the series field coil, increasing its magnetic field intensity. This larger magnetic field strength of the series coil will oppose the magnetic field of the shunt coil. This action permits the armature to increase its speed.

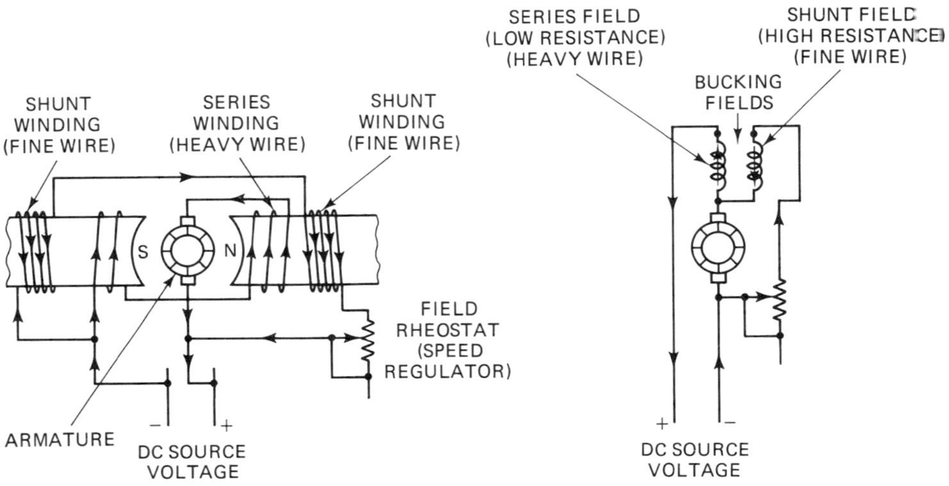

Figure 3-18 Wiring arrangement of the differentially wound compound motor (A). The series-wound and shunt-wound coils have opposing magnetic fields.

If, on the contrary, the load on the machine is decreased, exactly the opposite effect takes place. The flow of direct current through the armature and the series coil is reduced. This permits a greater current flow through the shunt field, increasing its magnetic strength, preventing the armature from rotating at excessive speeds.

The Cumulative Compound Motor

Both the differentially wound and the cumulative wound are two types of compound-wound motors. The cumulative-wound (Figure 3-19) has the high starting torque of the series-wound motor but because of the presence of a shunt field does not have its "runaway" characteristic. Unlike the differentially wound motor in which the current through the series field and shunt field moves in opposite directions (producing opposing magnetic fields), the flow of current in the cumulative-wound motor through the series and shunt fields is in the same direction. This results in magnetic fields which aid each other.

The circuit diagrams in Figure 3-19 are those of the cumulative-wound motor (A) and the differentially wound motor (B). Circuitwise the two are identical except for the direction of current flow through the series field winding. This is done by the way in which the series field coil is arranged physically. Here the direction of current flow is controlled by the way the series field coil is positioned with respect to the shunt field coil.

Compound-wound motors do not have the heavy load starting capabilities of the series wound DC motors but are better in this respect than shunt-wound types.

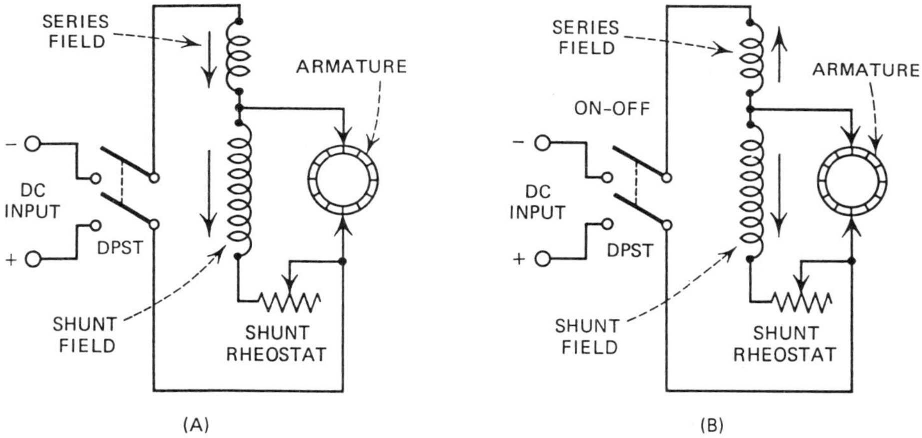

Figure 3-19 Two types of compound-wound motors: cumulative wound (A) and differentially wound (B).

STEPLESS MOTORS

DC motors of the types that have been described are constructed so that the shafts of these motors have continuous rotation. Modifications can be made in two respects. The motor, or its DC input source voltage, can be arranged to have armature rotation in one direction or the other. Another change can be made to provide control over the speed of rotation.

These motors, though, are placed under the general heading of stepless; that is, there is no control over the amount of angular rotation of the shaft. There are some motors, though, in which there can be precise control over the amount of shaft rotation, specified in degrees. They are known as stepper motors and are detailed in Chapter 8.

COGGING

Cogging, sometimes called slot-lock, is a magnetic detent phenomenon. It is caused by the fact that iron contains slots, into which the windings are inserted. These slots interact with the poles so that there are positions of minimum and maximum reluctance. The rotor detents to a position of minimum reluctance. There are various ways of minimizing cogging, but so long as the stator contains slots, cogging cannot be eliminated.

Skewing

Either the stator slots or the field magnets can be skewed with respect to one another so that the leading edge of the slot does not encounter the field magnet all at once but makes a gradual transition.

Fractional Slot Stators

By using a number of slots which is prime to the number of poles, the cogging effect under one pole is out of phase with the other poles, giving an averaging effect.

Latching Circuit

Some also incorporate a latching circuit to assure that they will turn on and stay on once they have seen a south pole and will not turn off until they see a north pole. This prevents them from turning off in the absence of a field in the interpolar space.

MOTOR NOISE

While heat is generally recognized as a motor energy loss, there is still another source of energy loss, and that is sound. Some motors are basically noisier than others, with the sounds produced representing a reduction in motor efficiency. It takes input energy, supplied by the voltage source, to have some portion of it converted to sound energy.

Sound can be produced in a number of ways, the most obvious of which is the movement of a substance, either alone or in contact with another. The size of the motor frame, the way in which it is constructed, the type of motor frame are all factors which give the motor its ability to resonate at some particular frequency. The resulting sound may be either audible or inaudible, but the fact that it cannot be heard does not mean there is no loss of energy in the form of sound. The resonance of a motor or any audible sound it produces can be emphasized by the way in which it is mounted. Insofar as motor noise can be reduced, the efficiency of the motor is increased.

HORSEPOWER

Horsepower (HP), like the inch, centimeter, ohm, or ampere, is a unit of measurement and is used for both electrical motors and generators. The input to a motor is electrical and is expressed in watts or thousands of watts (kilowatts or kw). The output of a motor, the turning power of the motor shaft, is in horsepower, and this measurement can be in while numbers or in fractions. Horsepower is a unit of mechanical power and is the capacity of a machine, such as a motor, to do work. One horsepower is the equivalent of raising 33,000 pounds one foot in one minute or 550 pounds one foot in one second. There is a relationship between horsepower and electrical power.

$$\begin{aligned} 1 \text{ HP} &= 746 \text{ watts} \\ &= 550 \text{ ft-lb/second} \\ &= 33{,}000 \text{ ft-lb/minute} \\ &= 550 \times 60 \text{ ft-lb/minute} \end{aligned}$$

Horsepower can also be indicated in terms of kilowatts.

$$\text{HP} = \frac{\text{kw} \times 1000}{746}$$

These formulas can be transposed in terms of power in watts or kilowatts.

$$1 \text{ kw} = \frac{\text{HP} \times 746}{1000}$$

$$1 \text{ w} = 1/746 \text{ HP} = 0.00134 \text{ HP}$$

Horses came before machines, and in the early days of the steam engine, some comparison was needed between the amount of work a horse could do and what a machine could do. Formulas provide such comparison.

As indicated previously, motors can be classified in many ways, but one of the most common is to designate them in terms of horsepower, either integral or fractional. The horsepower range of motors is substantial, with some rated at 1/100 HP (or less) to 1,000 HP (or more). Typical designations for fractional horsepower motors are 1/4, 1/3, 1/2, and 3/4, while integral sizes are 1, 1-1/2, 2, 3, 5, 7-1/2, 10, 15, 20, 25, 30, 40, and 50. While these numbers do supply a general idea of HP output, the list does not represent the large variety of HP outputs.

Horsepower Rating

This is the amount of horsepower a motor will develop at its shaft. A motor rated at 1 horsepower will deliver 1 horsepower of work at its shaft, or, in terms of power, the equivalent of 746 watts, based on the relationship

1 HP = 746 watts. This does not mean that 746 watts of electrical power is supplied by the power source, for there are electrical power losses due to friction, windage, I^2R (power) losses when a current flows through the motor windings, and eddy current losses in the laminations. The sum of these losses plus the horsepower delivered to the motor shaft is equal to the total amount of power supplied by the power source.

The element of time is not involved in these calculations. That is, the shaft horsepower is not the horsepower per hour, per minute, or per second. Thus, a 1 HP motor delivers 1 HP to the motor shaft at any moment.

Overload

The fact that a motor is rated at 1 HP does not mean this is the maximum that can be delivered to the shaft. As a general rule, a motor can tolerate a 25 percent overload for a period of about two hours. This does not mean the motor will operate more efficiently under these circumstances. As the demand for greater horsepower increases, so does the demand for greater power input. The motor losses due to the factors mentioned above may also increase.

EFFICIENCY

The efficiency of a motor is the ratio of its output to its input. To make this comparison the output and input must be in similar terms. If the input is in watts (and this is usually the case), the output must also be in terms of watts.

The efficiency of a motor can also be expressed as

$$\text{Efficiency} = \frac{\text{output}}{\text{input}}.$$

This may also be written

$$\text{Efficiency} = \frac{\text{input} - \text{losses}}{\text{input}}.$$

For all DC motors, horsepower output can be expressed as

$$P = \frac{\eta IE}{746}$$

where P = horsepower output, η = efficiency, I = line current, and E = line voltage. If the motor has no shunt field, the copper losses in that field can be omitted from the equation; losses in the series field can be handled similarly. If the motor has both types of fields, the losses for both must be included. Thus the losses for a compound-wound motor could be greater than those for either a series- or shunt-wound.

The efficiency factor can be combined with formulas involving power in terms of either watts or horsepower. The formula for horsepower can be stated as

$$HP = \frac{E \times I}{746}$$

which can be transposed to

$$746 \times HP = E \times I$$

$$I = \frac{746 \times HP}{E}.$$

If the efficiency is less than 100 percent, the formula becomes

$$I = \frac{746 \times HP}{E \times eff}.$$

Example

A motor used for operating a small lathe is rated at 1/2 horsepower. If the power input to the motor is 0.45 kw, what is the efficiency factor of the motor? What is its percentage efficiency?

Solution

$$1\ HP = 746\ watts$$

$$1/2\ HP = 746/2 = 373\ watts\ output$$

$$Input = 0.45\ kw = 0.45 \times 1{,}000 = 450\ watts$$

$$Efficiency\ factor = output/input = 373/450 = 0.83$$

$$Percentage\ efficiency = efficiency\ factor \times 100 = 0.83 \times 100 = 83\%$$

Example

A motor whose rated output is 60 HP is being used in building construction. This motor is to lift an elevator, whose total weight, with load, is 4 tons. The building has a height of 240 feet and the elevator is to travel from ground level to the top of the building in 1-1/2 minutes. Will the motor be able to handle this load?
Assume a ton = 2,000 lbs
4 tons = 4 × 2,000 = 8,000 lbs
HP = Number of ft-lbs per minute/33,000

Solution

$$HP = \frac{8{,}000 \times 240}{1.5 \times 33{,}000} = 39\ HP\ (approx.)$$

The motor will be capable of lifting the elevator the distance indicated.

Example

A motor working from a 220-volt DC line draws a current of 12 amperes and operates with an efficiency of 88%. What is the output rating of this motor in horsepower?

Solution

$$HP = \frac{E \times I}{746} = \frac{220 \times 12}{746} = 3.54 \text{ HP}$$

Efficiency factor = efficiency in percent/100 = 88/100 = 0.88
Output = input × efficiency factor = 3.54 × 0.88 = 3.12 HP

Example

How much current will be required by a 5 HP, 220-volt DC motor whose operating efficiency is 90%?

Solution

$$I = \frac{746 \times 5}{220 \times 0.9} = 18.84 \text{ amperes}$$

TORQUE

Torque is a force that in the case of motors is the force required to produce rotation of the armature. Torque is measured in inch-pounds, foot-pounds, or inch-ounces. In a motor, torque is proportional to the armature current and to the strength of the magnetic field. Thus, to produce greater turning power of the armature shaft requires an increase in armature current. The strength of the magnetic field is dependent on the number of turns constituting the armature coil and on the core. For the iron core, the greater its permeability, the stronger the resultant magnetic field.

Torque can be defined as a rotational force equal to the mass of a load times the radius of a pulley or length of a lever arm. Torque involves two factors. The motor must be able to start not only itself but its load as well.

How Torque Is Developed

When a current carrying coil is placed in a magnetic field, a force is produced which causes the coil to rotate. This is due to the reaction between the two magnetic fields with the force used to produce rotation known as torque.

Figure 3-20 shows an armature consisting of two turns. The coil will rotate until it is perpendicular to the flux lines. The force on conductor A pushes it up and that on B pushes it down. Since both forces are additive their effect is to make the coil turn. When one turn of the coil is in a position such that it is not affected by the permanent magnetic field, the other turn is in a position to receive maximum upward pushing on one side and maximum downward force

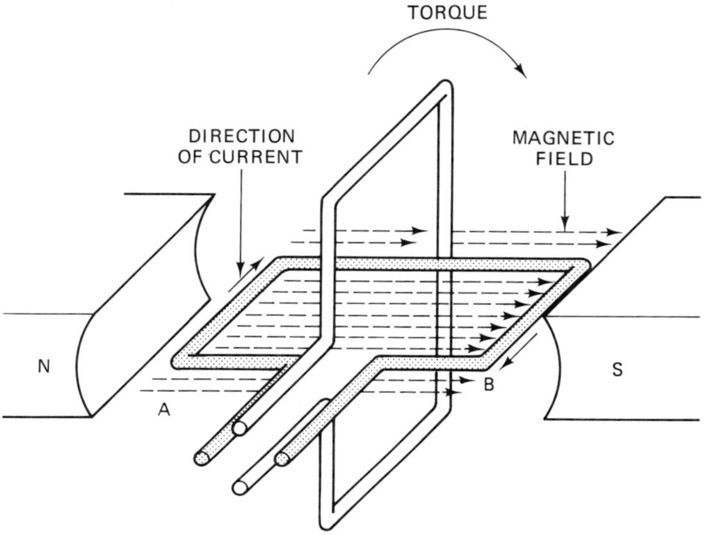

Figure 3-20 Torque is developed when armature conductors move across lines of flux supplied by the field magnet.

on its other side. Consequently, if the rotator consists of two turns, at least one of them is at work producing torque.

Actually there is only one position in which no torque is developed and that is in the case of a single turn coil occupying the position shown in B and D in Figure 3-21. At any other position there is always some development of torque and this, plus the inertia of the coil, keeps it rotating. The torque developed varies sinusoidally between its maximum and minimum positions (Figure 3-22).

For smaller motors, torque figures are in inch-ounces (in-oz), while for larger motors it is in foot pounds (ft-lbs).

Inch-Ounces and Foot-Pounds

An inch-ounce is the work done by a force of one ounce working through a distance of one inch. A foot-pound is the work done by a force of one pound working through a distance of one foot. Work requires time but the element of time is not involved. A man weighing 130 pounds, carrying a 5 pound bucket of paint up a 25 foot ladder does work against the force of gravity of $[(130 + 5) \times 25]$ or 135×25 or 3375 ft lbs, and this remains the same whether the ascent up the ladder takes 5 minutes or 10 or 20.

The torque of a motor can be expressed in a number of ways: full-load torque, starting torque, pull-up and pull-out torque, etc.

Full-load torque. The full-load torque can be expressed in terms of a formula as

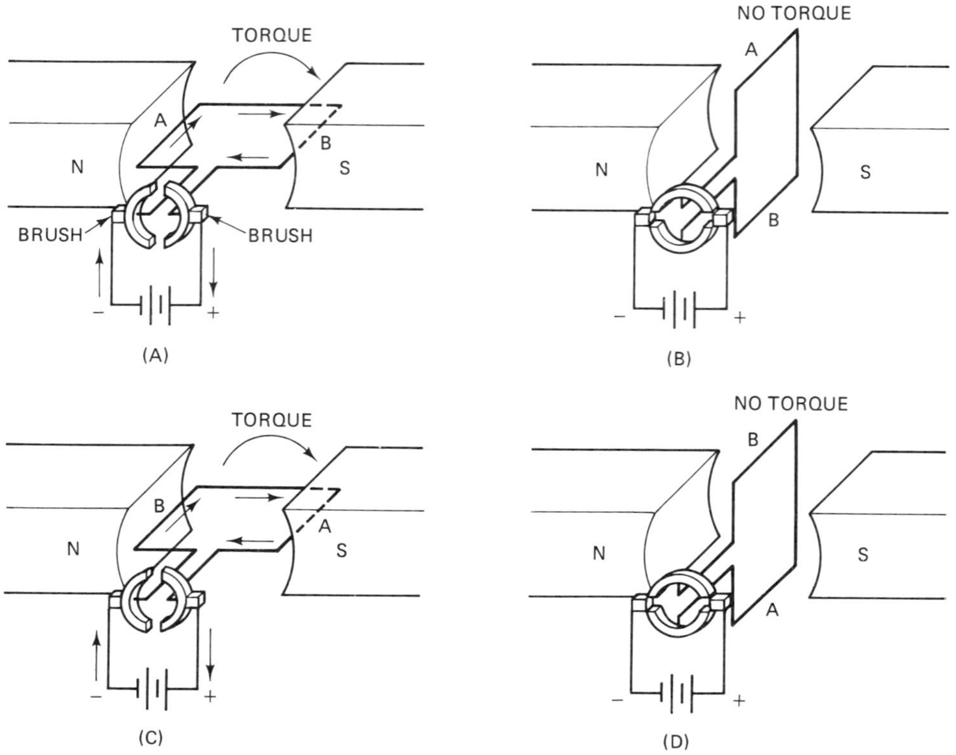

Figure 3-21 Development of torque: maximum torque (A and C) and zero torque (B and D).

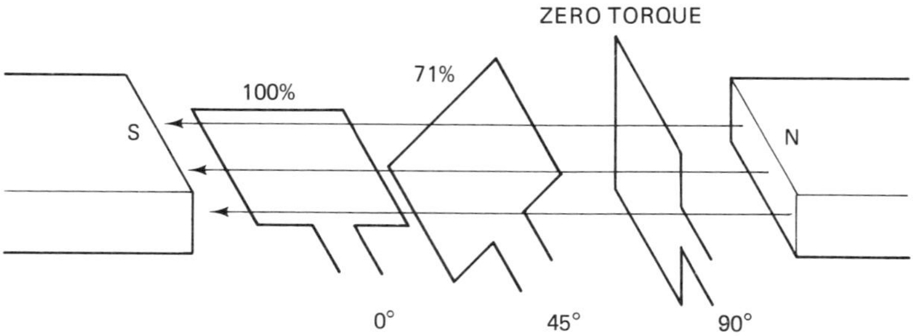

Figure 3-22 Variations in torque produced by an armature coil.

$$\text{Full-load torque} = 5252 \times \frac{\text{horsepower}}{\text{rpm}} (\text{in ft-lb}).$$

The torque is supplied in foot-pounds. The foot-pound is a unit of work that is equal to force multiplied by distance. If a 50 pound weight is moved a distance of 15 feet, the work done is $50 \times 15 = 750$ ft-lb.

Example

A motor having a HP rating of 20 has a running speed of 900 rpm. What is the twisting effect of its shaft, that is, its torque, at its rated load?

Solution

$$\text{Full load torque} = 5252 \times \text{HP/rpm}$$

$$= 5252 \times 20/900 = 116.7 \text{ ft-lb}$$

Starting torque. This is the torque that is developed with no rotation. It does not take into account the angular position of the rotor but the source voltage applied to the motor. No work is done in this initial rotor position.

Rated starting torque. This is the minimum torque produced by a motor geared for 1 rpm, at rated 60 Hz voltage, 25 degrees C ambient, when energized through a snap type switch. The motor may not reach rated synchronous speed when developing rated starting torque. The torque produced for motors other than 1 rpm will be inversely proportional to the rated motor speed.

Pull-up torque. Starting at zero, the rotor will accelerate until it reaches its operating speed. During this time the torque increases.

Running torque or stall torque. This is the amount of torque that a typical production motor will develop prior to stalling after having reached synchronous speed.

Typical synchronous torque. This is the amount of torque that a typical production motor will develop running at synchronous speed. Any additional torque load will cause the motor to slow down, run subsynchronously, and approach the running torque rating.

Rated breakdown torque. This is the torque supplied by the shaft at its rated input voltage and rated load without a drop in speed. It is an indication of the load that can be imposed on the motor without the rotor dropping to some slower speed.

Pull-in torque. This is a reference to a synchronous motor and is the constant torque supplied by the shaft at the point where the rotor functions at synchronous speed with the input voltage and frequency at rated amounts.

Ripple torque. This is the variation in torque due to the switching of windings; the magnitude is proportional to the output current. The frequency is based on the number of times the winding is switched per revolution multiplied by the revolutions per second.

The number of switching cycles per revolution is six times the number of pairs of poles for a bipolar motor and half that for unipolar.

Peak torque. This is a somewhat arbitrary value. It can be exceeded for very short periods of time, such as during acceleration. A more logical method of rating is based on the torque which can be produced for 15 seconds without exceeding a safe temperature.

Torque constant. This is the torque that will be produced for a given current input.

Shaft torque of a motor. While the armature is mounted directly on the shaft it does not follow automatically that the torque of these two, the armature and the shaft, are identical. Because of various losses, the actual amount of torque available at the shaft is always less than the armature torque.

Torque losses are the sum of a number of individual losses. These include friction in the bearings at both ends of the shaft, the friction of the brush against either slip rings or a commutator, windage due to the rotating armature, and if a fan is used, windage because of fan rotation.

Armature torque of a DC motor. Because of losses in a motor, armature torque is greater than the actual torque developed by the shaft.

Torque Requirements

In some applications the correct operating torque of a motor is determined by the manufacturer of the device operated by that motor. Thus, in the case of a refrigerator, an electric fan, a vacuum cleaner, and so on, the motor is an integral part of the component. However, there are other instances in which the torque of the motor requires a decision. If a separate motor is used to drive a lathe or a stand-alone sander, for example, the motor selected as the drive unit must have enough torque, and this is determined by the requirements of the load.

SOFT START MOTORS

A motor that is usually operated with a light load is sometimes referred to as a soft start type. Motors that fall into this category include shaded-pole induction, split-phase, and permanent split capacitor-induction units. These are all motors operated from an AC source voltage and are detailed in the following chapter.

In some instances the motor may be faced with a varying load ranging from soft to difficult. An electric drill is an example. If the drill piece is a thin gauge type and is being used to penetrate wood that is not dense, the torque required could be minimal. If the drill piece is one required to produce a half-inch hole in a metal such as steel, the torque required would be much more substantial.

RATED GEAR TRAIN LOAD

This is the amount of torque the motor's gear train (see Chapter 9) will carry independent of the performance torque ratings previously listed. There are two ratings, continuous duty and intermittent duty. Standard gear trains are rated at 20 in-oz continuous duty, where optimum life is desired, and 80 in-oz intermittent duty, which represents the minimum gear train strength beyond which damage may occur. Both ratings are independent of motor speed.

Since the available torque rating of a gear motor is inversely proportional to motor speed, the gear train rating should not be exceeded, even though additional torque is available within the motor. For example, a 1/4 rpm motor would be electrically capable of a 160 in-oz rated starting torque. However, the gear train may be damaged if torque loads exceed 80 in-oz.

MOTOR SPEED CHARACTERISTICS

There are a number of ways of classifying motors and their operating speed is just one of them. Motor speed, as described earlier in this chapter is the rotation of the shaft in revolutions per minute (rpm) without any consideration of gearing up or down.

SHAFT SPEED LIMITATIONS

Shaft speeds vary considerably depending on motor type. Shaft speeds are expressed in terms of maximum values. Thus, for so-called miniature motors, the shaft speed is about 10,000 rpm and is about half this amount for small motors of less than 1/2 HP. For some motors the maximum available speed depends on the frequency of the AC source voltage. For induction motors, for example, the maximum shaft speed is 3,450 rpm when the line frequency is 60 Hz but decreases to 2,875 rpm for 50 Hz. For AC generators that supply an AC output of 400 Hz, there is a wide range of available maximum motor speeds, such as 3,700 rpm, 5,400 rpm, 7,200 rpm, 10,500 rpm, and 21,000 rpm. For constant speed motors representative values are 1,140, 1,725, and 3,450 rpm.

The speed of rotation is often limited by the type of drive that is used, whether gear or belt, and the amount of HP output. These speeds are not supplied as single, specific values but rather over a range. The diameter of a machine pulley in inches, for example, is directly proportional to the motor rpm.

MOTOR SPEED COMPARATIVE ANALYSIS

The graph in Figure 3-23 illustrates the variations in speed of the different types of DC motors with load. Of these, the series-wound motor is most affected by load conditions. The shunt-connection type has an almost constant speed. The differential-compound motor is unusual in that its speed increases with load. The cumulative compound motor has a speed characteristic that is between that of the shunt- and series-wound types.

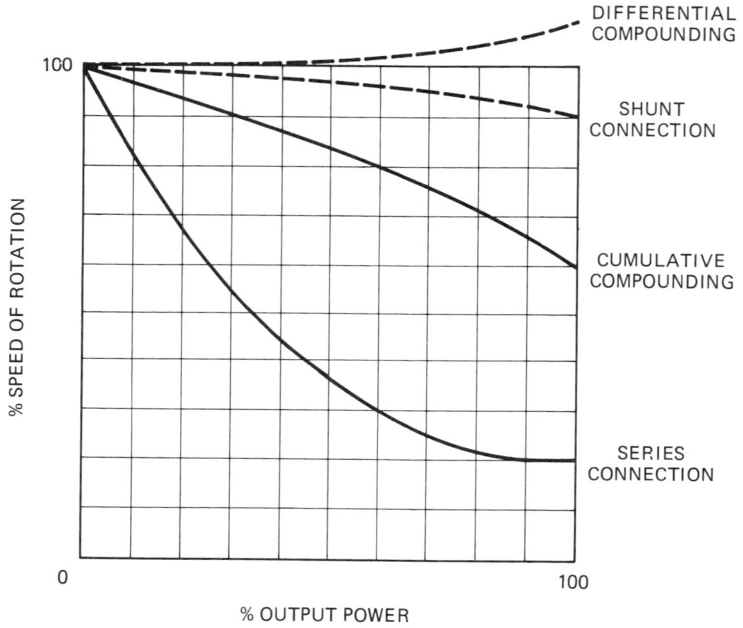

Figure 3-23 Effect of load variations on speed in different types of DC motors.

ENERGY LOSSES IN DC MACHINES

The input to these motors is DC power and the output is measured in terms of horsepower. Horsepower, though, based on the formulas supplied earlier, can also be expressed in terms of electrical power. When this is done, it will be noted that the electrical power output is always less than the electrical power input. This power loss is the sum of a number of individual losses in the motor, some of which are minor, others major. The efficiency of a motor is directly dependent on the extent to which motor losses can be curtailed, but since the motor is an electromechanical device, some losses would appear to be inevitable.

Copper losses. Whenever a current flows through a conductor, such as the field and armature windings of a motor, there will be losses of the current flow known as I^2R losses. While the resistance of a winding, such as the field winding of a series-wound motor can be very low, the amount of current is substantial. Further, the power loss is proportional to the square of the current. Since these losses occur in both the field and armature windings, the total copper loss is the sum of these two.

These power losses are dissipated in the form of heat. While heat is produced by other causative factors, the total heat may be substantial enough to require internal or external cooling. With internal cooling, fan blades are attached directly to the shaft; with external cooling, a separate blower fan is used.

Iron losses. These can be divided into two types: hysteresis and eddy current losses. Hysteresis losses are due to the fact that the armature revolves in an alternating magnetic field. As a result, its core becomes magnetized, but with a changing magnetic polarity. The magnetism of the iron or steel of which the armature core is made always has some residual magnetism, and it is the need to constantly reverse this residual magnetism which results in energy losses. Since the field magnets do not have polarity changes, they do not have hysteresis losses.

Eddy current losses. These are minimized through the use of laminations instead of solid iron cores. Eddy currents can heat the cores, and in turn these can cause damage to windings if their insulation cannot tolerate the temperature increase. Eddy currents, like currents flowing through windings, result in an energy loss. The thinner the laminations that are used, the lower this loss will be.

Losses due to friction. Even though the motor shaft has both its ends riding on a film of oil, there is still a certain amount of friction. There is also some friction between the brushes and the commutator bars on which they ride and some power loss due to current flow through the brushes.

Load loss. The heavier the load on a motor, the greater the amount of input power required. Some energy is used in getting the load started. If the motor is geared, there is some energy loss due to friction in the gear train.

Miscellaneous losses. There is some loss because of windage, motor noise, friction in the bearings, possible shaft eccentricity, motor vibration, and core loss due to flux distortion. Depending on the motor design, losses can be constant or variable, particularly in motors whose flux changes with load.

MOTOR OPERATING PROBLEMS

There are two basic types of motor problems: those external to the motor and those that require removal of the motor from its frame. Often enough the motor problem is so simple that it can be solved successfully by the user. In some instances, the motor repair may be easily accomplished, but because the motor is housed inside an appliance it may demand the services of an experienced repair technician, as for example, washing machine motors, refrigerator and oil burner motors.

CAUSES OF BRUSH SPARKING

Aside from improper setting of brushes, there are a number of other possible causes of brush sparking. The brushes are held in place by the action of a spring. The spring may have been pushed out of place or may have lost its spring action. In cheaper small motors a spring is not used, but the brush is held in place by a small piece of metal which is bent over, making contact with the brush and holding it in position. The use of this metal segment makes it difficult to replace the brush when it is worn, since it tends to break when it is bent backward and forward at the time a new brush is installed.

Not all brushes are alike, and some are softer than others, although all are made of carbon. In time, even with the best of motors, the brushes will need to be replaced. A worn brush will not only cause sparking but, if sufficiently worn, may permit the metallic brush holder to contact the commutator bars, scratching or even damaging them beyond repair.

The carbon brush can make the commutator segments dirty. This can often be seen as a black path on the commutator. The commutator can be cleaned with a small section of canvas, fine sandpaper, or a polishing paste made specifically for this purpose. If the commutator has not been cleaned for a long time, it may be difficult to remove carbon particles since some may have become imbedded in the commutator bars; consequently it may require repeated applications of polishing paste. Do not attempt to use a file or emery paper.

In time and with continued use, the commutator may become worn out of round. Its eccentricity can result in variations of brush pressure, and so the movement of the brush to and away from the commutator segments can result in sparking. Sparking can also result if an armature coil is either open or shorted.

Sparking can be caused if the motor is overloaded. Since a DC motor is started with the help of a variable resistor in series with the armature coils, an attempt to start the motor too rapidly (by reducing the resistance too quickly) will cause commutator sparking. Other sparking causes include brushes not properly set at their neutral point, mica insulation above the commutator segments (high mica), and an eccentric commutator.

PROBLEMS OF THE SELF-INDUCED EMF

The commutator of a DC motor can be regarded as a rapidly operating number of single-pole single-throw switches. It can be represented by the circuit in Figure 3-24, consisting of a single switch, a battery as the voltage source, and an iron-core coil. The coil is equivalent to an armature coil, the battery comparable to the DC supply of a motor, and the switch to one of the commutators of the motor.

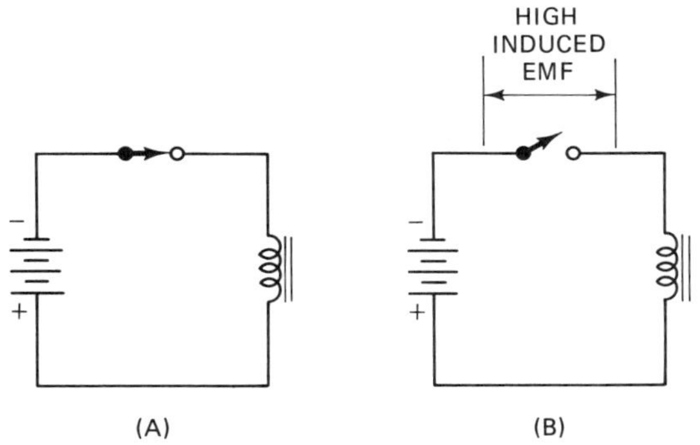

Figure 3-24 When the current reaches a steady state (A) the induced EMF is zero. When the switch is opened (B) an extremely high voltage caused by the rapid collapse of the magnetic field around the coil appears across the switch. The switch represents the commutator segments of a motor.

When the switch is closed, the flow of current produces a magnetic field around the coil. This magnetic field induces a voltage across the coil that opposes the applied battery voltage. The amount of induced EMF depends on the rate of change of the magnetic field, but the current gradually reaches a steady state, that is, the magnetic field ceases to grow and the induced voltage decreases to zero.

When the induced EMF reaches zero, the flow of current through the coil is maximum and so is the magnetic field around that coil. If the switch is now opened, the current drops to zero and the magnetic field collapses. In so doing, the magnetic field induces an EMF across the coil, but because the reduction of the magnetic field to zero is so extremely rapid, the induced voltage is extremely high and is often in kilovolts.

The thousands of induced volts appear across the open switch and, because the potential is so extremely high, can rupture the air across the switch terminals, ionizing the air and providing a conductive path. Given a high enough induced EMF, the result can be a spark across the open terminals of the switch.

In a DC motor, this effect of the induced EMF is shown as sparking at the brushes and, if continued over a period of time, can result in commutator and brush damage.

Techniques for Reducing Commutator Sparking

There are two techniques for reducing commutator sparking. One of these is to position the brushes along the neutral plane. This is a line drawn by joining two diametrically opposite points referred to as neutral points. These represent minimum induced EMF points, positions where the value of the induced EMF may not be strong enough to result in sparking. The neutral position or plane represents the points at which the brushes can be positioned so there is minimum sparking at the brushes.

When the motor is turned off, no current flows through the armature coils. The neutral plane can then be considered to be located at the point where an armature bar segment in contact with a brush is positioned half way between a pair of field poles. During rotation, a brush will short an armature coil. The large magnetic field around the coil will collapse suddenly, resulting in an extremely large induced EMF. It is this voltage that results in sparking at the commutator.

Positioning the brushes at the neutral points is a satisfactory solution, provided the load on the motor remains constant. However, as the load increases, the brushes need to be pushed back to avoid sparking.

BRUSH PROBLEMS

Like motor bearings, brushes should be examined regularly as part of a maintenance program. Brushes are subject to a number of different troubles. Brushes for motors should be exact replacements of those used originally. Problems can also be caused by worn brushes, insufficient brush spring pressure, some defect in the brush holder, a loose connection to the brush holder, and brushes in the wrong position on the commutator. Faults in the commutator are considered as a separate category.

If the brush does not press adequately against the commutator, the result will be excessive sparking. This can produce pitting in the shoe of the brush, causing the condition to worsen.

In some cases, the brush is not at its correct pressure position on the commutator. This results in a higher than normal amount of sparking. Move the brush holder to the electrical neutral on the commutator, a point at which little or no sparking takes place with the motor operating under its normal load. This does not apply to motors using interpoles, since it is their function to produce sparkless commutation.

Check the connections to the brush holders to make sure they are physically secure. The brushes should not be worn down to the point where it is possible for the brush holder to make contact with any commutator segments.

Identical brushes in identical motors will not necessarily have the same amount of wear. A motor that is frequently started and stopped will show more wear. Brush wear can be affected by temperature and humidity. Environmental operating conditions can also be a factor. The presence of chemical fumes and high levels of dust can reduce brush life. The brush should not be loose in its holder and should not be able to wiggle back and forth in that holder.

The brush holder itself may be part of a mechanical arrangement consisting of a brush-holder stud and a brush yoke. The brush-holder stud is a linkage between the brush holder and a yoke. The yoke can consist of various mechanical elements, a rocker arm, a ring, or some similar device for holding the brush-holder studs in their correct position with reference to the commutator.

COMMUTATOR PROBLEMS

Commutators and brushes probably develop more working problems than other parts of the motor. The commutator is subject to the centrifugal force of the motor and to the contact action of the brushes riding on its surface. While carbon and graphite do work as lubricants, there is still a certain amount of friction. As a result, the commutator can develop roughness along the brush path. The result of this is increased sparking.

The commutator can be smoothed by rubbing it with number 00 sandpaper wrapped around a wood block. Following the sandpapering, the commutator can be finished with a small length of canvas. If the surface area of the commutator has been neglected and deep cut grooves have developed, it may need to be turned down on a lathe or else replaced with one that is new.

Commutator wear by the brushes is not always uniform, and so there may be some eccentricity. When this happens, the brushes will seem to be riding up and down. The result is as though the shaft is off-center.

BEARING PROBLEMS

Bearing problems can stem from various causes. The most usual are due to lack of a proper schedule for lubrication, allowing the bearings to run dry. Another is the use of the wrong kind of lubrication. The lubricant may be so thin at the start that it gradually disappears with an increase in temperature. Finally, the shaft may have developed some end-of-the-shaft eccentricity causing undue friction. This problem may be small enough not to be physically noticeable but may still contribute to an increase in temperature. It may be evident by a gradual growth in motor noise level. The final result is a condition known as a "hot bearing," and it always happens when the motor is running.

When this becomes noticeable, do not stop the motor. Doing so will result in a loss of bearing temperature with the bearing contracting and gripping the armature shaft extremely tightly. This condition, known as seizing, may make it difficult or impossible to separate the two.

Instead, if the motor has a speed control, set it to its lowest operating speed. If a fan is at hand place it so a stream of air is directed against the bearing. If ice is available make a pack of it, wrapped in a cloth or towel, and hold it against the bearing.

Once the bearing temperature has dropped, flush the existing lubricant out with kerosene. The purpose here is to clean the bearing. Once the bearing seems to run clean, replace the lubricating oil, and then stop the motor.

COASTING AND BRAKING

Coasting is the number of degrees an armature will continue to rotate because of its momentum following the stoppage of the input power source. The load will also continue to operate, and in some instances this may not be desirable. The rotation can be less than a full revolution, in the case of motors driving a load or operating a gear train, or, as in the instance of ceiling fans, may continue for a large number of gradually slowing turns.

Braking Methods

There are various types of braking identified as resistive, dynamic, electromagnetic, and electronic, although some of these are duplicate names for the same function.

Figure 3-25 shows the connections for braking a compound-wound motor. The brake consists of a variable resistor placed in shunt with the armature field winding. The resistor is added to or removed from the circuit by a double-pole single-throw switch, S1. To achieve braking, power is disconnected from the voltage source lines connected to the motor. The armature, in continuing to rotate, will work as a generator, rather than a motor, and will discharge across variable resistor R, which will function as a load. A resistor used in this type of braking is sometimes called resistive but, because of the generator action of the armature, is also referred to as dynamic.

The variable resistance technique for braking can also be used for series-wound motors. The mechanical energy in the armature is dissipated in the braking resistance.

For some motors, braking is accomplished by the deliberate development

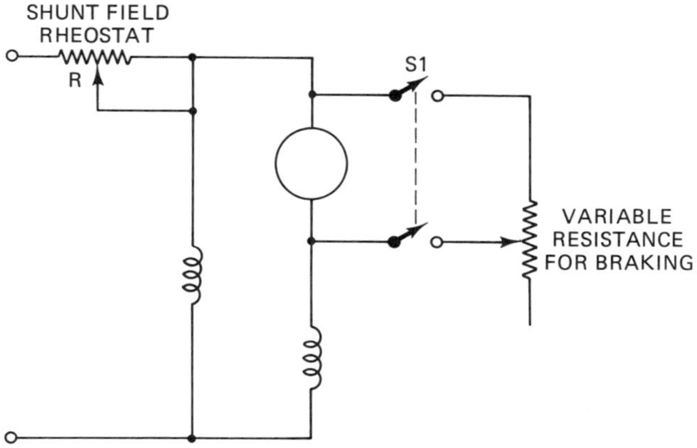

Figure 3-25 Method for braking a compound-wound motor.

of eddy currents. The magnetic fields surrounding these currents are in opposition to the magnetic field surrounding the armature and so supply the equivalent of an electronic drag.

FUSES FOR MOTORS

Since the starting current of some motors is substantially higher than the running current, using an ordinary fuse will result in problems. If the starting current of a motor is 20 amperes and the running current is 3 amperes, it would appear that a logical choice would be to use a 5-ampere fuse. Yet the 20-ampere starting current will destroy the fuse. For this reason it is better to use a time-lag type of fuse. This will permit a high starting current; it will open if the motor draws excessive current while working.

The drop in current of a motor from its start to its running mode is due to the electrical behavior of the motor. Motors and electric generators are quite alike, and as a matter of fact, while a motor is running it behaves like a generator and produces a voltage.

If the motor should jam or if for any reason the rotor stops turning, the counter EMF disappears and the motor behaves as if in the start position. With the loss of the counter EMF, maximum current flows into the motor. This current can damage the motor windings quickly, possibly starting a fire inside the motor by igniting the wire insulation and causing it to carbonize. The chart in Table 3-1 indicates motor running current and the maximum fuse sizes to use.

As the load on a motor increases, so does its armature current. The load can increase to the point where the current can cause damage. This can be the result of a jammed belt or gear drive. While this is external to the motor proper, an apparent increase in load can also be due to lack of or improper lubrication, a damaged shaft, or worn bearings.

TABLE 3-1 Fuse Sizes for Motors

Motor Amperes	Maximum Fuse Size
1 to 5	15
6	20
7 to 8	25
9 to 10	30
11	35
12 to 13	40
14 to 15	45
16	50
17 to 20	60
22	70
24 to 26	80
28 to 30	90
32	100
34 to 36	110

The recommended size of a motor fuse can be determined by the motor's running current and also by its horsepower output as listed in Table 3-2. The table is limited to DC compound-wound motors using 115 volts DC input.

TABLE 3-2 Fuse Size Determined by HP Output

HP	115 Volt DC Compound-wound Motors
1/8	3 amp.
1/6	3 "
1/4	4 "
1/3	6 "
1/2	8 "
3/4	10 "

There is a common misconception, fostered by the fact that fuses are rated in terms of current. It is not excess current that causes a fuse to open but rather the demand for excess power dissipation. Here power dissipation is in terms of $P = I^2R$. If the amount of current squared, in amperes, multiplied by its resistance, exceeds the rating of the fuse, the fuse will open. If it were possible to develop a fuse having zero resistance, no amount of current could possibly cause the fuse to open.

Prior to the time a fuse opens, the voltage across it, based on Ohm's law, $E = I \times R$, is very small. This is due to the fact that although the current is measured in amperes, the resistance of the fuse is extremely low. When the fuse opens, the voltage across the fuse terminals becomes equal to the line voltage. For this reason, removing an open fuse with unprotected fingers can easily result in an electrical shock.

In addition to fuses, motors can be protected by various types of thermal cutouts, resets, or circuit breakers. Their advantage is that they eliminate the need for replacement. Further, they remove the possibility of mistakenly using a wrong size fuse.

Resets

A reset is a current overload protection device. Its advantage is that, unlike a fuse, it does not need replacement. In the event of a current overload, the reset, like a fuse, will open and interrupt the line current.

There are two types of reset devices, manual and automatic. The manual reset is positioned outside the motor, often on the motor frame, but in some instances, as in the case of oil burner motors, the reset is placed on the oil burner housing. The reset consists of a device operated by a pushbutton, sometimes colored red. If the reset opens, it will need to be pushed to permit the motor to start again. If the motor does not start, it indicates some fault, either in the motor, or its load.

The reset, also known as an overload control, can be positioned inside the motor frame. Since it is not accessible without removing that frame, this is an automatic reset type. Both the manual and automatic resets function by means of a bimetallic strip. When this strip becomes hot, it triggers a switch, disconnecting input power to the motor. For automatic resets it is necessary to wait until the strip cools, thus closing its circuit.

Manual starting switch with overload control. The manual starting switch can be a push-button type housed in a metal box and equipped with an overload control. Its pushbuttons are clearly marked On and Off or Start and Stop. It contains a manual reset. The overload control can consist of a bimetallic strip or can be a magnetic type. The magnetic type uses a solenoid which keeps the line power to the motor active. In the event of a current overload, the solenoid opens the line switches.

Time delay fuses. Under some operating conditions the motor may have a temporary overload. This is not a motor or load fault but it can lead to an open fuse. Unlike an ordinary fuse, the time delay type has a double fuse element capable of handling the overload current. In the event of a short circuit, the fuse elements will open and will need to be replaced.

PERMISSIBLE TEMPERATURE RISE

Motors develop an increase in temperature during their conversion of electrical energy input into mechanical energy. The factor determining the permissible temperature rise is the kind of flammable materials used. Most of the motor, the frame, the commutator segments, the yoke, and the copper of the windings can tolerate high temperatures without damage. This does not apply to the insulating materials which are subject to charring and burning. These substances include cambric, silk, cotton, and sometimes paper, each having a different flammable point. Some of these may also be impregnated with chemicals that are fire retardant, and although they do not eliminate the possibility of fire, they can delay its onset. The so-called nonflammable materials, such as the mica used as an insulator between commutator segments and asbestos, can tolerate temperatures as high as 257 degrees F (125 degrees C).

The temperature rating for a motor does not assume 0 degrees F as the base or starting point, an impractical reference, but takes room temperature as the reference base; this is generally assumed to be 104 degrees F or 40 degrees C. This is rather high and is much above the general ambient. The permissible temperature rise is supplied as a number of degrees C above this arbitrary base.

FUSING CURRENTS OF WIRES

The fusing current of a wire is the amount of current at which a wire will melt. There are a number of factors involved in heat dissipation. Heat loss, for example, may be due to the type of motor frame used, that is, whether the frame is ribbed or finned or otherwise designed to dissipate heat or whether the motor is equipped with a blower and, if so, whether it is effective. Ambient temperature is another factor and the higher it is the less the ability of the motor to throw off heat.

DC MOTORS FOR ELECTRONICS

DC motors are at extremes. They are widely used in industry and transportation when heavy loads are to be handled with armature currents rated in amperes and power inputs often in hundreds of watts. At the other extreme are the small motors used in electronics and brushless DC motors, also known as permanent magnet motors or PM motors.

The only types of DC motors discussed in this chapter are those that use electromagnets as field coils. Such motors are intended for commercial applications. Electronics for motors is covered in Chapter 6 and brushless DC motors in Chapter 7.

4

AC MOTORS

The DC motor preceded the AC by about fifty years, for it was not until the latter part of the 19th century that AC became available. With the gradual introduction of AC into the home, a number of different types of motors that would operate from household outlets were gradually introduced and became popular because they were labor saving devices.

As a result, in general, DC motors were largely used for industrial applications, AC motors in the home. Subsequently, DC motors became popular for in home and portable use. This was due to two factors: the considerable reduction in motor size and the improvement in batteries.

AC MOTOR TYPES

While the listing of AC motors would appear to be extensive, many of these are simply variations. Thus, capacitor motors could be specified as capacitor-start motors, split-capacitor, and two-value capacitor motors. Another general heading would be AC single-phase motors. Under this heading, for example, we have squirrel-cage motors, but these could also be designated as shaded-pole motors or as split-phase induction types. Motors could also be described as reversible or nonreversible types.

While DC motors have relatively few listings, AC types have a large variety. A partial grouping could include AC series, autotransformer capacitor, universal, hybrid, polyphase types such as single-phase, two-phase, and three-phase, repulsion start, squirrel-cage, induction, motors with built-in electronics, motors using external electronics, reluctance synchronous, repulsion, repulsion-start induction, shaded-pole induction, split-phase capacitor-start, split-phase permanent capacitor, stepper types, multispeed, dual voltage, low and high frequency, servomotors, geared motors, fractional horsepower, subfractional horsepower, split-phase induction, capacitor-start unidirectional bidirectional, self-starting induction-reaction synchronous, permanent-split-capacitor, induction hysteresis synchronous, and so on.

One of the confusing aspects of AC motor designation is the fact that the same motor may have several different names. In some instances the name change is slight and sometimes there are slight electrical modifications in the motors.

SYMBOLS

There are two basic types of symbols used for motors, whether AC, DC, or universal, operating from either AC or DC. Symbols can be generalized (Figure 4-1) and such symbols are applicable to all motors. A motor symbol can also be specific and in that case is applicable to a single motor type. In both instances, whether generalized or specific symbols are used, there is no standardization, and so a number of symbols are available for both applications. To avoid confusion, the type of motor is sometimes lettered alongside the symbol or is described in the accompanying caption or text material.

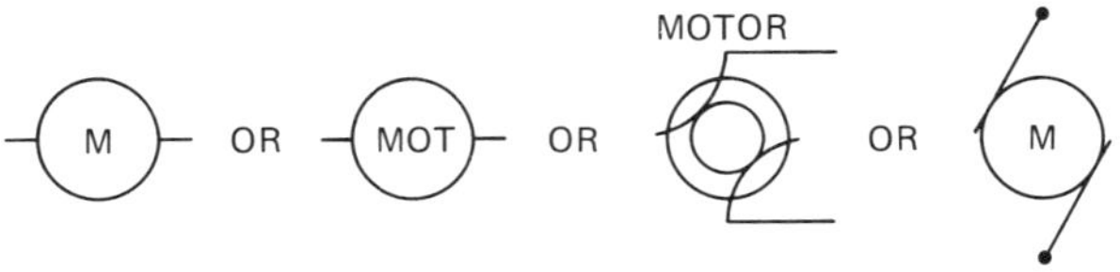

Figure 4-1 Generalized motor symbols.

LINE VOLTAGE VARIATIONS

The AC source voltage for motors, the line voltage, is variously listed as an amount between 115 and 121. There is no standardization and the data plate on appliances indicating AC line voltage can be any number selected by the manufacturer.

The AC line voltage at any moment depends on a number of variables: the distance of the run between the fuse box or circuit breaker box and the motor-connected outlet; the gauge of the connecting wire; the number of appliances connected to this particular power line branch; the current requirements of those appliances; and the amount of the overall load imposed on the utility's power lines.

Consequently, the amount of AC voltage present at the outlet to which the motor is connected can vary. For multiphase AC the line voltage can range between 220 and 240 volts. The amount of line voltage indicated in the motor circuits which appear in the following pages observe this pattern of inconsistency. To have selected a single specific voltage might make that appear as a standard.

Unlike the voltage, the power line frequency, 60 Hz, is remarkably constant and will seldom fluctuate more than a half to one cycle. As a result, the average frequency over a given time period is either 60 Hz or quite close to it.

MOTOR CURRENT FEED

Current can be brought directly into field and armature coils from a voltage source, as in the case of DC machines such as the series, shunt, and compound types, or in AC series motors, or into the stator of AC machines by conduction and the rotor by induction. Typical examples are the AC induction and repulsion motors.

THE BASIC AC MOTOR

The basic AC motor, shown in Figure 4-2, is essentially the same as the fundamental DC motor, but there is an important difference. Instead of a commutator, the AC motor uses a pair of slip rings. Since the input to this motor is AC, no commutation is required. Unlike the commutator, slip rings consist of

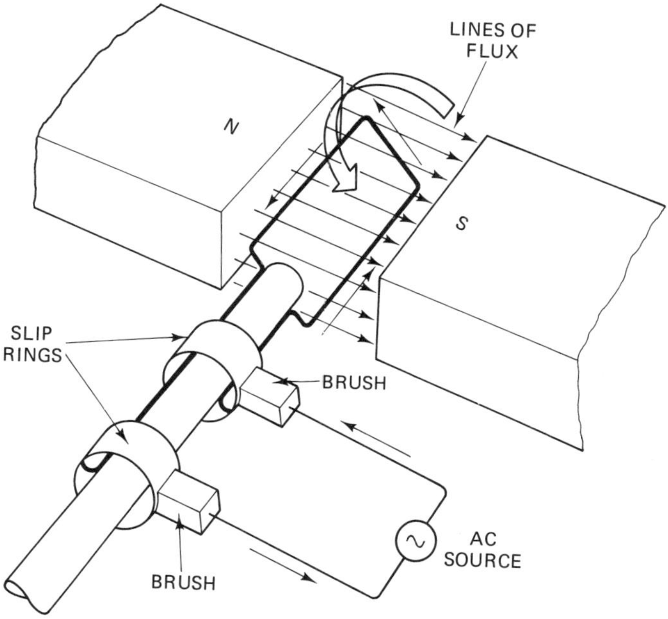

Figure 4-2 Basic AC motor.

separate conductive segments insulated from each other. The two rings are continuous metal bands, have no physical contact with each other, and are insulated from the shaft which supports and rotates them. Slip rings are necessary to supply the AC source voltage to the rotating armature.

As in the case of DC motors, the brushes can be any one of the carbon or metallicized types.

Armature rotation is obtained due to the fact that the magnetic polarity of the armature keeps changing in step with the input AC. This produces forces of attraction and repulsion between the armature and the field poles. The field, as shown in this illustration, is supplied by a permanent magnet, but a field coil supplied by DC could also be used.

PHASE

Motors and AC generators are quite similar in the way they function. For a motor the input is electrical energy; the output is mechanical energy. A generator is just the reverse. It takes in mechanical energy; its output is electrical energy.

Figure 4-3 is a simplified drawing of an AC generator. It consists of a

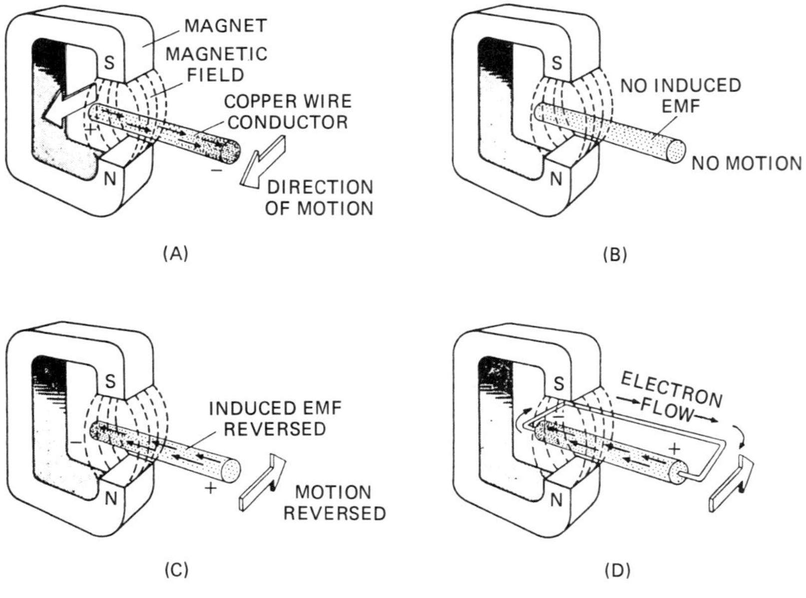

Figure 4-3 Method of producing an induced voltage (EMF). In A, the wire is moving through the magnetic field of the magnet and a voltage is induced across the conductor. In B, there is no induced voltage since the conductor is not moving. When the wire moves again, as in C, there is another induced voltage. Drawing D shows the direction of current flow.

permanent magnet field which supplies a north and south magnetic pole. Positioned between these two poles is an armature shown here as a single-turn coil; this would have a large number of turns in an actual machine. Not shown in this drawing is a motor whose drive shaft rotates this armature.

The armature, rotating in the magnetic field supplied by the field magnet, has a voltage induced across it. This induced voltage is brought to a pair of slip rings mounted on the armature shaft. This voltage could then be led from the slip rings to an AC outlet. The AC voltage supplied by this simple generator will have a frequency determined by the speed of rotation of the armature, and the amplitude of the voltage will depend on the strength of the magnetic field supplied by the field magnet and by the number of turns of the armature winding. The output AC will be single-phase sine wave and will resemble the waveform in Figure 4-4.

If another coil, similar to the first, is mounted on the same shaft as the first, but with both coils at right angles to each other (Figure 4-5A), a voltage

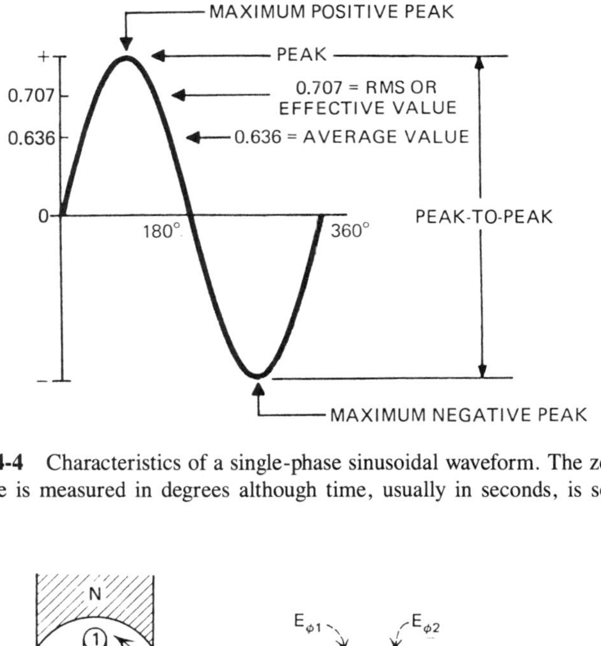

Figure 4-4 Characteristics of a single-phase sinusoidal waveform. The zero reference line is measured in degrees although time, usually in seconds, is sometimes used.

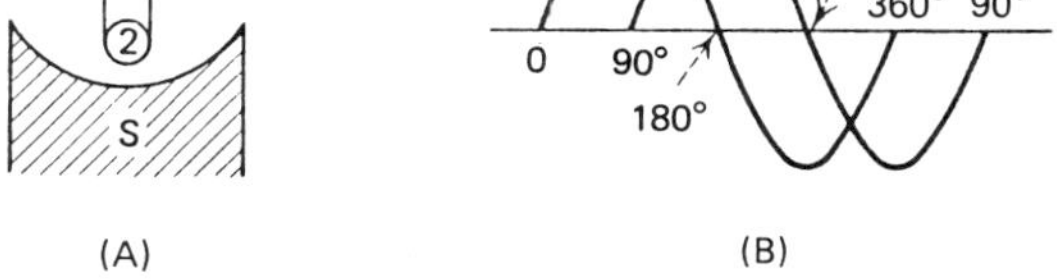

Figure 4-5 Independent armature coils spaced 90° apart (A) produce a pair of sine wave voltages having a phase separation of 90° (B).

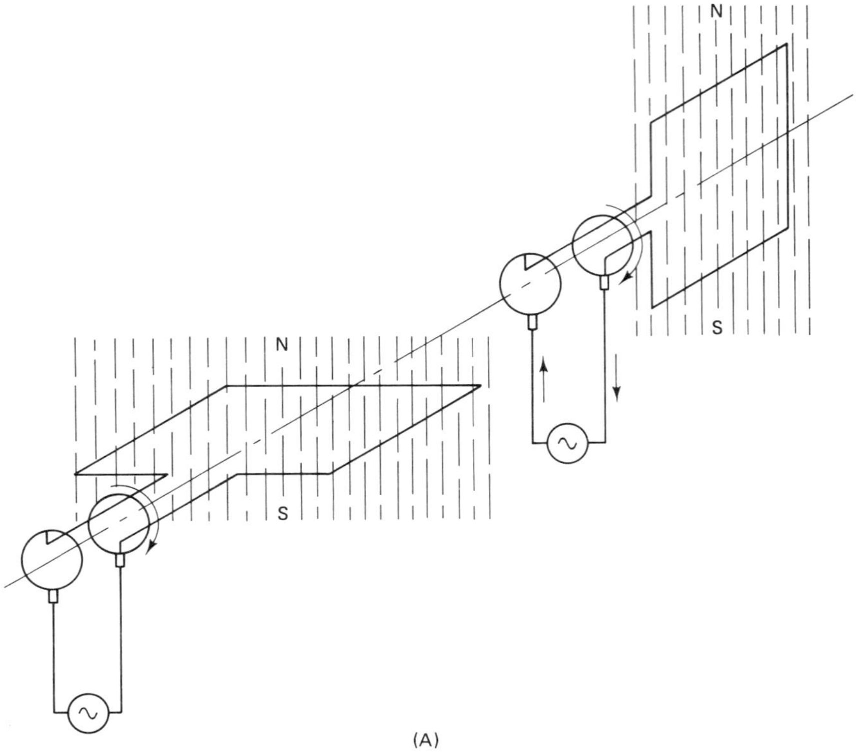

(A)

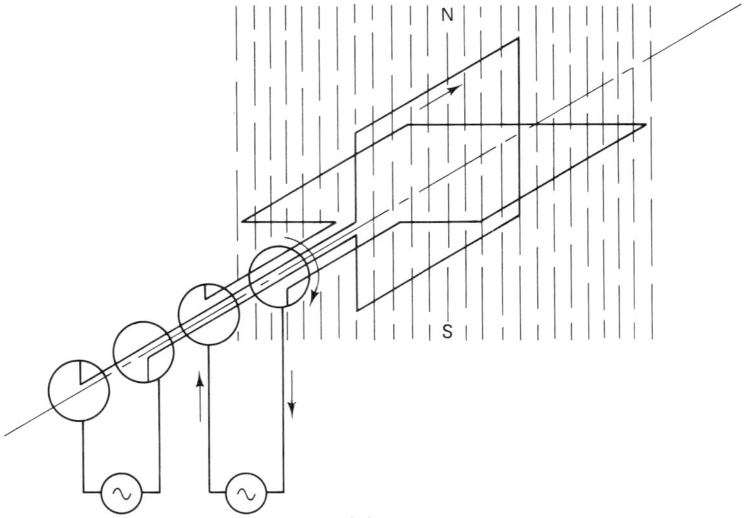

(B)

Figure 4-6 The two single-phase generators (A) can be combined into a single generator supplying two independent AC voltages that are out of phase by 90° (B).

will be induced across both. We now have a pair of identical voltages with the same amplitude and frequency, both brought out of the generator via slip rings. Since these two voltages are independent, the first can be identified as phase 1 ($E_{\phi 1}$) (Figure 4-5B), and the second as phase 2 ($E_{\phi 2}$). The two voltages are alike in every way except that of time.

Because of the way the armature coils are positioned with respect to each other, separated by 90 degrees, their induced voltages will start and finish at different times, as indicated in the graph in Figure 4-5B. Because the phase-1 voltage starts first, it is said to lead the phase-2 voltage. Conversely, the phase-2 voltage can be said to lag voltage phase 1. And if both armature coils are identical, each of the two voltages will have the same amplitude. Since they are mounted on a shaft and rotate at the same speed, their frequencies will be identical. These two voltages (Figure 4-6A) are led out of the generator on individual slip rings and so the two out-of-phase voltages can be used independently (Figure 4-6B).

This basic concept of developing two independent voltages can be carried a step further as indicated in Figure 4-7A. Here the number of separate armature coils has been increased to three, with each of these coils separated by 60 degrees. The voltage produced by each of these coils is brought out to three pairs of independent slip rings. The slip rings are made of continuous rings of a metal (like hard copper) and are insulated from each other and from the shaft on which they are mounted. Not shown in any of these drawings is the motor shaft that drives the shaft of the armature of the generator.

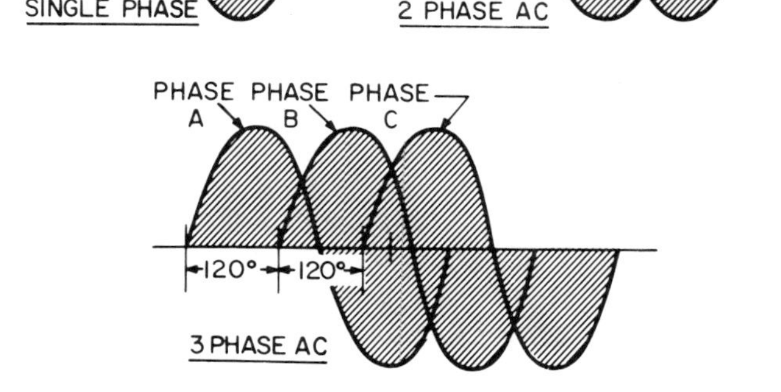

Figure 4-7A Single-phase, two-phase, and three-phase AC voltages.

The graph in Figure 4-7B shows the three out-of-phase voltages that have been produced. They all have the same amplitude and the same frequency, but each has a different starting and completion time. This generator can supply single-phase AC, two-phase AC, and three-phase AC. A motor that uses just one of these phases is known as a single-phase motor; one that uses two or more is referred to as polyphase. Practically, motors are either single-phase or three-phase types. Single-phase AC is the type available at most in-home power outlets and is used for motors in food mixers, electric fans, and small air conditioners.

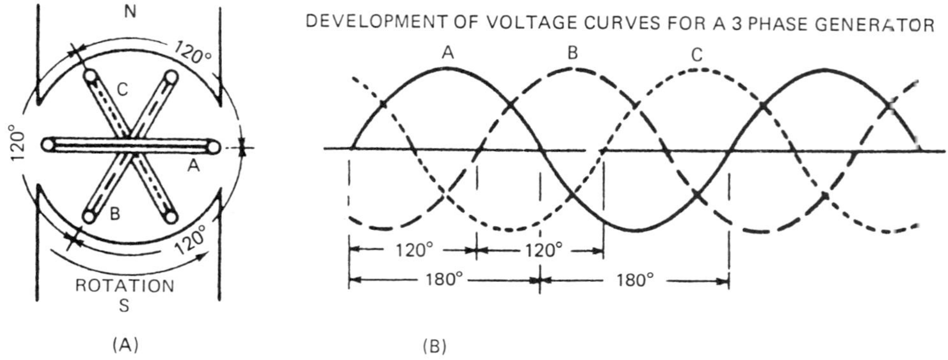

Figure 4-7B Independent armature coils, A, B, and C, are spaced 60° apart (A) and produce three independent AC voltages, also spaced 60° apart (B).

Two-Phase Connections

A single-phase and a polyphase motor can be connected to an AC power line depending on the wiring arrangement of that line. As an example, consider the two-phase AC power arrangement shown in Figure 4-8. The polyphase motor has a pair of armature coils at right angles to each other, indicated by drawing the symbols for the coils at 90 degrees. Each of these coils has its own pair of slip rings, and so the output from each is 120 volts. The illustration shows the use of two motors: one a 120-volt single-phase type, the other a 120-volt two-phase. The single-phase motor has just two AC input terminals; the two-phase motor has four. Additional loads, such as lamps, can be connected across either or both of the 120-volt lines. An advantage of this polyphase arrangement is that the load current is distributed between two pairs of power lines.

Delta and Wye Connections

The drawing of the two-phase system in Figure 4-8 shows that the armature coils are not internally connected. However, there are two ways of connecting them. The one shown in Figure 4-9A is referred to as a delta connec-

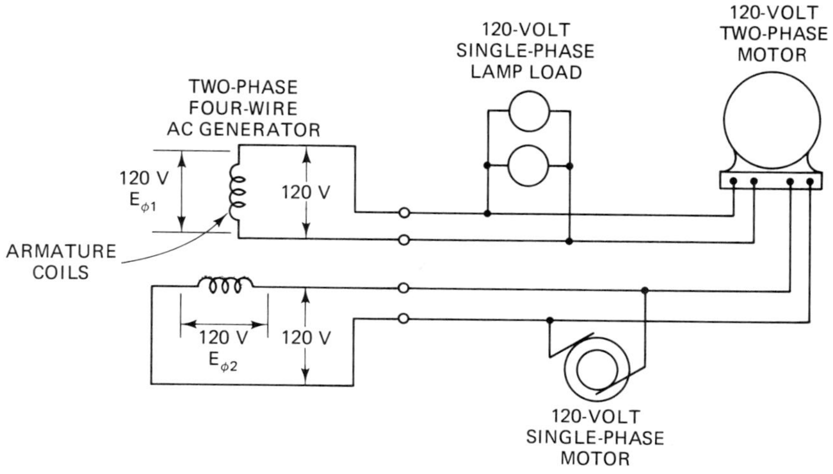

Figure 4-8 A two-phase AC generator supplies operating voltage for single-phase and two-phase motors and can operate other loads as well.

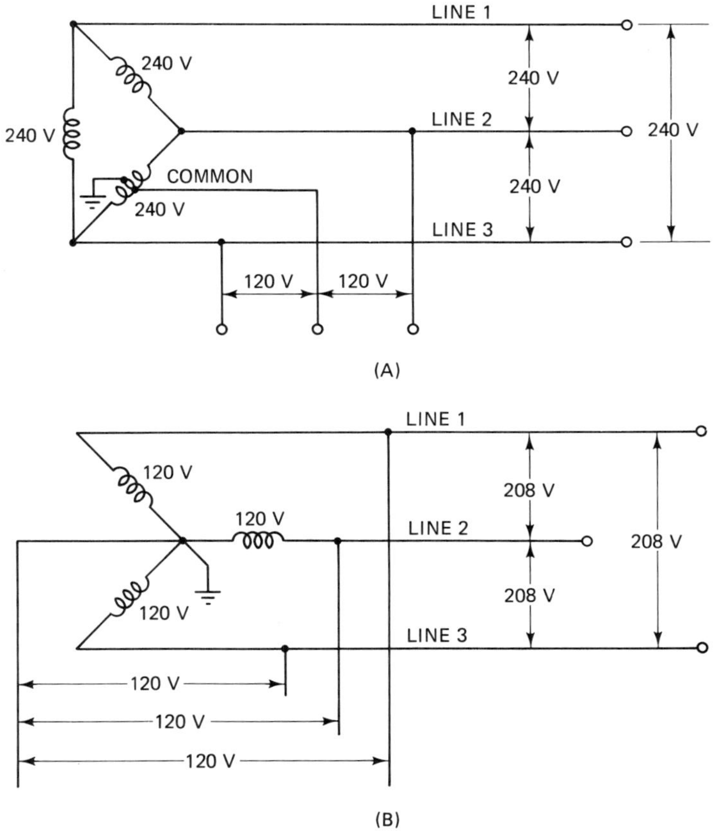

Figure 4-9 Delta connection (A); wye connection (B).

tion because of its resemblance to the Greek letter delta, while the other, in Figure 4-9B, is known as a Y, usually spelled wye, because it resembles that letter.

Note that in Figure 4-9A, which uses a delta-wired armature, that each of the armature coils develops 120 volts, and so this is the voltage available for a three-phase motor. Other 120-volt loads can be connected across two of the output leads. With the delta connection, the maximum available output voltage is that produced across one of the armature coils.

Figure 4-9B is the drawing of a wye system. Unlike the delta system in which the voltage output is from any one of the three armature coils, voltage output in the wye is the voltage from across two of the armature coils that are connected in series. However these voltages are not directly additive. Thus the individual armature voltage is 120 volts and the output from across two of these coils is not 2×120 but is 1.73×120 since the two voltages are not in phase but are out of phase. The output voltage is approximately 208. This wye arrangement is known as a four wire system. Note this is done by connecting a lead to the junction at which all three armature coils are connected. The advantage of this arrangement is that the AC generator can supply two different output voltages. One of these is 120 volts, used for operating appliances requiring this input voltage, possibly small units such as fans and mixers, while the higher voltage can supply the motor of a device such as a washing machine. The higher voltage load could be a three-phase motor; the lighter loads are single-phase types.

Single-Phase Power

While single-phase AC is commonly used in the home and industry, single-phase AC generators are rare. In a single-phase system only one AC voltage is involved. The power delivered to home power outlets is single-phase and most home motorized electrical appliances are designed to work from single-phase AC, but this power is generated by a polyphase source.

The generation of polyphase power is preferable since such power is less pulsating than single-phase power. Further, polyphase motors are more efficient, and less copper is needed in a three-wire power distribution line than in a two-wire line. Two-phase is the most elementary of the polyphase systems, but electric power distribution most often uses three-phase.

Three-Phase Power

Polyphase rotating-coil or rotating-field AC generators are equipped with three coils with a separation of 60 degrees or 120 degrees between coils, depending on how the coils are wired.

AC Power Delivered to the Home

Electrical power delivered to homes may be three-wire AC, single-phase with 120 volts each phase. This system also supplies 120 + 120 or 240 volts. However, if a three-phase system is used, consisting of a neutral and three other feeders, the voltage delivered may be 120 volts, single-phase, or $120 \times 1.732 = 208$ volts.

SINGLE-PHASE vs. POLYPHASE MOTORS

AC motors can be placed in two basic categories depending on the type of line voltage input. This input can be either single-phase or polyphase. In turn, polyphase motors comprise two-phase and three-phase units.

Single-phase motors. Single phase motors include repulsion, induction, series, capacitor, and synchronous. There are also some variations under these general headings.

Polyphase motors. A polyphase motor is one whose AC input source voltage consists of more than one phase. This polyphase voltage can be supplied by a power line, that is, the power line may be able to deliver a multiphase voltage, or else single-phase input power may be modified in some way so that it assumes the characteristics of polyphase power. Thus, the fact that a motor is connected to a single-phase voltage source does not necessarily mean that the motor is a single-phase type.

If the AC input to a motor is from a polyphase power line, this input is usually three-phase, since two-phase AC power is rare. Despite this, as indicated in the preceding paragraph, there are two-phase motors with single-phase power converted to two-phase by a reactive component, such as a coil or capacitor, that is part of the motor and is external to the power line.

POWER FACTOR

The fact that a component is connected to an AC power outlet does not necessarily mean the component accepts or uses the power delivered to it. If a coil or a capacitor is connected to an AC outlet very little of the AC power supplied to it will be used, and almost all of it will be returned to the source. Coils and capacitors do have some resistance associated with them, and so some power is used by these components.

Like resistors, coils and capacitors do have an opposition to the flow of alternating current, described as reactance, which, like resistance, is measured in ohms. The ratio of the resistance of a coil or capacitor to its impedance is called the power factor, abbreviated as pf. Pf can be expressed as

$$pf = \frac{R}{Z}$$

where R is the resistance of the component and Z is its impedance, with both expressed in ohms.

Power Factor in Resistive Circuits

For either single- or polyphase power, the power factor in resistive circuits is 100 percent, sometimes called unity power factor. Since 100 percent power factor = 1, this digit is omitted in resistive power calculations.

Example

What is the power used by a 100-ohm resistor carrying a current of 5 amperes? The resistor is connected across two wires of a single phase voltage.

Solution

$$P = I^2 \times R \times PF$$

$$\text{Since } PF = 100\% = 1$$

$$P = I^2 \times R = 5 \times 5 \times 100 = 25 \times 100 = 2{,}500 \text{ watts}$$

$$2{,}500/1{,}000 = 2.5 \text{ kilowatts}$$

Real Power vs. Reactive Power

Power that is used and not returned to the power source is called real power. Power returned to the source is known as reactive power. For real power only resistance (R) is involved.

$$P = I^2 \times R$$

For reactive power only inductive or capacitive reactance is involved.

$$P = I^2 \times X$$

For apparent power involving reactance and resistance

$$P = I^2 \times Z$$

The ratio of real power to apparent power supplies the power factor (pf).

$$pf = \frac{\text{real power}}{\text{apparent power}} = \frac{I^2 \times R}{I^2 \times Z} = \frac{R}{Z}$$

Wattmeters measure real power only, also known as true power. The power factor is a comparison between real power and apparent power and can also be written as

$$pf = \frac{E \times I \times \cos \theta}{E \times I}$$

or

$$pf = \cos \theta = \cos \frac{R}{Z}$$

VA and KVA

VA is the product of volts and amperes. Thus, VA = V × A where V is in volts and A is in amperes. KVA is used when large voltages are involved and is convenient for avoiding large numbers. Thus, 3,000 volts is equal to 3 kv. KVA is the product of voltage, measured in kilovolts, multiplied by the amount of current in amperes.

VARS

A var is the reactive power of one reactive volt-ampere.

$$var = E \times I_{reactive}$$

A kilovar = 1,000 vars. Var is an acronym made of the first letters of the words volt-ampere reactive. While a wattmeter measures only true or real power, reactive power can be measured by an electrodynamometer. The instrument measures reactive power only for sine waves having a frequency for which the instrument is adjusted.

Need for a High Power Factor

If a factory has an average working requirement of 200 amperes at 220 volts, it uses a total of 200 × 220 or 44,000 volt-amperes. 44,000/1,000 = 44 kilovolt amperes (kva). If this factory has a 100 percent power factor, kva is equivalent to true watts.

If a similar factory also requires 200 amperes at 220 volts but has a 50 percent power factor, the kva is still 44, but the real power is only half this amount.

In both examples the same amount of energy is delivered, but in the second case the factory uses only half. For this reason electric utilities charge extra where a low power factor is involved. A factory might have a low power factor if it uses reactive components extensively, such as motors.

For home use the power factor is close to 100 percent since most home appliances are resistive—boilers, electric ranges, toasters, heating pads, wall heaters, lamps, etc. Some appliances, such as electric fans, turntables, cassette decks, and compact disc players, do have a reactive component, but this is generally small.

The formula for kilovolt-amperes in single-phase circuits is

$$KVA = \frac{E \times I}{1{,}000}$$

The formula for kilovolt-amperes in three-phase circuits is

$$KVA = \frac{E \times I \times 1.732}{1{,}000}$$

For single-phase motors, the horsepower output can be calculated from

$$HP = \frac{\eta IVpf}{746}$$

For three-phase motors the formula is

$$HP = \frac{\sqrt{3}\,\eta IEpf}{746}$$

where HP is the output in horsepower, η is the efficiency of the motor, I is the line current, E is the line voltage, and pf is the power factor.

If the AC input to a motor is from a polyphase power line, this input is usually three-phase, since two-phase AC power is rare. Despite this, there are two-phase motors with single-phase power converted to two-phase by a reactive component such as a coil or capacitor that is part of the motor and is external to the power line.

Power in Three-Phase Circuits

The power formulas for three-phase circuits are the same as those used for single-phase, except that the $\sqrt{3}$ factor or 1.732 is involved.

The formula for a single-phase resistive circuit is

$$P = E \times I$$

If the circuit is three-phase resistive, it is

$$P = E \times I \times 1.732$$

Finally, if the three-phase circuit is reactive, that is, is either capacitive or inductive, the formula is

$$P = E \times I \times pf \times 1.732$$

Although this looks like a new formula, it is the same one used previously for power, but now taking pf and polyphase voltage into consideration.

Example

A three-phase star-connected motor is connected to a 208-volt circuit. It draws a current of 10 amperes and has a 50% power factor. How much input power is required by this motor?

Solution Since the motor windings are inductive, the motor, as a load on the power supply, is regarded as a reactive element.

$$P = E \times I \times pf \times 1.732$$

$$P = 208 \times 10 \times 0.5 \times 1.732 = 1{,}801.3 \text{ watts} = 1.8 \text{ kilowatts}$$

The power factor of a three-phase motor can be calculated from

$$pf = \frac{P}{\sqrt{3}(E_L)(I_L)}$$

where pf is the power factor, P is the power delivered to the motor, E_L is the line voltage between any two phases, I_L is the line current, and $\sqrt{3} = 1.732$.

Current in Three-Phase Circuits

The formula for finding the current in a three-phase circuit can be developed in the same way.

$$I = \frac{P}{E \times pf \times 1.732}$$

Horsepower, Efficiency, and pf in Three-Phase Circuits

As in the case of single-phase circuits, or in working with DC motors, efficiency is a factor in electrical calculations for three-phase work. If the HP, operating voltage, efficiency and pf are known, the amount of current can be calculated by

$$I = \frac{746 \times HP}{E \times eff \times pf \times 1.732}$$

This formula is similar to those used previously, except that the $\sqrt{3}$ factor (1.732) is now involved in the calculations.

Example

How much current is required by a 5 HP, 220-volt, three-phase motor whose operating efficiency is 90% and power factor is 80%?

Solution

$$I = \frac{746 \times HP}{E \times eff \times pf \times 1.732}$$

$$I = \frac{746 \times 5}{220 \times 0.9 \times 0.8 \times 1.732} = \frac{746}{44 \times 0.9 \times 0.8 \times 1.732}$$

$$= \frac{746}{54.9} = 13.59 \text{ amperes}$$

Efficiency

Efficiency, previously considered in connection with DC motors, is also a factor in AC machinery. If the horsepower is known in a single-phase circuit, the current can be calculated by

$$I = \frac{746 \times HP}{E \times eff \times pf}$$

This formula is much the same as the formula for current for DC motors. The only difference is that in AC circuits the power factor must be included if it is less than unity. Efficiency is sometimes represented by the Greek letter η.

Example

An AC motor rated at 3 HP is connected to a single-phase 110-volt AC power line. The efficiency of this motor is 90%. What are the current requirements of this motor, rated as having a 90% power factor?

Solution

$$I = \frac{746 \times HP}{E \times eff \times pf} = \frac{746 \times 3}{110 \times 0.9 \times 0.9} = \frac{2,238}{89.1} = 25.1 \text{ amps}$$

Efficiency is also a factor in electrical calculations for three-phase work If the HP, operating voltage, efficiency, and pf are known, the amount of current can be calculated by

$$I = \frac{746 \times HP}{E \times eff \times pf \times 1.732}$$

This formula is similar to those used previously, except that the $\sqrt{3}$ factor is now involved in the calculations.

Example

How much current is required by a 5 HP, 220-volt, three-phase motor whose operating efficiency is 90% and power factor is 80%?

Solution

$$I = \frac{746 \times HP}{E \times eff \times pf \times 1.732}$$

$$I = \frac{746 \times 5}{220 \times 0.9 \times 0.8 \times 1.732} = \frac{746}{44 \times 0.9 \times 0.8 \times 1.732}$$

$$= \frac{746}{54.9} = 13.59 \text{ amperes}$$

Example

A three-phase motor whose efficiency is 90% works with a pf of 80% from a 208-volt power line. It has a 2-HP rating. What is the current drain of this motor?

Solution

$$I = \frac{746 \times \text{HP}}{E \times \text{eff} \times \text{pf} \times 1.732} = \frac{746 \times 2}{208 \times 0.9 \times 0.8 \times 1.732}$$

$$= \frac{1{,}492}{259.38} = 5.75 \text{ amperes}$$

Power Delivered to a Three-Phase Motor

The formula shown below is applicable to wye and delta wound stator windings.

$$P = \sqrt{3}(E_L)(I_L) \cos \theta$$

where P is the total power delivered to the three-phase motor, E_L is the line voltage measured between any two phases, I_L is the current in a single line, θ is the angle between voltage and current in any phase, and $\cos \theta$ is the power factor and is equal to R/Z.

Two-Phase Motor Operating from a Three-Wire Line

It is possible for a two-phase motor to work from any two-phases of a three-phase power line. The voltage relationships are expressed by

$$E_d = \sqrt{2}(E_p)$$

where E_d is the voltage across the stator winding and E_p is the voltage measured from the center tap of the stator winding to either side of the line.

Summary of Characteristics of the Three-Phase Motor

Load: Difficult starting load

Starting Torque: 200% to 300% of the running torque

Starting Current: Three to six times as much as the running current

Horsepower: 1/2 to 400 HP

Phases: Three

Source Voltage: 120 to 240 AC, 240 to 480 and higher

Speed: 900, 1,200, 1,800, 3,600 rpm
Reversibility: Yes
Uses: Industrial machinery, elevators, pumps

ROTATING MAGNETIC FIELD

If a bar magnet is suspended above a compass and if the magnet is rotated horizontally, the pointer of the compass will follow the movement of the magnet. Depending on the initial position of the bar magnet and the compass pointer, the effect of the two magnets, the bar and the compass, will be one of attraction or repulsion. In the case of repulsion it will be as though the bar magnet pushes the compass pointer, and in the case of attraction the effect will be as though the bar magnet pulls the compass pointer.

For a polyphase motor, the magnetic field around each field coil increases and decreases, but in a specific successive time frame. First one magnetic field reaches a peak, then the next, and finally the remaining field. This succession of magnetic fields produces an effect comparable to a single rotating magnetic field. It is this rotating magnetic field which induces a current in the armature, which produces its own magnetic field. The reaction between this field and that of the rotating magnetic field of the field coils is such that the armature is compelled to rotate. This rotation can be caused by attraction or repulsion between the two magnetic fields, that of the field coils and that of the armature, and is the basis of AC motor operation.

The same effect takes place in DC motors. The DC supplied to the armature is changed to AC by the action of the commutator. However, there is also a time displacement of the currents flowing through the armature and so the result is that of a rotating magnetic field.

THE UNIVERSAL MOTOR

While it is convenient to classify motors as AC or DC, there is one type that will operate from either DC or single phase AC (Figure 4-10) and so is referred to as universal. It is widely used for home appliances such as sewing machines, vacuum cleaners, fans, electric razors, mixers, food grinders, and so

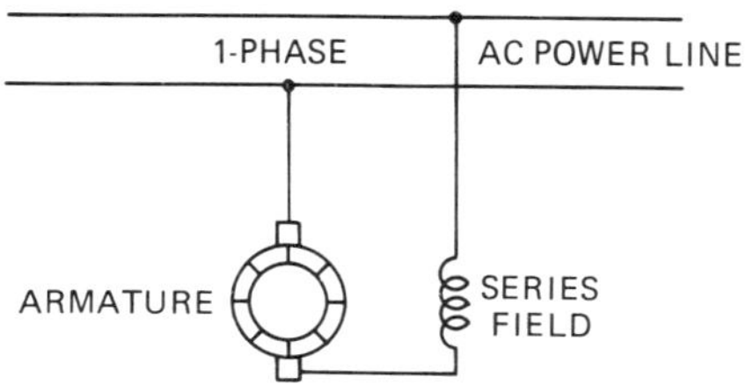

Figure 4-10 AC single-phase power connection for the universal motor.

on. Since the use of line powered DC has long since been replaced by AC, the fact that a universal motor can operate from a DC power line is a touted feature of no great value.

Figure 4-11A is a cross-sectional view showing a pair of electromagnetic type field poles, adjacent to a curved shoe surrounding the armature, the rotor, as closely as possible. Figure 4-11B is the symbol for this motor. Sometimes in using this symbol, as in drawing 4-11C, the motor frame or housing is shown as a circle while the field coil is represented by a jagged line. The difference in symbols arises from the fact that the drawing in 4-11B is taken directly from electronic symbols; that in 4-11C is from electrical symbols.

The universal motor has a series type winding, that is, the same current flows through both field coils and the armature and, like DC motors, is equipped with a commutator and brushes. There are some variations in construction of the stator winding. The straight series type is wound around the stator pole. The compensated series-wound type has the field coils positioned in slots. This results in better speed regulation and a higher starting torque. The straight series type is simpler in its construction and is used in sizes that range up to 1/3 HP while the compensated type is usually found in higher rated motors.

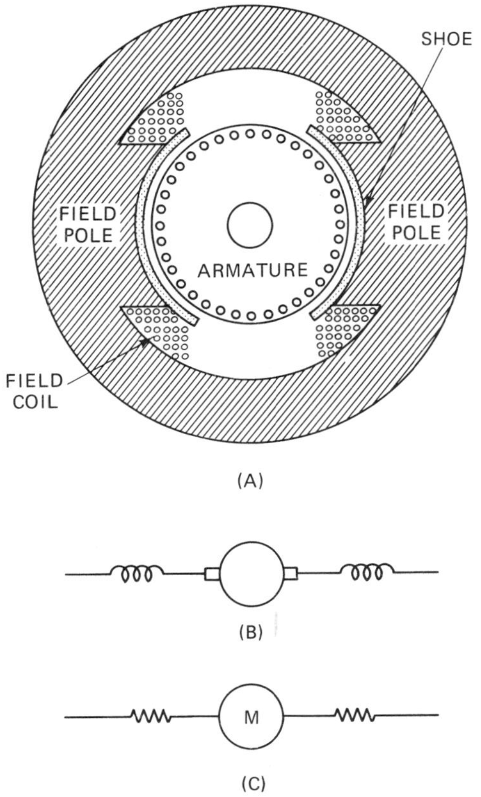

Figure 4-11 Universal motor (A); symbol (B); alternate symbol (C). A commutator is used instead of slip rings to enable the motor to operate on DC.

The universal motor has certain advantages over other AC types. It has a higher operating speed which can be controlled, a feature that is desirable for certain types of home appliances, such as electric fans, sewing machines, and food mixers. The speed can be governed by putting a resistor in series with the line connected to the motor. The resistor, built into the appliance, can be a single fixed unit having a number of taps, permitting the selection of a number of predetermined speeds, or can be a variable resistor (Figure 4-12) with speeds ranging from close to zero to maximum. This feature is found, for example, in electric drills, with speed selection governed by the trigger. The torque of the motor is maximum at its lowest speed and decreases as the speed is permitted to increase. If, as in the case of an electric drill, the drill piece is removed from the work, the motor speed will rise. In the absence of a load the rotational speed of the armature may become greater than 10,000 rpm, possibly from 12,000 to 18,000 rpm. Representative full load speeds are from 5,000 to 10,000 rpm.

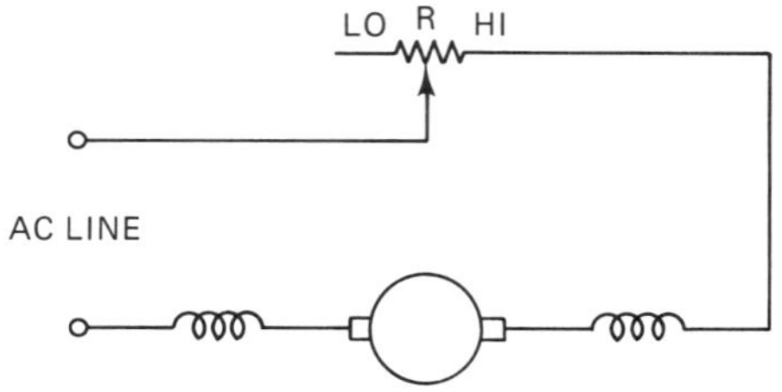

Figure 4-12 Speed control for a universal motor. The rheostat, R, is simple and effective, but there is a power loss in this component.

Another method of speed control of the universal motor is the use of a tapped field winding (Figure 4-13). The taps are arranged so that more or less of the field winding is used, thus governing the strength of its magnetic field. This arrangement has two advantages over the series resistor method. The resistive arrangement can have a substantial power loss and, further, can result in the production of considerable heat. Thus, the tapped winding method reduces operating costs.

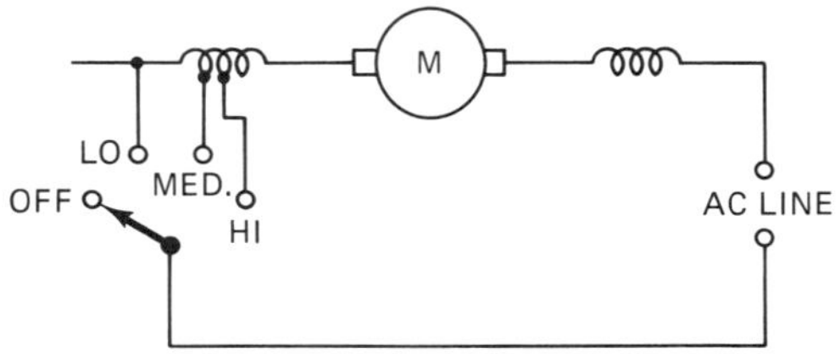

Figure 4-13 Universal motor with tapped field winding.

The direction of rotation of the armature cannot be changed by transposing the line cord plug in its socket. A change in direction can be obtained by reversing the direction of current flow through either the armature or the field coils, but not both. This does not imply that the motor will operate as efficiently. Appliance manufacturers select that rotational direction which will permit the appliance to supply best performance.

Construction of the Universal Motor

The two field windings of the small universal motor (Figure 4-14) such as those ranging from 1/20 to 1/8 HP consist of two coils of wire held in position in a field frame made of thin sheets of steel laminations with the field coils wrapped with insulating tape to keep the two from shorting. The armature is also made of laminated steel but is slotted longitudinally with the armature winding inserted into these slots. The entire armature assembly is mounted on a steel shaft on which the commutator is also mounted with the armature leads connected to it. In some universal motors, a set of fan blades is positioned on the shaft. Current is brought into the armature coils via soft carbon brushes which are spring loaded against the commutator segments.

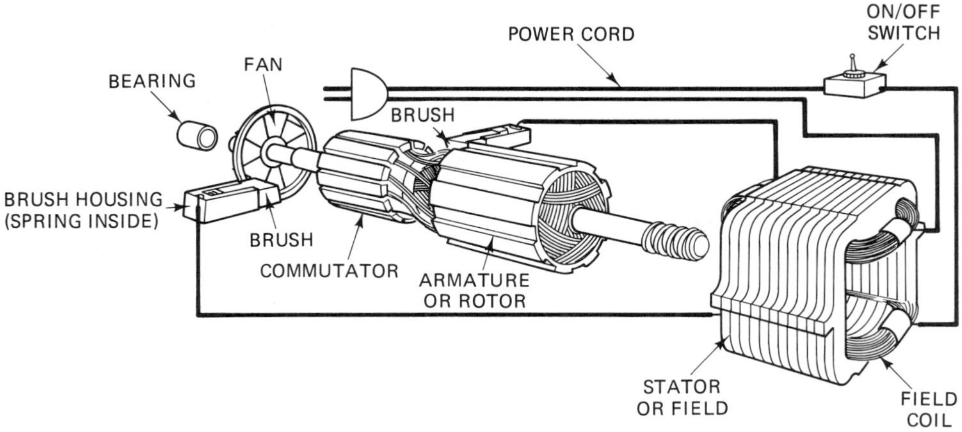

Figure 4-14 Construction of a universal motor.

Basic Operating Characteristics

The universal motor is characterized by high speed and high torque. It is sometimes used with a reduction gear in applications where a slower speed is desirable. Because the universal motor is series-wound, its behavior is similar to that of the series-wound DC type. While its no-load speed is quite high, it is not enough to damage the motor, something more likely to happen with large size DC units. In practice, these motors are usually operated at 3,500 rpm, but sometimes higher. In addition to changing its speed with the load, the speed can be varied by changing the frequency of its AC voltage input or its amplitude. The speed can be changed over wide limits by voltage control. For the typical in-home line voltage, the frequency is remarkably constant and is usually 60 Hz, although 50 Hz is sometimes used.

For a given size and weight, this motor has high power output. However, the motor does have some disadvantages. The fact that it is capable of high speeds means that its thrust bearings at each end of the motor axle must be well made for they can show substantial wear due to high armature speeds. Universal motors are not as quiet as AC motors that work without the need for commutators. Brush sparking can cause ratio or television interference and the brushes may require frequent attention.

Types of Universal Motors

There are two basic types of universal motors: the concentrated-pole noncompensated motor, and the distributed field compensated motor. Usually motors rated at 1/4 HP or less fall within the first category; those having a higher power rating in the second. Universal motors are most often fractional HP types, in the range of 1/200 to 1/3. In terms of electrical power input this is from about 4 watts to approximately 250 watts.

The fact that motors are identified as universal does not mean they are all identical. One of these, the concentrated-pole noncompensated type is less costly to manufacture since the field-pole laminations, the pole pieces, and the yoke are the result of a single machine punch. However, this technique has resulted in a lower performance capability than the distributed field compensated type since its HP rating is lower. Another way of categorizing universal motors is by the number of field poles or by the way in which these poles are constructed. Quite commonly there are two field poles diametrically opposite each other and mounted on the interior of the motor frame (Figure 4-15). The second type has the field coil winding mounted in slots cut in the iron core of the coil. Since the field and armature windings are in series, just as they are in the series-wound DC motor, the same amount of current flows through both.

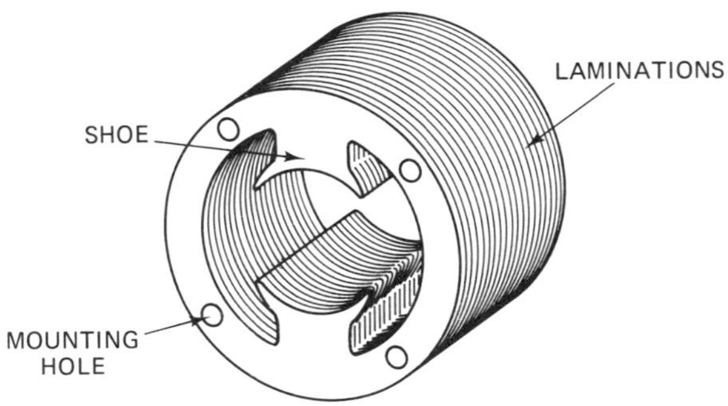

Figure 4-15 Laminated structure of the core for the field coils of a universal motor.

The DC/AC Feature

Offhand, it would seem highly desirable to be able to use either DC or AC as the input voltage, but this feature is of no great value. Practically all homes are equipped with AC, and while some factories have both AC and DC, both have about the same voltage value. Further, the frequency of the AC line must be about 60 Hz. It is of some help to the motor servicing technician that the word universal used on the motor's name plate is an immediate indication that the motor is a series-wound type.

Universal Motor Brush Problems

Since the universal motor will often operate at high speeds it is to be expected that the brushes may show substantial wear, especially since they contact commutator segments instead of slip rings. Excessive brush wear or inadequate brush spring tension can result in a completely inoperative motor. Typically, the length of a brush is about 1/2 to 3/4 inch.

Manufacturers supply brush and spring replacements for their universal motors and it is always best to use an exact substitute for this repair. If a repair manual is available the correct brush, brush holder, and associated spring can be identified by a manufacturer's part number. An incorrect replacement can cause erratic motor operation. The correct spring tension should be about 1.5 to 2 pounds per square inch of brush surface.

The commutator has a circular contour and it is necessary for the brush surface to have the same shape. This can be done by wrapping the commutator with fine sandpaper and then, with the brush contacting the sandpaper, rotating the armature by hand. The abrasive side of the sandpaper should face the brush.

If the commutator has worn spots, it can be serviced with a lathe. This should be done if inspection reveals grooves. If commutator wear is not excessive it can be sandpapered manually.

Mica is used to separate and insulate the commutator segments. The mica should have the same rate of wear as the commutator segments but this is not always the case. If the copper wears more, the mica will show above the surface of the commutator bars, a condition known as high mica. If a lathe has been used to true the commutator surface, the mica and the commutator segments will be flush. The mica can then be undercut using a hand-held hacksaw blade. The mica should be cut until it is about 1/32 inch below the surface of the commutator.

Armature Problems

There are four possible armature defects: three are electrical and one is physical. Electrical defects include a ground fault in the armature or an open or a short circuit. The armature and field poles are very close, and so an eccentricity in the shaft, loosened mounting screws, or excessive bearing wear can produce physical contact.

The grounded armature. The insulation of the armature winding may be defective, worn, or may have been subjected to a hostile environment. The wiring of the armature may contact the edge of one or more laminations and may do so intermittently.

Under these circumstances, the motor may produce a thumping noise during rotation, possibly at a specific speed. The noise may vary in intensity. Other possible symptoms include smoking or the motor frame being hot to the touch. To check, use an ohmmeter and do a continuity test between each commutator bar and the armature shaft. Repeat this test for each commutator segment. The ohmmeter should read open circuit for each of these tests. When doing this test do not depend only on the on/off switch. This switch should be in its off position and the power plug should be disconnected from the power source.

Open circuit. An open circuit in one of the armature coils could be caused by a break or by separation of the end of the armature coil from its associated commutator bar. The separation may be permanent or intermittent. The symptoms may be the same as for a grounded armature coil. To check, measure the resistance between the commutator segments to which the suspected coil is connected. The ohmmeter should read the resistance of the coil. If it shows an open circuit or gives an infinite resistance reading, the problem is that of an open in that armature coil circuit.

Physical contact between rotor and stator. This is evidenced by noise from the motor. Under severe conditions the motor will get hot. The problem may be due to poor seating of the motor shaft in its bearings or a bent motor shaft.

Troubleshooting the Universal Motor

Motor does not start. Power may not be reaching the motor. Trace the power loss as far back as the fuse block. Check fuses or circuit breakers. Test the power outlet with a test lamp. Examine the plug and make sure the power cord from the motor is properly connected. Try a different power outlet. If DC is used as the energy source, try a different battery. Test the battery under load. A voltage check of battery is useless since even a practically discharged battery will show potential across its terminals. Put the battery on charge if it is a rechargeable type.

The motor's on/off switch may be defective. Check connections of power cord at the motor's power input terminals. Make sure armature brushes are properly seated, are not worn out, and make good contact with armature's commutator. Brushes may be sticking or the commutator may be dirty. If arcing appears at commutator, the power cord and power outlet are working.

Check to make sure armature is free to rotate. It may be rubbing against the field poles. Check bearings to make sure they have not seized. Resistor speed control, if one is used, may be open or may be set for maximum resistance. If motor uses reduction gears these may be jammed.

Either of the field coils may be open.

Motor vibrates or is very noisy. The motor bearings may be so worn that the shaft has too much play in its end thrust. Bearings may be too dry due to lack of lubrication. The motor may have loose parts. Check all screws, nuts, and bolts. Make sure interior of motor is free of lint and dirt. If the motor has a cooling fan make sure its blades do not strike or rub against adjoining motor parts. The fan should be securely fastened to the motor shaft. Support for the motor should not be loose.

Motor overheats or smokes. One of the armature coils may have a shorted turn. This could be indicated by excessive sparking at the commutator. The field coils could also have a shorted turn. This is a possibility if the field coil is very hot to the touch after the motor has been on for a very short time. The same is true of the armature coils. If the armature and field coils are covered with low grade insulation and motor is operated in a damp environment, excessive current leakage is a possibility. Dry the motor windings using a heat lamp but do not overheat and char the insulation.

Motor sparks or causes excessive RFI. Excessive arcing at the commutator can cause RFI (radio frequency interference). This can result in a crackling sound in AM radio receivers, much less so or not at all in FM sets. It may not affect television sound. RFI reaches radio and television receivers by feedback through the power line and via radiation. Connect radio and TV sets to a different branch of the power line and use a line noise filter. Minimize the cause of commutator sparking. Check brush springs, try a new set of brushes, and clean the commutator. Examine the commutator for high mica and file it down if necessary. Polish the commutator bars if pitted. Replace armature if bars are loose.

CONNECTING WIRE SIZE FOR SINGLE PHASE MOTORS

The wire connecting the AC power source to the motor must be capable of carrying the starting current. Since this is so much larger than the full load current, it is used for the determination of wire size. Wire size is referenced to the American Wire Gauge (AWG).

TABLE 4-1. **Circuit Wire Sizes for Individual Single-phase Motors**

Horsepower of Motor	Volts	Approximate Starting Current Amperes	Approximate Full Load Current Amperes		Length of Wire Feet							
					25	50	75	100	150	200	300	400
1/4	115	20	5	AWG	14	14	14	12	10	10	8	6
1/3	115	20	5.5	AWG	14	14	14	12	10	8	6	6
1/2	115	22	7	AWG	14	14	12	12	10	8	6	6
3/4	115	28	9.5	AWG	14	12	12	10	8	6	4	4
1/4	230	10	2.5	AWG	14	14	14	14	14	14	12	12
1/3	230	10	3	AWG	14	14	14	14	14	14	12	10
1/2	230	11	3.5	AWG	14	14	14	14	14	12	12	10
3/4	230	14	4.7	AWG	14	14	14	14	14	12	10	10
1	230	16	5.5	AWG	14	14	14	14	14	12	10	10
1 1/2	230	22	7.6	AWG	14	14	14	14	12	10	8	8
2	230	30	10	AWG	14	14	14	12	10	10	8	6
3	230	42	14	AWG	14	12	12	12	10	8	6	6
5	230	69	23	AWG	10	10	10	8	8	6	4	4
7 1/2	230	100	34	AWG	8	8	8	8	6	4	2	2
10	230	130	43	AWG	6	6	6	6	4	4	2	1

The best arrangement is to have the motor placed as close to the AC power source as possible. A smaller distance means using connecting wires having a larger gauge and these are not only less expensive but are easier to handle. But even when the correct wire gauge is used there is always a voltage drop (a voltage loss) across the wire. The voltage that appears at the motor's power input terminals is equal to the voltage at the fuse box minus the drop across the connecting wires. If this voltage drop is excessive, the motor may not start or may start and run sluggishly. It is inadvisable to use ordinary lamp cord for connecting motors to power sources. Use an extension cord; extensions have the disadvantage of seldom being of the correct length.

If a motor requires a large amount of current, supplying that current can be a problem. There are two factors involved. One is the size of the wire to use and the other is the distance from the power service entrance to the motor. This distance will often be less than 130 feet.

The listing in Table 4-1 supplies wire gauge numbers for connecting wires and these are determined by the horsepower of the motor and the distance. The on/off switch, if installed external to the motor, must be rated for the maximum current to be carried by the power line.

SQUIRREL-CAGE MOTORS

As indicated earlier, the armature of a motor consists of a laminated electrical sheet steel core. The core winding may consist of heavy copper bars joined to each other or else put into core slots, with the bars connected to each other by equally heavy rings of copper.

Known as a squirrel cage, this works like a step-down iron-core transformer having a shorted single-turn secondary winding (Figure 4-16). In the case of the transformer, the magnetic field around the primary winding induces a large current flow in the secondary. In the case of the motor, the magnetic field surrounding the field coils induces a large current in the squirrel cage. In neither the transformer nor the motor is a physical connection required by the secondary winding or the squirrel cage to an outside voltage. The energy is supplied by the primary winding (transformer) or the field coils (motor).

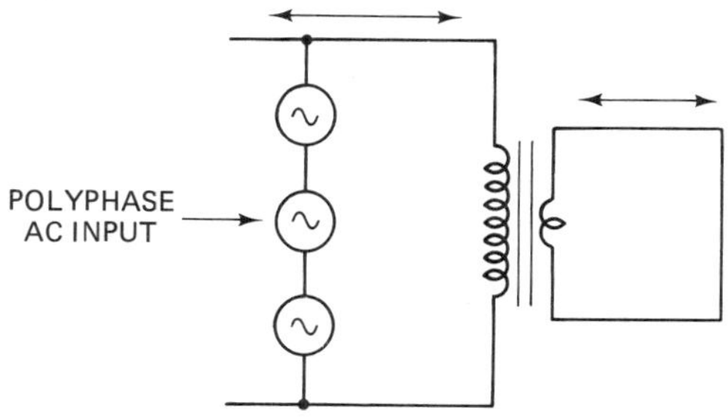

Figure 4-16 A squirrel-cage motor behaves like a transformer having a single-turn shorted secondary.

The electrical energy in the squirrel-cage rotor is obtained from the windings of the field coil, even though there is no physical connection between the two. As indicated, this energy transfer is via induction. The varying magnetic field around the field coils cuts across the squirrel cage, inducing a voltage across it, which in turn results in a flow of current.

This electrical action is equivalent to imposing a load on the field coils and because the squirrel cage is a shorted turn winding, the load is a heavy one. As a consequence, a strong current flows in the field coils whose windings must be capable of carrying this current.

Because of the way the current is produced in the squirrel cage, motors of this type are sometimes referred to as induction motors.

Rotation of the squirrel cage is obtained since the polarity of the rotor is always opposite that of the field coils (the stator). Note that the squirrel-cage induction motor does not have a commutator nor is there any need for carbon brushes. The AC power line can be connected directly to motor terminals which then lead to the field coils.

Squirrel-Cage Motor Speed

The speed of a squirrel-cage motor is determined by the number of pairs of poles, the frequency of the AC supply source, and the load to some extent.

The drawing in Figure 4-17 is a simplified version of the squirrel cage (sometimes referred to as an amortisseur winding), consisting of two horizontal bars; the magnets are formed by current flowing through the coils of wire

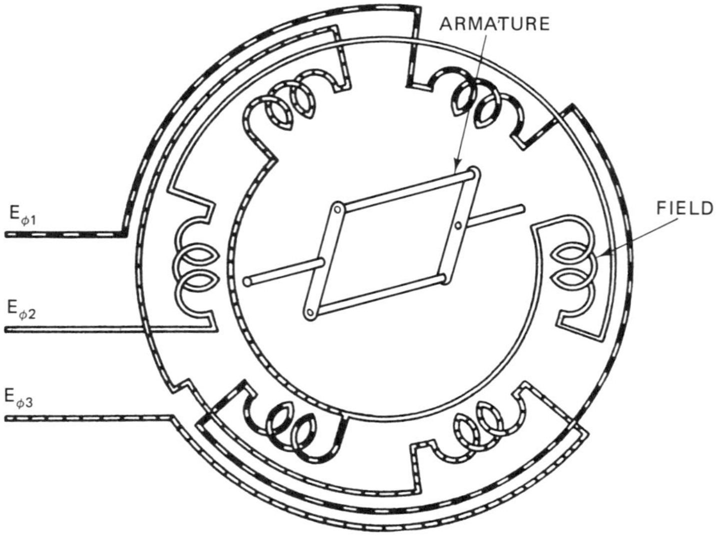

Figure 4-17 Basic structure of a squirrel-cage armature. Just a single element is shown here.

surrounding these two bars. The polarity keeps changing since an alternating current is used by the field coils. It is the varying magnetic field accompanying this current that induces a voltage across the bars, resulting in a current through them.

The maximum voltage is induced across the horizontal bars since these cut across most of the flux lines produced by the field coils. Since the vertical end bars have a motion that is parallel to the flux, they move through very few of the lines and so do not add much to the induced current in the squirrel cage.

Figure 4-18 supplies a view of the squirrel cage rotor. While the cage is usually formed of copper or aluminum rods, in some motors flat strips of copper are used. The advantage of this arrangement is that no end-short-circuiting rings are required. This type of assembly cuts manufacturing costs, but since the volume of copper used is less than with the rods, the amount of torque developed is less. Figure 4-19 illustrates the end pieces, rods and shaft of one type of squirrel cage.

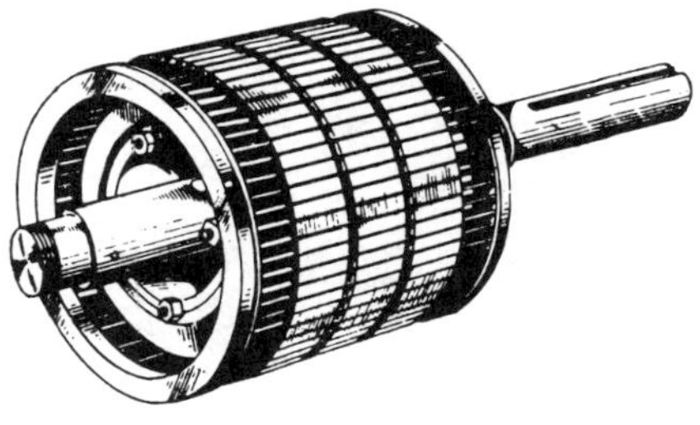

Figure 4-18 Squirrel-cage rotor. Copper strips are sometimes used instead of rods in the squirrel-cage rotor.

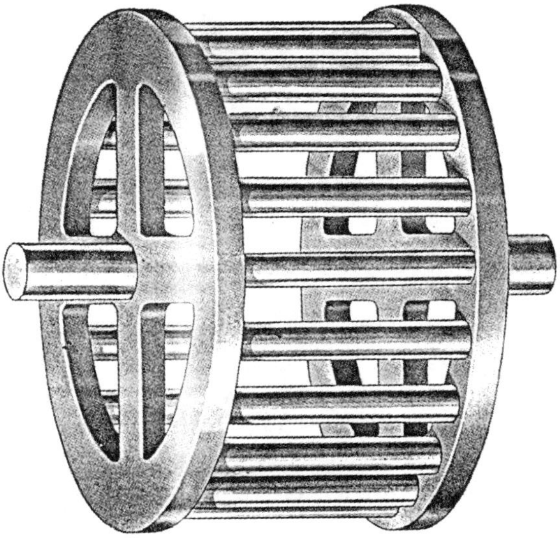

Figure 4-19 View of squirrel-cage rotor showing the end pieces, the copper connecting rods, and the shaft.

Some motors are equipped with a double squirrel cage. Ordinarily, the single squirrel cage has very low resistance and high inductance. The second cage, positioned on top of the first cage, but separated from it, has a much higher resistance and a lower inductance. Figure 4-20A is the symbol for this motor.

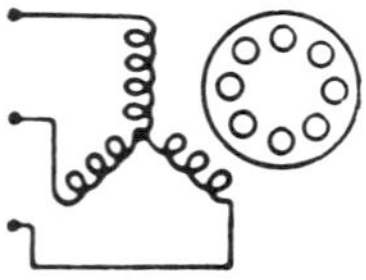

(A)

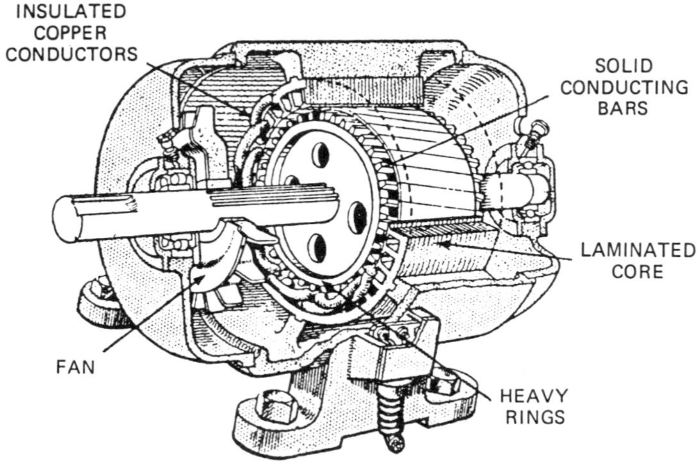

(B)

Figure 4-20 Symbol for a three-phase squirrel-cage motor (A); Induction motor using squirrel cage rotor (B).

Squirrel-Cage Motor AC Input

The wye structure for the field coils indicates that this motor type is a polyphase unit using three-phase AC input. The rotor is squirrel cage, a designation used to differentiate it from other rotor types that use coils and which are sometimes described as wound rotors.

Squirrel-Cage Motor Usage

The squirrel cage motor can be used wherever three-phase AC is available. Commercially, it can be designed to have a high starting torque for applications such as hoists and heavy door operation.

Squirrel-Cage Motor Characteristics

Because of its form of rotor construction, this motor type is possibly the simplest motor available. It is noted for its high starting torque and good efficiency. Its field coil structure can be either a wye or delta arrangement, a setup that results in a rotating magnetic field that surrounds the rotor. However, it does have some characteristics that could block its usage. The three-phase input requirement could exclude it from most in-home usage. Commonly, three-phase is used for motors rated at about 3 HP, but there are some single-phase motors in the high HP range. It is not built single-phase and so requires the AC output from a three-phase power line to operate. It draws a lagging current.

Power is the product of voltage and current, that is, $P = E \times I$. To keep rotor current within reasonable limits, input voltages far in excess of in-home line voltages are used, and so there is some relationship between wanted HP and AC voltage input. Thus, fractional HP motors generally have a line voltage input of 120. Small motors, such as those used for washing machines, may operate at 240 volts. Motors requiring a large HP output may use 480 volts, and this puts them into the industrial category. Motors rated at 100 HP or more can have AC inputs of 2300 volts.

INDUCTION MOTORS

The induction motor is comparable to a transformer. In the transformer, the current flowing in the primary winding induces a voltage across the secondary winding and this voltage, in turn, results in current flowing through the secondary. In the induction motor, a voltage is induced across the armature by the magnetic field surrounding the current in the field coils. No commutator is required, and the field coils are directly connected to the AC power line. The varying line voltage produces the equivalent of a rotating magnetic field. This has the effect of pulling the magnetic poles set up in the armature by the current flowing through it.

Induction motors are polyphase types and can be subdivided into two major groups: the wound-rotor and the squirrel-cage-rotor (Figure 4-20B).

Wound-Rotor Induction Motor

The rotor of this motor consists of groups of coils and are designed so they have the same number of poles as the stator windings. While such motors are often polyphase types they can be designed to work from a single-phase AC line. In all instances the purpose is to produce a rotating magnetic field. Whether the rotor consists of a squirrel cage or is a wound type winding, a strong current is produced by induction. The magnetic field accompanying this induced current reacts with the rotating magnetic field of the stator, thus causing the rotor to turn. This turning motion is slightly slower than that of the rotating magnetic field. To achieve the maximum induction effect the air gap between the stator and the armature is made to have minimum clearance.

If the input voltage is single-phase AC there are two stator windings, one of which is a starting winding, the other a running winding. To achieve the necessary phase difference, either a large amount of inductance or capacitance is connected in series with the starting winding. Because of the presence of these reactive elements, the currents through the two windings, the starting winding and the running winding, are out of phase, and so are their magnetic fields. It is this phase difference which results in a rotating magnetic field. When the rotor reaches its operating speed, a switch which operates centrifugally opens the connection to the starting winding.

The three-phase induction motor has several advantages over the single-phase type. Since it has three equally spaced windings, spaced 120 degrees apart, the result is the induction of magnetic fields around the rotor that are out of phase with each other, producing the equivalent of a rotating magnetic field. The reaction of these fields with those of the stator windings produces armature rotation. No starting winding is needed and the centrifugally operated switch is not required. This motor is self-starting.

SYNCHRONOUS SPEED

The speed of the rotating magnetic field of the stator winding is referred to as its synchronous speed.

ROTOR SLIP

The difference between the synchronous speed and the running speed of the rotor is known as rotor slip or sometimes more simply as slip. The rotor speed of an induction motor is expressed by

$$N_2 = N_1(1 - s)$$

where N_2 is the rotor speed, N_1 is the synchronous speed (speed of the rotating field), and s is the slip expressed as a decimal. The slip of the rotor in an induction motor is

$$s = \frac{N_1 - N_2}{N_1}$$

The rotational rate of the magnetic field is proportional to the frequency of the AC input and is inversely proportional to the number of poles. In terms of a formula it can be expressed as

$$N_1 = \frac{120f}{P}$$

where P is the number of poles, f is the frequency in hertz, and N_1 is the speed of the rotating field. The number of poles is their number per phase. A two-phase motor will have four poles and a three-phase motor will have six.

The formula supplied above shows that the motor is independent of its load and is directly dependent on the frequency of the source voltage. Consequently, this type of motor can be regarded as a constant speed device, assuming, of course, that the load does not exceed the breakdown torque of the armature.

The answers to problems supplied by these formulas is in decimal form. It can be changed to a percentage by multiplying the result by 100. In some instances the number 100 is included in the formulas and so the result is directly in percentage.

Table 4-2 supplies the synchronous speed of induction motors operating from a line frequency of 60 Hz.

TABLE 4-2. Synchronous Speed

Poles	60 Hz
2	3,600
4	1,800
6	1,200
8	900
10	720
12	600
14	514.2
16	450
18	400
20	360
22	327.2
24	300
26	277
28	257.1
30	240
32	225
34	212
36	200
38	189.5
40	180
42	171.5
44	163.5
46	156.6
48	150
50	144
52	138.5
54	133.3

Example

An induction motor operating from a three-phase, 60 Hz power line, operates at a speed of 1,730 rpm. What is the percentage slip of this motor?

Solution The speed of the rotating magnetic field can be calculated by multiplying the line frequency by 120 and dividing the result by the number of poles.

$$\text{Speed} = \frac{60 \times 120}{4} = 1{,}800 \text{ rpm}$$

The slip is equal to the rpm minus the rotor speed divided by the rpm.

$$\text{Slip} = \frac{(1{,}800 - 1{,}730)}{1{,}800} = .039$$

$$0.39 \times 100 = 3.9\%$$

There are two speeds involved here: the speed of the rotating magnetic field and the actual speed of the armature. The fact that there is a difference between these two does not mean the motor does not operate at a constant speed. In this example the motor speed is 1,730 rpm and it remains at this value for its load range. However, there can be a modest difference between its full load and no load speeds.

Example

A four-pole induction motor operating two-phase from a 60 Hz power line has a no load speed of 1,790 rpm and a full-load speed of 1,750 rpm. What is the synchronous speed of this motor, the percentage of rotor slip under no-load conditions, and the amount of slip with the motor working at full load?

Solution The synchronous speed is

$$\frac{120 \times f}{P} = \frac{120 \times 60}{4} = 1800 \text{ rpm}$$

The slip at no load is

$$\frac{1800 - 1790}{1800} = 0.0055 \text{ or } 0.56\%$$

The slip at full load is

$$\frac{1800 - 1750}{1800} = .02778 = 2.78\%$$

REPULSION MOTORS

There are various kinds of repulsion motors, including repulsion induction, repulsion, repulsion start induction, and induction wound rotor. The differences among these are based on motor structure.

The Repulsion Motor

When the brushes are aligned in the vertical plane, as shown in Figure 4-21, they are in their electrically neutral position. Reversal of armature rotation is obtained by shifting the position of the brushes to the left or right of electrical neutral, as indicated by the dashed lines. The motor is equipped with a brush shifting mechanism which may be marked to indicate the neutral point and the number of shift degrees, left or right. The brush shifting device may also be equipped with a limit stop.

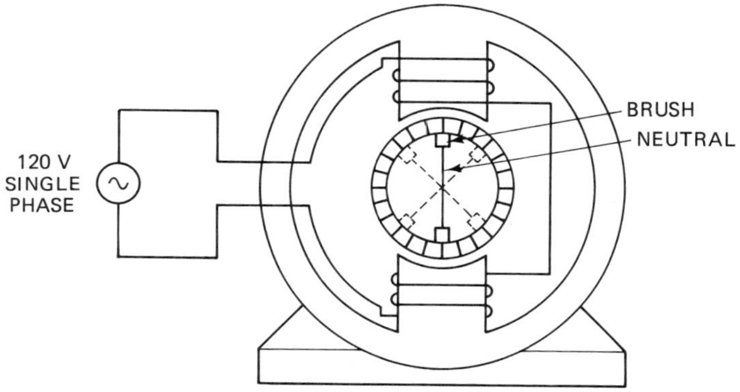

Figure 4-21 Repulsion motor.

An interesting feature is that the brushes are connected to short circuit the armature, comparable to the technique used in squirrel-cage motors. The single-phase 120-volt AC input is led directly into the field coils which are connected in series. There is no external physical connection to the armature, with current flow induced by the magnetic field around the field coils. The motor functions like a step-down power transformer having a shorted secondary winding.

Repulsion-Start Induction Motor

This motor is characterized by having a wound rotor, as opposed to a squirrel cage; it also has a commutator and brushes. The function of the commutator and brushes, unlike those used in DC motors, is not to change the input from DC to AC, but rather as an aid in starting the rotor. When it reaches about 75 percent of its running speed a centrifugal governor lifts the brushes up from the commutator and puts a short-circuiting ring around the commutator. The motor then runs as though equipped with a squirrel cage rotor.

This rotor, whose symbol appears in Figure 4-22, is characterized by a high starting torque comparable to that of a series DC motor but unlike that motor its starting current is not excessive. Unlike other AC power motors it is suitable for speed control and reversal of direction of rotation of its armature by adjustment of the position of its brushes.

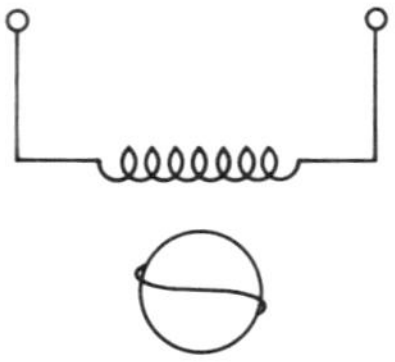

Figure 4-22 Repulsion motor symbol.

Summary of the Characteristics of the Repulsion Start Induction Motor

Load: Difficult starting load

Starting Torque: Very high and can be as much as four times the running torque

Starting Current: About three times the running current

Horsepower: 1/6 to 20 HP

Phase: Single

Source Voltage: 120 to 240 AC

Speed: 1200, 1800, 3600 rpm

Reversibility: Yes

Uses: Industrial machinery, conveyor belts

Repulsion-Induction Motor

The repulsion induction motor is somewhat unusual in the sense that it has two armature windings. One of these is a squirrel cage type and the other is a wire wound. As shown in Figure 4-23, the squirrel cage winding is positioned beneath the slots intended for the wound armature. The coils of the multiple coil armature windings are connected to a commutator, mounted on the same shaft with the armature, with carbon brushes riding on the commutator. (Figure 4-24). The wound armature is often a lap wound type and is cross connected.

The repulsion-induction motor has a number of advantages. It has a dual voltage input capability, with operating characteristics comparable to those of

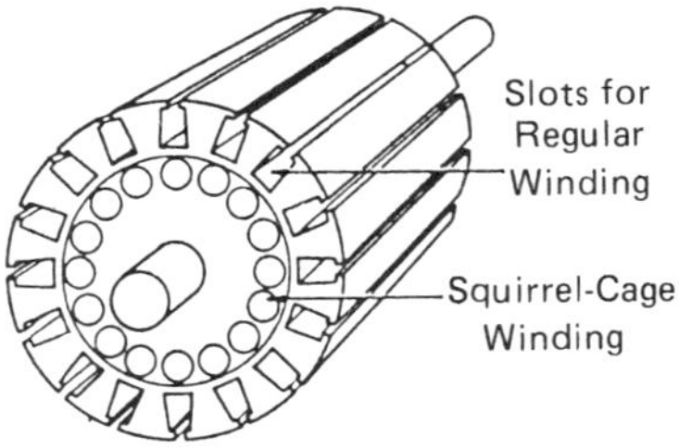

Figure 4-23 Structure of the repulsion induction motor.

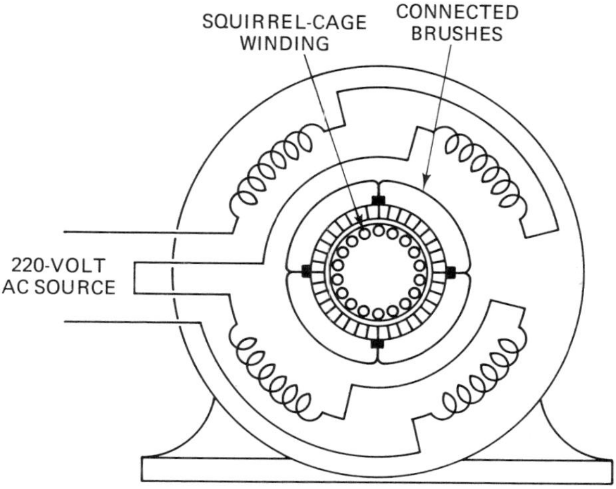

Figure 4-24 The repulsion induction motor may have a cross-connected lap-wound winding.

the compound-wound DC motor. The motor has a high starting torque and its operating speed remains fairly constant. The direction of armature rotation can be reversed by shifting the brushes by approximately 15 degrees. The amount of brush shift and its direction may be indicated on either the brush holder or on the motor frame. In some motors there are limit stops to make the positioning of the brushes easier. This motor does not require a centrifugal switch.

Some repulsion-induction motors are equipped with compensating windings (Figure 4-25) for increasing their power factor Table 4-3 supplies the fuse current ratings for fractional HP repulsion-induction motors.

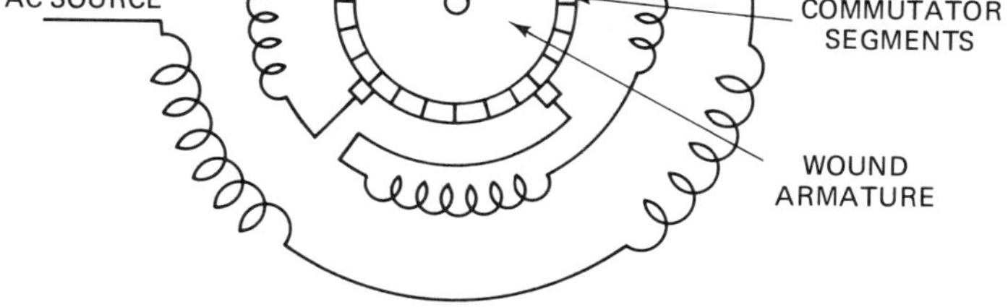

Figure 4-25 Use of compensating winding.

TABLE 4-3. Fuse Current Ratings for Repulsion-Induction Motors

Horsepower	Current Rating
$^1/_8$	4 amperes
$^1/_6$	4 amperes
$^1/_4$	6 amperes
$^1/_3$	8 amperes
$^1/_2$	10 amperes
$^3/_4$	15 amperes

SHADED-POLE MOTOR

There are a number of different types of motors that use a squirrel cage as the armature, and the shaded-pole motor is one of these. This motor is very much like the single-phase AC induction motor except for the modification of one or more of its field poles. The purpose of the change is to make the motor self-starting. At one time, induction motors were used in electric clocks and had a thumb screw on the back plate to enable the user to give the armature a starting twist. With the shaded-pole feature this is no longer necessary.

The rotor of the shaded-pole is not a wound form but uses a squirrel cage instead, and is an induction type. The stator consists of a winding whose input is single-phase AC. The notched laminations of the stator core are fitted with a single heavy ring or band of copper, referred to as a shading coil. Because of its structure it is one of the simplest types of self-starting motors (Figure 4-26).

The shaded-pole motor, sometimes called a shaded-pole induction type, does not use a commutator, nor does it have any need for a brush assembly. As

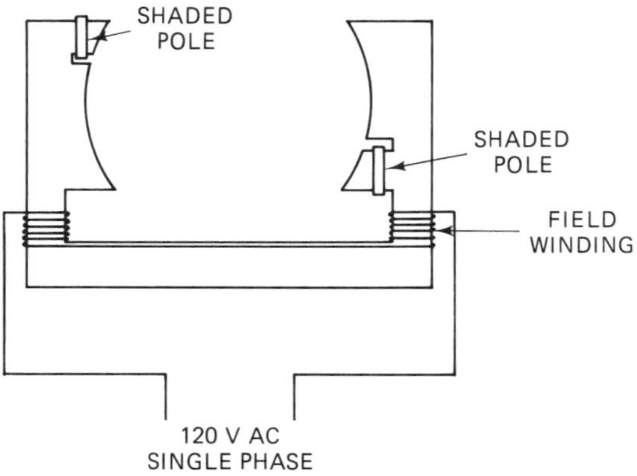

Figure 4-26 Shading poles.

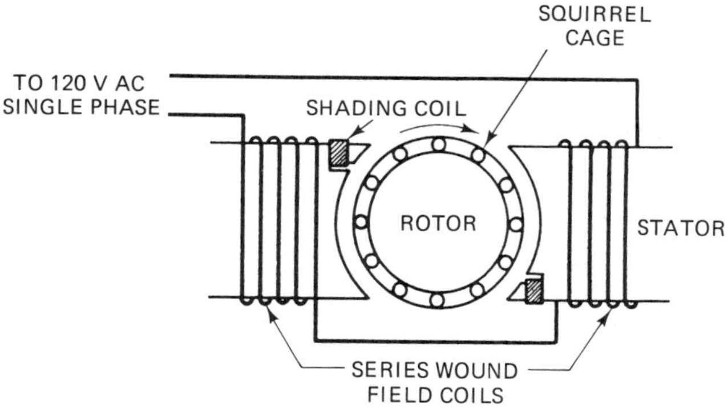

Figure 4-27 Two-pole shaded-pole induction motor.

shown in Figure 4-27, the AC line is brought directly into the series wound field coils. There are no physical or electrical connections to the squirrel-cage rotor.

The shaded-pole motor can be regarded as a step-down transformer, with the field coils working as the primary, and the shading coils and squirrel-cage rotor as shorted turn secondaries. Because of the step-down turns ratio and the fact that the secondary windings are shorted, a substantial current flows in these windings, producing a corresponding substantial magnetic field. Unlike the field coils, both the squirrel-cage rotor and the shading coils are bare copper, that is, they are not covered with any insulating material.

Shaded-pole motors are inefficient, have a low starting torque, poor speed regulation, and are generally not considered as reversible types, although reversing is possible. Starting torque is in the range of 50 to 80 percent of the full-speed torque at the moment the motor starts. The torque then rises momentarily to 120 percent at 90 percent of full speed, dropping to normal torque at operational speed. These motors are fractional horsepower types with 1/25 HP common. However, because of their structure their manufacturing costs are lower in comparison with other motor types having an equivalent HP rating.

The Salient Pole

A salient pole is a modification in a field pole and consists of one or more through-notches in its laminations. A short-circuited copper strap is fitted on the salient pole and is known as a shading coil.

This type of motor is usually quite small and is manufactured in horsepower ratings up to about 1/20. Figure 4-28 shows a four-pole motor. These motors get field coil excitation from the AC power line. The squirrel-cage rotor has no physical connections to any outside voltage source. Both motors, the

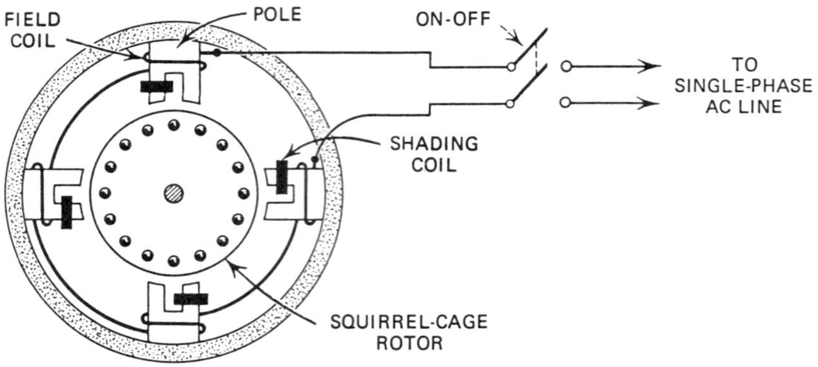

Figure 4-28 Four-pole shaded-pole motor.

two-pole and the four-pole, have their field windings connected in series, with the two leads representing the coil ends connected to the AC power line.

At the beginning of the sine wave AC cycle the magnetic flux around the field coil increases. The shading coil (or coils, if there is more than one) is in this flux field, and as a result an EMF is induced across the shading coil. The presence of the shading coil disturbs the distribution of magnetic flux around the field pole so that most of the flux is around that area of the pole that is not close to the shading coil.

When the sine wave AC input voltage reaches its peak value, at the 90-degree point of its cycle, its rate of change becomes zero. At that moment the magnetic field neither increases nor decreases, and so the induced EMF across the shading coil and the current through it become zero. As a consequence, at this time the magnetic flux becomes more equally distributed over the entire face of the field pole.

This action which is repeated is equivalent to a movement of the magnetic field across the pole faces. The conductors of the squirrel cage are cut by this moving field, resulting in a force that turns the rotor in the direction of the moving magnetic field. When the motor reaches its operating speed the shading coil (actually a solid copper ring) has minimum effect. This means that the shaded pole motor does not need external assistance to get started.

The advantage of this motor is that its manufacturing cost is low. It does not need a commutator and has no brushes. It has a low starting torque, low efficiency, and is used in small fans and clocks.

Most shaded-pole motors are single edge salient types, that is, just one edge of the pole is split, and as a result these motors are not usually reversible, but there are some that can be reversed. Shaded-pole motors are often used in small fans, hair dryers, humidifiers, motor operated toys, electric clocks and advertising displays. There are also some that have both the leading and trailing edges of their poles split to accommodate shading coils. Motors of this kind are designed for special applications.

Cogging

The induction motor may be confronted with a cogging problem, a condition in which there is a substantial variation in torque and is dependent on the position of the rotor. At this time the salient pole of the rotor is as though it were nonexistent and has no effect on the operating characteristics of the motor. Skewed slots on the armature minimize cogging at lower speeds.

Shaded-Pole Motor Symbols

Figure 4-29 illustrates the symbol commonly used to represent this motor. It shows two coils with the larger representing the field coils, the smaller the shaded pole winding. The circle in the center enclosing a number of much smaller circles is the squirrel cage rotor.

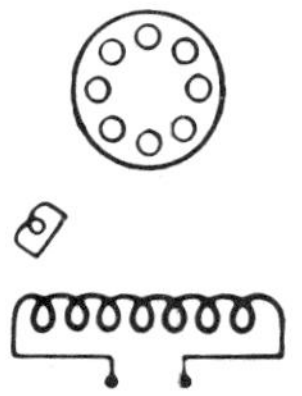

Figure 4-29 Symbol for a shaded-pole induction motor.

Summary of the Characteristics of the Shaded-Pole Induction Motor

Load: Designed for easy starting

Starting Torque: Low. Can be less than or equal to running torque

Horsepower: Ranges from 1/20 to 1/4 HP

Phase: Single

Source Voltage: 120

Starting Current: Low

Speed: 900, 1,200, 1,800, 3,600 rpm, constant

Reversibility: No

Noise and Vibration: Medium

Efficiency: Low to medium

Uses: Motor operated toys, hair dryers, small fans, electric clocks, advertising displays

Speed Control: Can be obtained by series resistance or iron-core inductor

Commutator: None

Brushes: None

Slip Rings: None
Field Coils: Series wound

Troubleshooting the Shaded-Pole Motor

Motor will not start. No sound from motor. Check AC power input. Check power outlet with test light or lamp. If no power, examine the fuse in the fuse box or reset circuit breaker. If fuse or circuit breaker keep opening there is a short circuit in the motor's line cord or the motor. If the outlet has power examine the blades of plug. Bend blades if necessary for better outlet contact. Examine the line cord for breaks. Make sure the line cord is securely connected to the motor input terminals. Check on/off switch. If you do not hear a click as switch is tried it may be defective. Test by shorting switch.

Motor will not start. Humming sound from motor. The motor bearings may be bound. Try turning the rotor manually. Lubricate bearings. Field coil may be open; test for continuity.

Motor runs but gets hot. There may be an excessive load on the motor. Motor bearings may need lubrication. Field coil may have a short circuit in one of its turns. Touch the line cord leading to the motor and the metal extensions of the line plug. They should not feel more than moderately warm. If hot, the motor may have a short.

Motor has poor torque. Motors of this type have low starting torque. Line voltage can fluctuate and may be very low. There may be a shorted turn in the field coil. The motor may give evidence of burning or smoking. Bearings may be bound or require lubrication. The squirrel-cage rotor may be rubbing against the field coil. Motor parts may be loose; check for loose screws, mounting bolts, cooling fan. Interior of motor may have accumulated dirt and lint.

Motor vibrates or is too noisy. Bearings may be worn or require lubrication. The motor shaft may be defective. Motor parts, screws, and bolts, may have worked loose. Screws holding the motor to the base support may no longer be tight. The motor frame should be free and clear of any adjacent metal parts.

Motor smokes. One or more turns in field coil may be short circuited. Motor bearings may require lubrication. Moisture may have collected inside motor. Make sure motor fan, if motor is so equipped, is not damaged and rotates freely. If the motor has vent slots, make sure these are clear of any obstructions and are not clogged with dirt or lint. If the motor is housed in some appliance and is not out in the open, it should have access to ventilation. Keep motor away from other heat generating electrical devices.

SELF-STARTING CLOCK MOTORS

Some self-starting clock motors are built with a hard-steel rotor which is not laminated and becomes a permanent magnet. Its poles "lock in" with the rotating magnetic field supplied by the stator, resulting in synchronous rotation. Motor power using this design is lower than that of squirrel-cage type; this is not significant when the structure is used as a clock.

The self-starting clock may be equipped with a gear train to reduce shaft speed with that speed set at 1 rpm. This is used for the sweep second hand while a reduced speed gear train is used for the minute and hour hands.

SYNCHRONOUS MOTORS

A synchronous motor is one in which the armature rotates in step with the rotating magnetic field of its field coils, and it is from this behavior that the motor gets its name, since its armature is synchronized to some multiple of the power-line frequency. It maintains its speed except when it is heavily overloaded and so is regarded as a constant-speed motor.

The speed of rotation of the armature in rpm can be determined by the number of pairs of poles and the line frequency.

$$\text{rpm} = \frac{\text{f} \times 60}{\text{p}}$$

Here rpm represents revolutions of the armature per minute, f is the frequency of the sine wave AC input voltage, p is the number of poles and is always an even number like 2, 4, 6, 8, or more and is a constant. Thus, the rpm for a four-pole motor operating from the 60 Hz power line (with Hz representing cycles per second) is

$$\text{rpm} = \frac{60 \times 60}{4} = \frac{3600}{4} = 900 \text{ rpm}$$

Figure 4-30 illustrates a synchronous motor of the type commonly used in clocks. The motor structure consists of hardened steel comprising a set of E and I laminations. The AC input is supplied to an inductor wound around the I laminations forming the field or exciting coil. The few turns shown in the illustration are just representative and do not indicate the actual number.

The field in this example consists of a pair of salient poles with each equipped with a copper ring working as shading coils with the shading coils making this motor self starting. The rotational speed of the synchronous motor

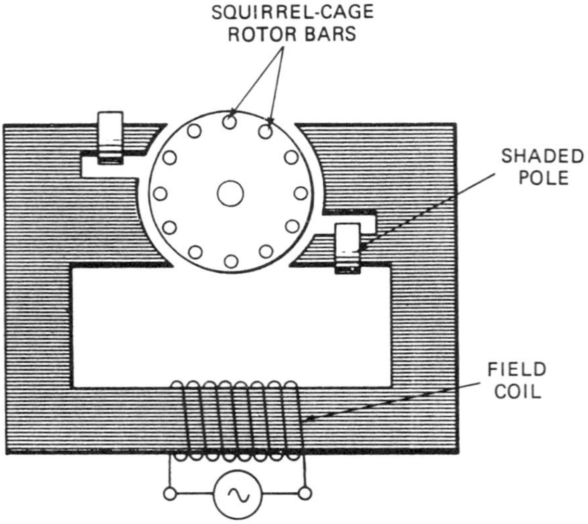

Figure 4-30 Fractional HP hysteresis-synchronous two-pole motor.

is independent of the amplitude of the line voltage or of any input voltage fluctuations. Thus, line voltage changes caused by load changes will not affect the motor. It is however directly dependent on line voltage frequency, and while that frequency may change by plus or minus one cycle throughout the day, its average remains remarkably constant. As a consequence the fractional horsepower synchronous motor is widely used for clocks and for transcription turntables in high-fidelity audio systems. They are desirable in these applications because of their self-starting feature and because of their ability to maintain a constant speed whether loaded or unloaded. Larger synchronous motors may be used in industry when machinery must be operated simultaneously. With industrial type synchronous motors the rotational speed will be maintained within the designed load operating range of the motor but will decrease speed when heavily overloaded or when the AC voltage of its power line input drops substantially.

Typically, the speed of a motor will decrease with an increase in load, but with the exceptions indicated above, the synchronous motor is the only true constant speed AC motor.

Synchronous Motor Variations

Although a motor may appear in the listing of a particular type, there are usually a number of variations in design in an effort to meet specific operating needs. One example is the use of a variable frequency oscillator (vfo) in place of line power. Such motors are usually small fractional HP types. The frequency setting of the vfo determines the rotational speed of the motor, but the strength of the voltage output of the oscillator is kept constant. If the power output of the vfo is not adequate, it can be followed by a power amplifier. The power amplifier (pa) is untuned and so strengthens the amplitude of the signal without altering its frequency. The combination, known as a master oscillator power amplifier (mopa) is equivalent to that of a small transmitter.

Self-Starting Induction-Reaction Synchronous Motor

This is a combination of induction and synchronous motors. It features a shaded-pole field that is slotted, with six slots typical. It is these salient poles which supply the reaction effect. The armature is not a wound type but is a squirrel-cage arrangement.

Hunting

Under correct operating conditions the rotor will reach its synchronous speed and will maintain it. However, the armature can hunt, a condition in which the rotor exceeds or does not reach synchronous speed. This speed variation above and below synchronous speed is referred to as hunting. It can be caused by an unexpected and quick change in the operating load, or a rapid change in the amplitude of the input AC voltage.

DC Synchronous Motors

As a general rule the field coils of a motor are fixed in position while the coils or squirrel cage rotate. From an electrical point of view there is no reason why this arrangement cannot be reversed, that is, the field coils can rotate while the armature coils are fixed.

The field coils are mounted on an assembly whose center structure is the motor shaft. The field coils are supplied from a DC source and since they will be rotating require the use of slip rings. These are needed since there must be some way of delivering current to coils that are in rotary motion. The slip rings, also known as collector rings, are pairs of copper rings with carbon brushes riding on their outside surfaces. One of these brushes is connected to the minus terminal of the DC supply, the other to the plus terminal. The slip rings are simply convenient connecting devices and do not commutate or rectify the input voltage. They are part of the shaft assembly and rotate with it.

As in AC only types, the speed is determined by the AC line frequency of the current supplied to the stator coils and the number of rotor poles. When the rotor field coils are connected to their DC source, the N and S poles they produce move into synchronization with the rotating magnetic field of the alternating current supplied to the stator windings. The AC stator winding is operated from polyphase AC.

Polyphase Synchronous Motors

Synchronous motors can be single phase or polyphase with the latter type having a three-phase input. The AC operated stator has either a wye or delta formation and supplies a rotating magnetic field. While the illustration in Figure 4-31 shows the stator windings connected to a three-phase line, the input could also be supplied by the three phases derived from a reactance network.

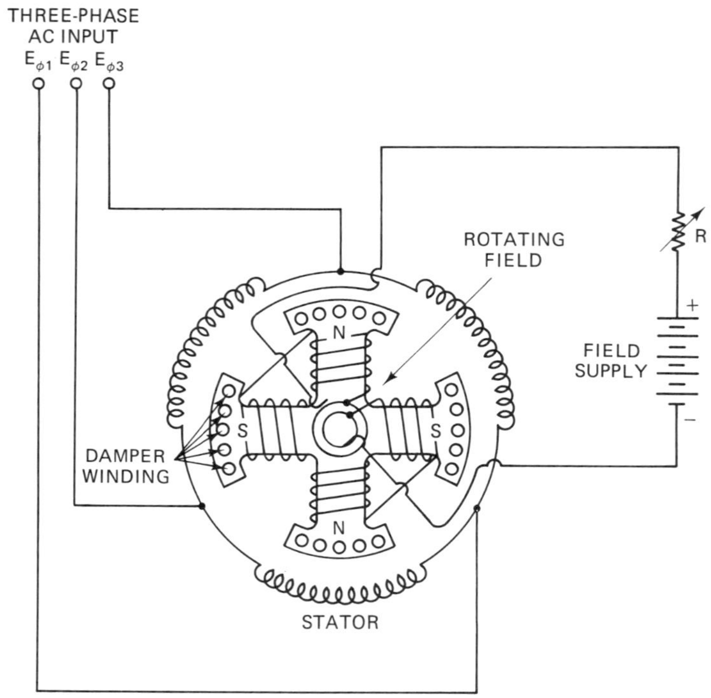

Figure 4-31 Three-phase synchronous motor.

The rotor of this motor gets its excitation from a DC voltage source and this can be either a battery supply as indicated in the illustration or from a power supply. In this motor it is the field windings that rotate with the magnetic poles of the field coils remaining unchanged. In some of these motors, permanent magnets are used in place of coils, and in that case no external DC supply is required. If an external DC supply is used, the current from that supply is delivered to the rotating field coils via slip rings. Still another variation of this motor uses an iron rotor, not equipped with any windings; nor is this rotor initially magnetized. Magnetic poles are induced in the iron rotor by the magnetic fields around the stator windings.

The torque of a synchronous motor is produced at synchronous speed and the motor is not self-starting. This is true for both single- and polyphase types.

Hybrid Motor

To overcome its nonautomatic starting feature, the hybrid motor is equipped with part of a squirrel-cage structure in the pole faces of the rotor. The motor is a hybrid since it is both a synchronous and an induction type. The squirrel-cage segments are identified as damper windings, but are also known as amortisseur windings, from the French for the word killer.

A rheostat, R, is put in series with the DC field supply and is used to control the amount of current delivered to the field windings. This supply does not affect the synchronous speed of the motor but is used to control its torque output.

Synchronous motors are characterized by having zero slip, that is, the speed of rotation is directly determined, without variation, by the frequency of the input.

Hysteresis Synchronous Motors

Fractional horsepower motors can be made to operate with current supplied to a field coil but with no electrical energy delivered to a rotor which has some residual magnetism; these are referred to as hysteresis synchronous.

The motor (Figure 4-32) has a pair of salient field poles using shading coils and so this motor is self-starting. The rotor is a squirrel-cage type and is also cut for salient poles. Input AC power to the field coils produces a rotating magnetic field and this field, sweeping across the squirrel-cage rotor causes it to turn. The synchronous speed is determined by the number of poles machined into the rotor and the frequency of the AC input to the field coil. The squirrel-cage rotor in this example has two poles while the AC power source has a frequency of 60 Hz. Under these operating conditions the rotor speed will be 1800 rpm.

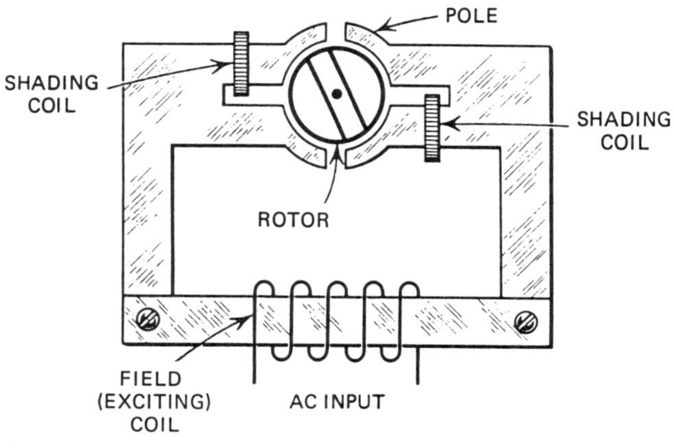

Figure 4-32 Two-pole hysteresis synchronous motor.

Another type of hysteresis synchronous motor has a rotor made of a number of ring-shaped thin plates of hard magnet steel. Prior to assembly, the steel is magnetized and then subjected to an accelerated aging process. While this reduces its magnetic strength, the magnetism that does remain will do so at a fairly constant level for a long time.

A rotating magnetic field is obtained by AC voltage excitation of the field coil. The rotating magnetic field induces magnetic poles in the rotor that follow after the stator poles. The rotor will then start turning, increasing its rotation until it reaches synchronous speed.

The AC input to this motor can be single phase or three phase. Its advantages with both types, single or three phase, include constant speed over its designed torque range, ability to accelerate high inertial loads, low noise and vibration at its load point, and low starting current.

The rotor used in these motors contributes to its high cost if made of samarium cobalt. It is also inefficient when compared to an induction motor.

SPLIT-PHASE INDUCTION MOTORS

Not only voltages but currents as well can be out of phase. Since it is a current that supplies a magnetic field, consequently a pair of magnetic fields can also be out of phase. As a result, it is possible to produce the equivalent of a two-phase motor with just a single AC voltage input.

The drawing in Figure 4-33 shows the arrangement of a split-phase induction motor. There are two stator windings, a main or running winding and an auxiliary winding or starting winding.

There are two factors that determine the phase angle of the current through a coil with respect to the voltage across it. If a coil could be developed

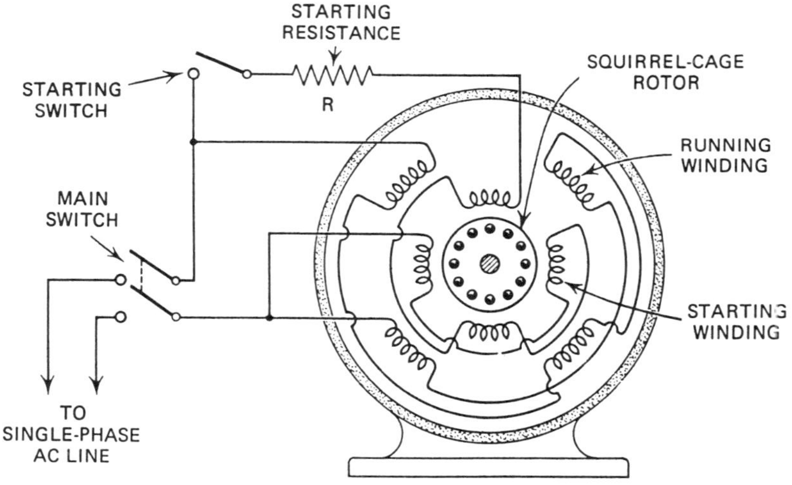

Figure 4-33 Split-phase induction motor.

with zero resistance, the current through it would lag the voltage across it by a maximum of 90 degrees. The greater the amount of resistance compared to the reactance of the coil, the smaller the phase angle. If a coil could be developed with resistance only and no reactance, the phase angle between the current through it and the voltage across it would be 0 degrees. The phase angle, then, can be determined by the ratio of coil resistance to its inductive reactance.

If the main winding of a motor and that of an auxiliary winding use different gauges of copper wire and a different number of turns, they will not have the same resistance, and so the magnetic fields across each of these coils will be out of phase with each other. The effect obtained will be the same as that of a two-phase motor. The result is a rotating magnetic field which causes the armature to turn.

Although the running and starting windings are shown drawn at right angles to each other, this does not mean to imply that the phase angle of their magnetic fields is 90 degrees, for it is actually somewhat less than this. The difference in phase angle is obtained by winding the starting winding with a smaller gauge wire than the running winding and reducing its inductance by placing it higher in the slots in the laminated field structure.

After the rotor is in motion, the starting winding is disconnected by means of a centrifugal switch, usually when the motor has reached about three-fourths of its synchronous speed. This is done by the switch S, shown in the illustration.

Split-phase induction motors are fractional horsepower types with fairly good speed regulation. Horsepower ratings are in the range from 1/50 (14.9 watts) to 1/3 HP (248.66 watts). The starting torque is low and is just a fraction of its normal full speed torque. This motor is suitable for easy-to-start light loads.

The problem with the split-phase motor is the large amount of its starting current and this can be as much as six to eight times its full speed running current. This current drain is heavy enough to cause momentary light dimming when the motor is first turned on.

These motors are used for oil burners, office appliances, fans, and blowers.

The split-phase fan motor. The split-phase fan motor in Figure 4-34 has two windings, a main or running winding, and an auxiliary or starting winding, L1. Unlike other types of split-phase motors, no centrifugal switch is used and the starting winding always remains in the circuit.

The coil identified as L2 is in series with the applied AC voltage. This coil works as an autotransformer and determines the amount of voltage applied to the main windings. When the switch, S1, is in its high position, the voltage step-up ratio is increased, raising the main winding current and the running speed. The voltage is reduced when the switch is in its low position. The transformer coils are mounted in the base of the fan and so is the speed control

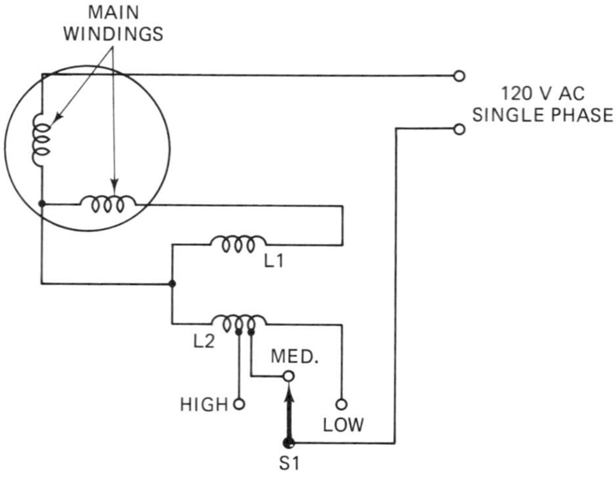

Figure 4-34 Split-phase fan motor.

switch. The setting of this switch is operated by a knob control on the outside of the base.

Troubleshooting the Split-Phase Induction Motor

Motor does not start. This problem has two basic possible causes. The trouble may be external to the motor, or internal. The male plug may not be inserted in its outlet or may have become loose. Check the outlet to make sure it is live, using either a line voltage tester or a test lamp. If there is no power, connect the motor plug to an extension cord and try some other outlet, first making sure that the outlet is live. For a dead outlet, check the fuse box for an open fuse or circuit breaker. Make sure the line cord is connected to the male plug and check the cord for breaks and to be sure it is securely connected to the motor. If an extension cord is being used make sure the extension has not become disconnected from the main line cord. Check the on/off switch and make sure connections to it are secure.

Some split-phase motors have a resistance in series with the starting winding to limit the amount of starting current. The motor will not start if this resistor is open. The motor has two starting windings. Make sure neither is open. Check to make sure the main windings (field windings) are not open. The squirrel-cage rotor may be touching field pole (or poles) of main winding, or the squirrel-cage rotor may not be able to turn freely. Try spinning the rotor manually. Make sure the bearings are lubricated and that the interior of the motor is free of dirt and lint.

Motor overheats or does not reach running speed. The motor may be overloaded. If a belt drive is used, loosen or remove the belt and check it. Check the load to determine if it revolves freely; the main winding may have a shorted turn. The motor may be equipped with a cutout switch that is not functioning.

Motor vibrates or is excessively noisy. Check for loose parts on the motor frame or inside the motor; the motor data plate may be rattling. Make sure the motor has no loose screws or plates. Motors mounted on wooden boards tend to emphasize motor sounds. The motor may be causing adjacent non-motor-related parts to vibrate. Vibration can be caused by worn or inadequately lubricated bearings or by a shaft which has become bent. The gear train or belt, driven by the motor shaft, may be loose, or the motor may be subject to sympathetic vibration if an adjacent part is shaking.

Excessive bearing wear. The gear train or belt may be too tight or the motor shaft may be out of alignment. If the bearings were not lubricated on a regular schedule, they may have run dry. Check for dirty bearings and be sure the lubrication used is correct. The shaft may be out-of-round.

Motor smokes. The starting windings or main windings (or both) may have one or more shorted turns. Moisture inside motor may be supplying a leakage path between current carrying windings. The motor may be operating under conditions of very high humidity or it may be in the path of a water drip and have a motor frame unable to keep water out.

The cut-out switch, if motor is so supplied, may be faulty. Turn the rotor by hand to determine if it is bound and unable to turn or cannot turn freely. Squirrel-cage rotor may be making contact with poles of the main or starting windings. The inside of motor may have accumulated dirt or lint.

Summary of Characteristics of the Split-Phase Induction Motor

Load: Designed for easy starting

Starting Torque: Equivalent to its running torque or as much as 50% greater

Starting Current: 300% or more of full speed current

Horsepower: Ranges from 1/20 to 3/4 HP

Phase: Single

Source Voltage: 120 AC

Speed: 900, 1,200, 1,800, 3,600 rpm

Reversibility: Yes

Uses: Fans, small machine tools, office appliances, oil burners

Speed Regulation: Good

Split-Phase Motor Symbols

Figure 4-35 shows two symbols used for split-phase induction motors. Another symbol appears in Figure 4-36, but this is for a two-speed type.

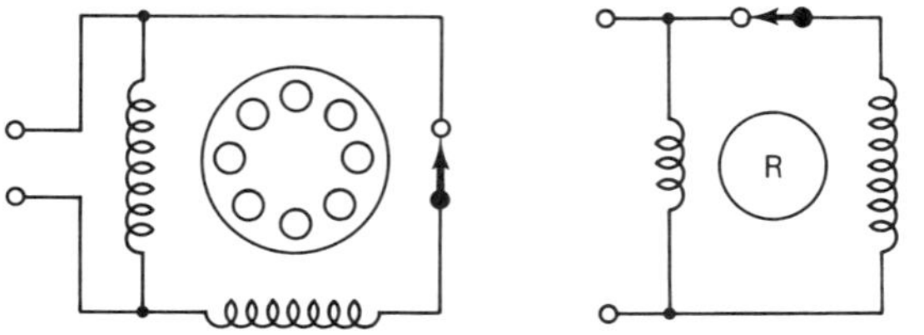

Figure 4-35 Symbols for the split-phase induction motor.

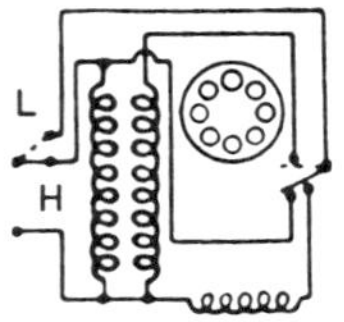

Figure 4-36 Symbol for a two-speed split-phase induction motor.

CENTRIFUGAL SWITCHES

A centrifugal force is one that tends to impel an object outward from the center of rotation. Motors like the split-phase and capacitor types can use such a switch to open the connections to a starting winding when the armature is close to approximately 70 percent of its final running speed.

There are a number of different types of such switches, but one of the more commonly used consists of a pair of ball weights kept in position by a spring. The spring keeps a pair of metal contacts in position so that a starting winding is part of the motor circuit. The magnetic field around the starting winding is phase displaced from that of the running winding, with the result that these two supply a rotating magnetic field. The starting winding, essential for getting the armature into motion, adds little to the motor's torque.

Figure 4-37A shows one type of centrifugal switch. At the time that the source voltage is applied, the spring holds the switch contacts together, with the conductor supplying a path for current flow through the starting coil. When the armature reaches a predetermined running speed, centrifugal force exerted on the metal ball weights becomes sufficient to compress the spring, pulling the conductor and its contacts outward, thus breaking the connections for the starting winding.

Another type of centrifugal switch appears in Figure 4-37B. The drawing at the left indicates the condition that exists at the moment the motor switch is turned on. The starting winding is connected in the circuit since the contacts of the switch are closed. The flyweights shown in the drawing are held in place by a spring, thus keeping the contacts closed.

As the motor approaches its operating speed, centrifugal force drives the flyweights outward, as indicated by the drawing at the right. The bars holding the switch contact closed move inward, overcoming the force of the springs

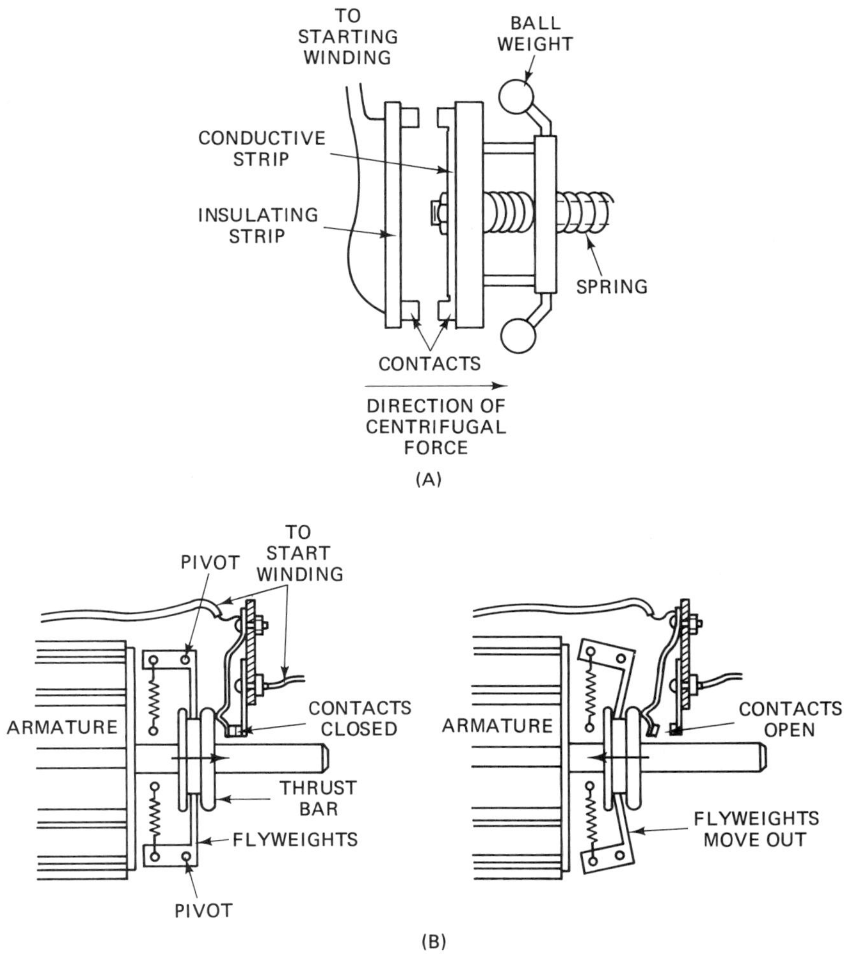

Figure 4-37 Centrifugal switches: ballweight type (A), and flyweight type (B).

and opening the contacts. This interrupts the circuit of the starting winding and it is effectively removed.

CAPACITOR MOTORS

When a pair of resistors are connected across a single phase AC line, the voltages across each of the resistors will be in phase with each other and also with the applied voltage. If a capacitor is substituted for one of the resistors, the voltage across the capacitor will be 90 degrees out of phase with that across the resistor and the applied AC. This assumes the capacitor is a perfect unit and has no internal resistance. For practical circuits the phase separation is somewhat less than 90 degrees.

If a coil is substituted for the other resistor, the result will be an L-C type, consisting of a circuit made up of a series connected coil and capacitor. The coil and the capacitor are reactances and are 180 degrees out of phase. If the reactance of the coil is small compared to that of the capacitor, the effect will be the equivalent of some small reduction in the total reactance. The presence of the capacitor will still produce a substantial amount of phase shift.

Since the voltage across the capacitor and that of the AC input are out of phase, this arrangement can be used to produce the equivalent of a two-phase setup and so is suitable for two-phase operation.

There are three basic types of these motors: capacitor-start, split-capacitor, and two-value capacitors. Capacitors are commonly used to supply a substantial starting torque since their use supplies the equivalent of a rotating magnetic field. Two windings are used, one connected to a capacitor and the other not so connected. As a result, there is a phase difference between the two arrangements, and it is this phase shift that permits the motor to develop a stronger starting torque.

CAPACITOR-START MOTORS

The capacitor-start motor, shown in Figure 4-38, is also a split-phase motor, but with the required phase shift obtained by using a capacitor in series with an auxiliary winding.

With the auxiliary winding remaining permanently in position, the motor behaves as though the input was two-phase AC. The capacitor also functions as an energy reservoir, smoothing the pulsing effect that is the result of two-phase operation. As a consequence, the motor operates more quietly. However, a change must be made in the type of capacitor used. Electrolytics are not suit-

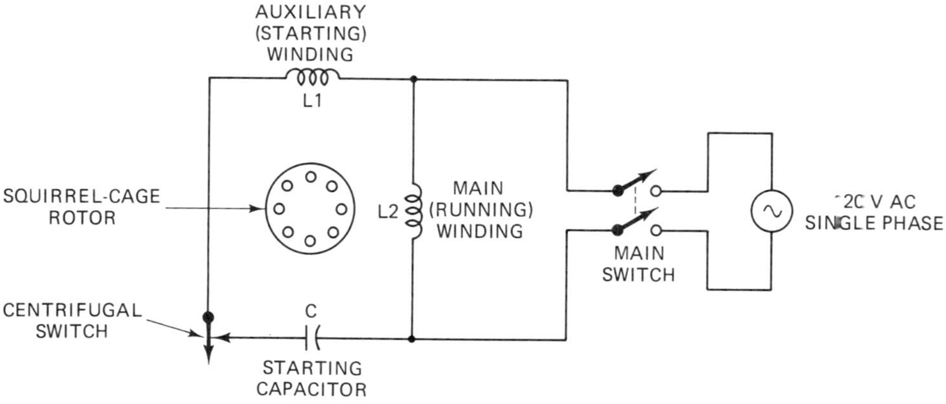

Figure 4-38 Capacitor-start motor.

able for permanent use and so must be replaced by an oil-impregnated sealed-paper type.

With the permanent capacitor motor, the starting torque is not ideal, for the torque is a compromise between the starting and running torques. This can be overcome by using two capacitors.

When the double-pole single-throw switch is closed, the motor is connected to the AC, 120-volt power line. The starting capacitor, the centrifugal switch, and the auxiliary (starting winding) are all in series and are connected to the voltage source. At this time the centrifugal switch is closed. When the motor gets close to its operating speed, the centrifugal switch opens and the starting capacitor and auxiliary winding are disconnected from the line voltage.

Summary of Characteristics of the Capacitor-Start Motor

Load: Difficult starting load

Starting Torque: 300% to 400% of the running torque

Horsepower: 1/6 to 10 HP

Phase: Single

Source Voltage: 120 to 240 AC

Speed: 900, 1,200, 1,800, 3,600 rpm

Reversibility: Yes

Uses: Machine tools, grinders, blowers, ventilating fans

Capacitor-Start Motor Symbol

Figure 4-39 illustrates the symbol for the capacitor start motor. Most of the symbol is fairly standard, such as the squirrel-cage rotor and the running coil. However, there are several variations for the capacitor and the centrifugal switch.

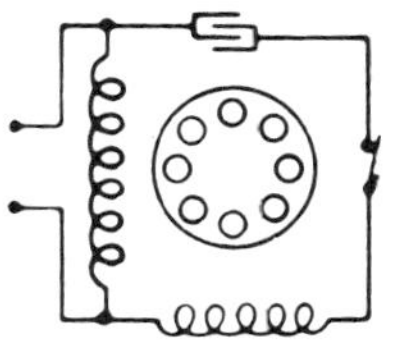

Figure 4-39 Symbol for the capacitor-start motor.

TWO-SPEED MOTOR

Figure 4-40 is the circuit of a two-speed motor, capacitor-start type. The switch shown at the top left has two positions, low speed and high speed. The switch at the top right is a centrifugal unit; the letter S indicates starting

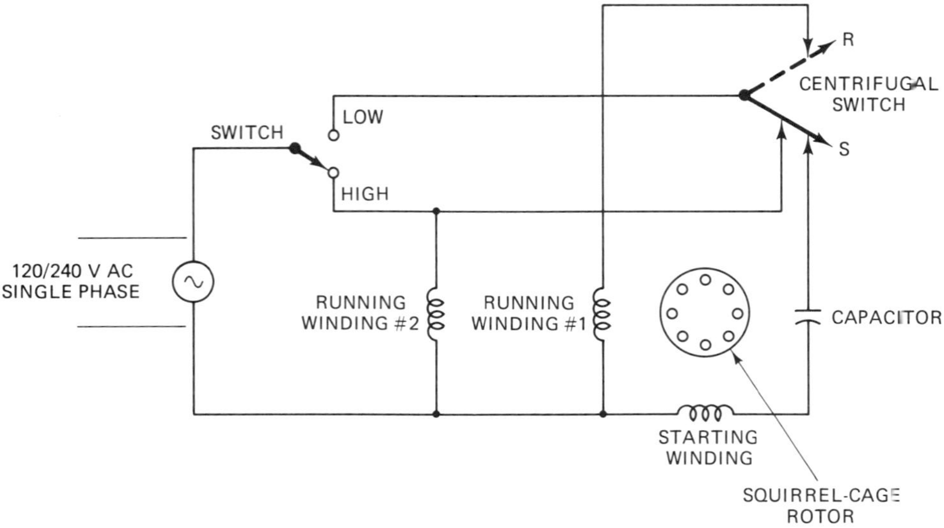

Figure 4-40 Two-speed capacitor-start motor.

torque, the letter R is an abbreviation for running torque. There are two running windings, #1 and #2, with running winding #1 used for low speed and #2 for high speed. The capacitor is in the circuit when the motor starts but is disconnected when the motor approaches its running speed. This could be the type of motor used for a fan where two speeds are wanted for a specific, unvarying load.

THE TWO-CAPACITOR MOTOR

The two-capacitor motor, as its name implies, uses a pair of capacitors, with the value of the capacitors selected for best starting and running torque. As shown in Figure 4-41, the two capacitors are in shunt with one of them in series with a centrifugal switch. When capacitors are wired in parallel, the total capacitance is the sum of the individual capacitances. This is the condition that prevails when the motor is first turned on. After the armature gets close to its final running speed, the switch opens and only the running capacitor remains.

The two capacitors are different types. The running capacitor, the one that remains permanently connected, is a sealed paper type; the starting capacitor is an AC electrolytic.

Two-Capacitor Motor Symbol

Figure 4-42 shows the two-capacitor motor symbol.

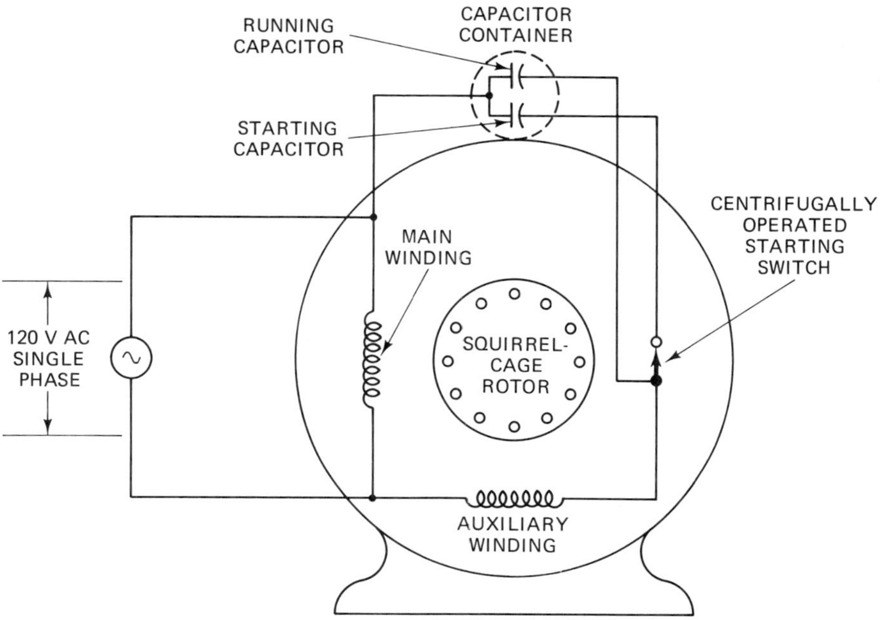

Figure 4-41 Two-capacitor motor.

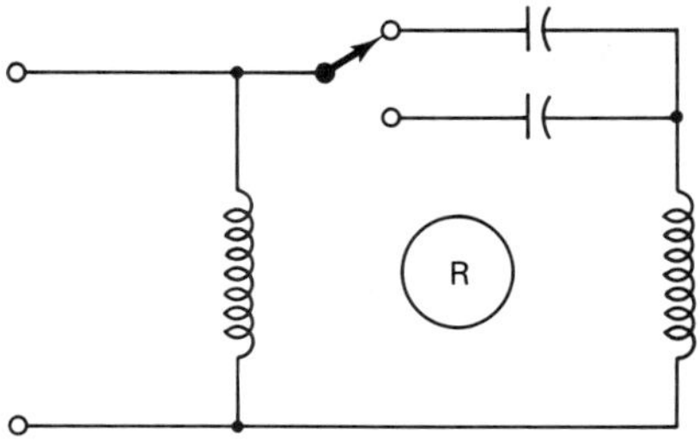

Figure 4-42 Symbol for two-capacitor motor.

PERMANENT CAPACITOR MOTOR

For some applications it is not necessary for the motor to have a centrifugal switch and so the start capacitor and the auxiliary winding remain permanently connected. This can be done for motor operations in which the load remains constant. The torque is not as good as in motors using a centrifugal switch.

The capacitor is wired in series with the auxiliary winding and so behaves like a series tuned L-C circuit. The inductive reactance of the coil and the capacitive reactance of the capacitor are 180 degrees out of phase and have a cancelling effect. If both reactances are exactly equal they will cancel completely, and so the net reactance of the combination will be due entirely to the resistance in that circuit. As such, the resistance will be 90 degrees out of phase with the reactance of the main winding. However, this phase difference will be this amount only at a specific AC input frequency to the motor. The circuit is equipped with a centrifugal switch, and when the armature is close to synchronous speed will open, removing the capacitor.

The capacitor, usually known as a starting capacitor, is often a dry AC electrolytic. Because the operating frequency of the motor is fairly low, 60 Hz, large values of capacitance are required, and this can be most conveniently supplied by this type of capacitor. The capacitance tolerance of such capacitors is fairly large, and so obtaining a full 90 degree phase shift is seldom practical, but fortunately is not a necessity, and so in practice lower values of phase shift are used.

One of the chief advantages of the permanent capacitor motor is its high starting torque, much higher than that of the split-phase motor. Consequently, this motor is desirable when the load is such that a higher starting torque is necessary. Further, the starting current is low and motor operation is quiet. The motor also does not interfere with radio or television reception, and so it is favored as a motor for use in refrigerators.

Autotransformer Effect

Unlike other power transformers which have individual primary and secondary windings not connected physically, an autotransformer consists of a single tapped winding (Figure 4-43). A voltage step-up is obtained because of the turns ratio: the secondary portion has more turns than the primary.

An autotransformer can be used in conjunction with the starting capacitor of the capacitor motor. The autotransformer does not change the AC line input frequency, but does step up the voltage applied to the capacitor above that supplied by the line. This produces an effect equivalent to increasing the capacitance. As a result, this makes possible the use of a capacitor having a smaller value of capacitance. The capacitor, however, must have a working voltage rating capable of withstanding the higher value of voltage.

Split-Capacitor Motor

This version of capacitor motor is diagrammed in Figure 4-44. The capacitor consists of two units, one of which is a starting type, the other a running unit; both are housed external to the motor frame in a metal container. When the motor reaches its running speed the starting capacitor is disconnected by the centrifugal switch.

The motor also has a pair of windings, one of which is the main, the other the starting winding. The starting winding is also disconnected from the circuit when the centrifugal switch opens.

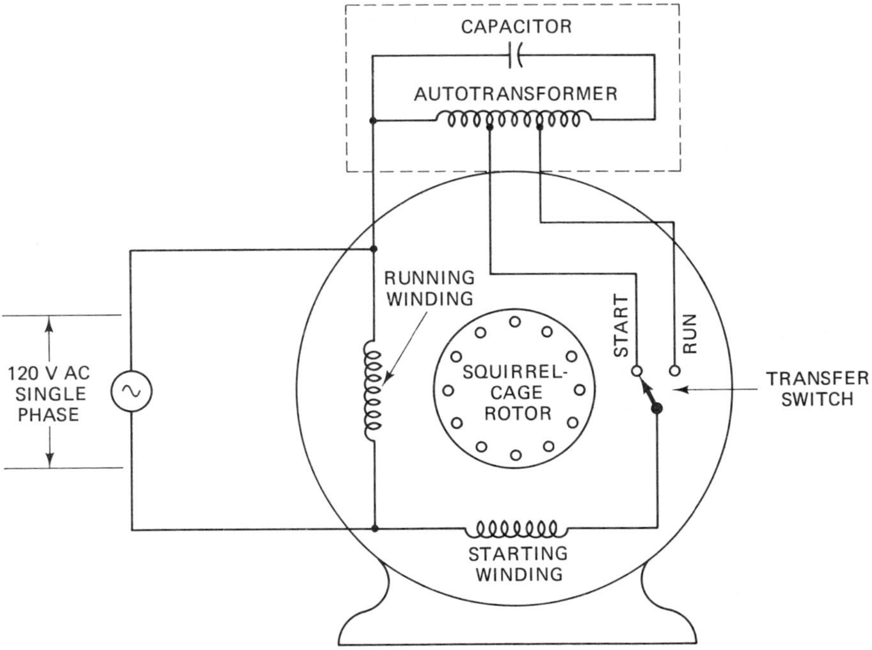

Figure 4-43 Motor using starting capacitor and autotransformer.

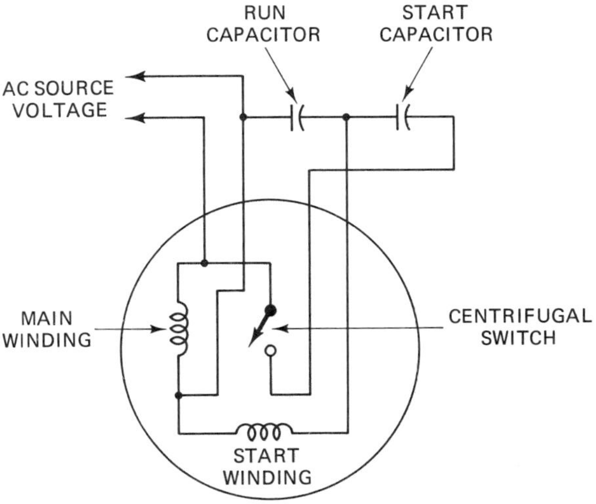

Figure 4-44 Split-capacitor motor.

Summary of Characteristics of the Split Capacitor Motor

Load: Designed for easy starting
Starting Torque: Low. It can be 50% to 100% of the running torque
Starting Current: Low
Horsepower: 1/20 to 1 HP
Phase: Single
Efficiency: High
Source Voltage: 120 or 240 AC
Speed: 900 to 1,800 rpm, variable
Reversibility: Yes
Noise and Vibration: Low
Uses: Fans, compressors

THE PERKEY CONCEPT

The Perkey concept is a technique for starting a motor having a very high horsepower output that is used for driving a heavy load. Figure 4-45 shows the working arrangement.

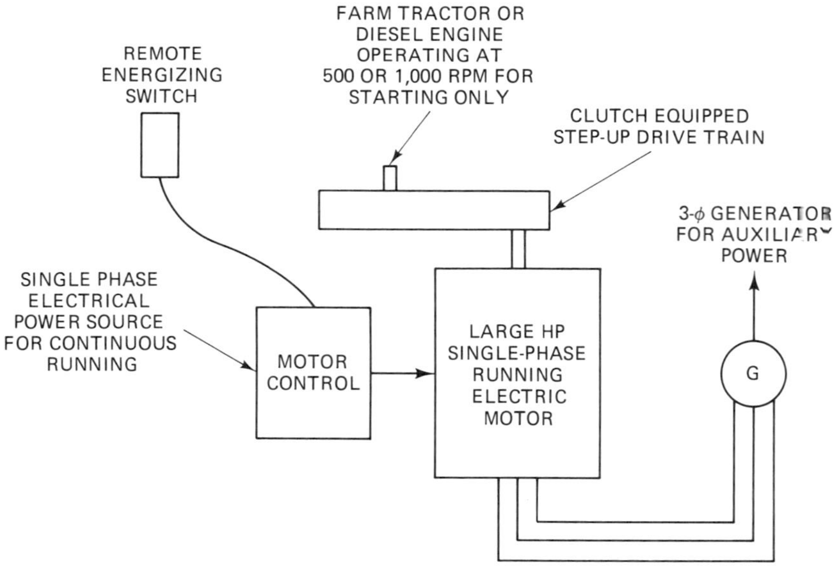

Figure 4-45 Perkey method for starting a motor driving a heavy load.

For continuous running, the motor is operated from a single-phase AC power source. Auxiliary power is supplied by a three-phase AC generator. To start the motor, the drive shaft is connected via a step-up drive train to a tractor operating at 500 to 1,000 rpm with the tractor's engine used for starting only. Once the motor reaches its operating speed, it is disconnected from the tractor's drive train.

DIRECTION OF ROTATION

In some instances, the direction of rotation of an AC motor can be changed, the technique depending on the type of motor.

Single-Phase Capacitor Motors

This motor (Figure 4-46) consists of a pair of main windings and also a pair of auxiliary windings. These windings are in series and are connected to the AC input source. Rotation is obtained through the use of a starting capacitor which supplies an out-of-phase working condition. The change of direction of rotation is obtained via a single-pole double-throw switch marked FOR (forward) and REV (reverse). The change of armature direction is via a capacitor which is switched from one main winding to the other.

Single-Phase Series Motors

Some of these can have the rotation changed by shifting brushes. Others can be reversed by transposing the series field connections. Do not reverse connections to the armature when doing this. It may not be possible to make changes if the motor windings are internally connected and cannot be reached for reconnections.

Split-Phase Motors

These are characterized by having a starting winding and a running winding. The starting winding consists of many turns of fine wire, while the running winding has fewer turns of a much heavier gauge wire. To reverse the running direction, transpose the connections of either the starting winding or the running winding. Do not transpose the leads to both coils.

Two-Phase Motors

To reverse direction of rotation, transpose the connections of either one of the phase windings.

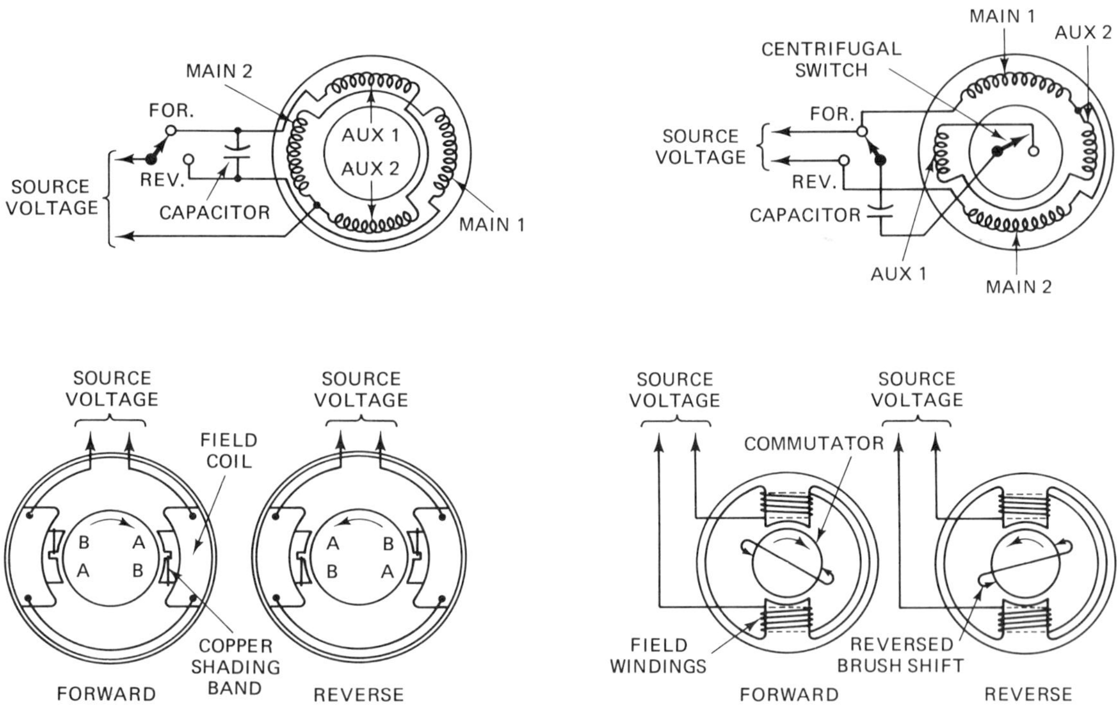

Figure 4-46 Methods of reversing the direction of rotation.

AC MOTORS FOR HEAVY LOADS

A number of motors have been designed for use with heavy loads; such motors are characterized by having a strong starting torque. These motors include types such as the capacitor-start induction-run; repulsion-start induction-run; capacitor-start, capacitor-run; three-phase general purpose; Perkey concept motors; and repulsion-start capacitor-run.

Loading

Motors vary in their ability to start a load. A split-phase motor has the capability of starting a load that is 150 percent of its running load. A capacitor-start motor has a 300 percent starting overload capacity and a repulsion-start motor can handle a load that is 400 percent of its running load. The starting load is a short time effect.

BRUSH USE

Commutators are usually associated with DC motors and slip rings with AC. Brushes are always used with commutators and slip rings. There are some AC motor types that use a commutator; these include series, repulsion-start, induction-run, and repulsion start-capacitor run. However, the use of a commutator always brings along some problems. These include radio frequency interference (RFI) when arcing takes place between the commutator segments and the brushes. Repulsion-start motors arc only when they start rotation. The use of brushes always means more maintenance, and while this applies more often to the brushes because of the effects of wear, maintenance is also needed for the commutator, the brush springs, and the brush holder. Consequently, the fact that a motor is AC operated does not necessarily mean a reduced maintenance schedule.

EFFECT OF THE MOTOR ON THE POWER LINE

Because of the large amount of starting and running current they require, motors in industry may be supplied with their own power line and will be operated independent of other electrical loads. That is usually not the case in the home where a motor will be connected to an outlet that serves other devices. If the motor has a high starting current, an electric lamp connected to the same outlet may become dim momentarily. As the motor increases its speed, it develops a larger counter EMF and so its current requirement drops substantially. If the lamp does not brighten, if it remains dim, the motor is in trouble, and its rotor may either be turning very slowly or not at all.

5 SYNCHROS

GENERATOR-MOTOR COMBINATION

Synchro, an abbreviation of the word synchronization, refers to a combination generator-motor. Synchro is known by a number of other names, some of which are trade names, but the one most commonly used is selsyn. Selsyn is an acronym for self-synchronization. Other names that are used include selsyn generator, selsyn motor, motor control transformer, differential generator, and differential motor. Although a synchro system is a generator-motor combination, its electrical behavior is close to that of a transformer and it is for that reason that the word transformer is sometimes used in describing it.

The generator portion of the synchro is a mechanical generator that delivers signals to its associated motor (sometimes more than one motor is used) and in this sense can be regarded as a transmitter. The data that is transmitted is the angular position of the rotor, and in a synchro system, the rotor in the generator and its associated motor are made to have the same angular displacement. These data are sent via wires connecting the generator and its following motor.

The stators of both units are three coils wound in a wye arrangement. The salient-pole rotor is a single coil and is connected to a pair of slip rings for AC input. While its construction is similar to that of a motor, the generator is not required to deliver mechanical power and so its output is not measured in terms of horsepower. However it is rated in terms of torque, for we are interested in the behavior of the rotor.

Basic Arrangement

Figure 5-1 shows the basic arrangement of the synchro. The generator in this circuit diagram is at the left, the controlled motor at the right. The three coils form the wye shape of the stator. The single coil shown in the center of the partial circle is the rotor and the two dots, one at the top and the other at the bottom of this coil, are the connections to the slip rings. It is via these slip rings that the line voltage, 120 volts AC, is brought into the rotor.

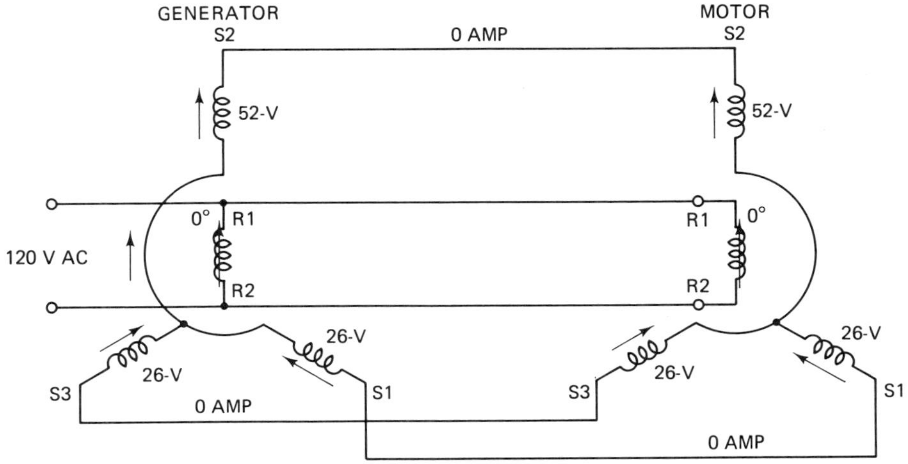

Figure 5-1 Voltage distribution in synchro when rotors of the generator and motor are at zero degrees. Since the voltages are equal there is no current flow between the two.

Note that the rotors of the two components, the generator and the motor, are wired in shunt and that at the present moment they are in the zero degree position. The three coils that form the wye all have a common connection and the corresponding coils are connected at their other ends.

One of the coils of the generator is identified as S1, the other two coils as S2 and S3. The same identification numbers are assigned to the coils in the motor. As a wye, the angles of separation between the coils in both the generator and the motor are 120 degrees.

Current Flow in the Synchro System

When both rotors are in the zero degree position no current flows from the transmitter to the motor (or receiver). The reason for this is that although there is a voltage drop across each of the coils in the generator, as well as in the motor, there is no potential difference between the generator and motor coils. The illustration shows an EMF of 52 volts across S2 of the generator and exactly the same voltage across S2 of the motor. There is no voltage difference between them, consequently there is no current flow. This analysis also applies to the other two coils.

The Generator as a Control Unit

If the rotor of the generator revolves through some angle, such as 30 degrees, the equilibrium that had existed will be upset. The voltage distribution across each of the stator coils of the wye will change and could be as shown in Figure 5-2. There is now a potential difference between coil S2 of the generator and the corresponding coil of the motor, and so we now have a current flow between them. The voltage across the other coils of the wye of the generator has also changed and so there is also a current flow between the remaining coils of the generator and the motor. There is no longer zero torque on the rotors of these two components. The torque on the motor shaft will be such that it will move in the direction of balance, that is, it will also rotate 30 degrees. When it reaches this angular displacement, the voltages across the coils of the stator in both generator and motor will once again be in balance; there will be no potential difference and no current flow between generator and motor. As a consequence, the rotor of the motor, having reached an angle of 30 degrees, will stop.

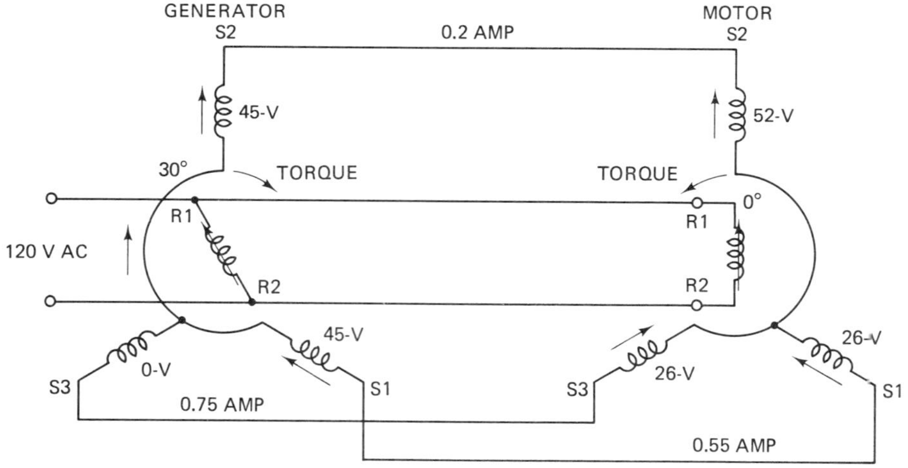

Figure 5-2 Voltage and current distribution when generator moves to 30-degree position. Motor rotor is at zero degrees.

This technique permits the use of the generator as a control unit. If its rotor is made to turn a certain number of degrees, whether electronically or manually, the motor at some distance away, possibly located in a hostile environment, will also turn the same number of degrees. But since it is working as a motor, its shaft will have the horsepower required to perform some useful work. The shaft of the motor can be made to rotate either clockwise or counterclockwise. Unlike other motors, the rotation is not continuous but is deliberately limited to a certain number of degrees.

Induced Voltage in the Stator Windings

The relationship between the rotor and stator windings is similar to that of a transformer with the rotor connected directly to the AC input, while the EMF induced across the wye windings of the stator depends on the turns ratio of the rotor to the stator.

The input to the rotor is single-phase line voltage and is assumed, in this instance, to be 120 volts. This number often ranges between 110 and 121 and is dependent on the amount selected by the local power utility. If the load on the generators of the local utility is a heavy one, the line voltage could drop below 110.

The 120-volt line input is not the peak voltage, but is the effective or rms value of 70.7 percent of the peak. The voltage induced across the stator coils is 52 volts rms and, as shown in the waveform of Figure 5-3 is 52 volts. This is the maximum effective value (not peak). To calculate the peak voltage, multiply the rms amount by 1.414 ($52 \times 1.414 = 73.528$ volts). This induced voltage occurs at two points on the waveform: when the rotor is at its 90 degree position and when it is at its 270 degree position. It is at these positions that the magnetic coupling between the stator and rotor windings is maximum.

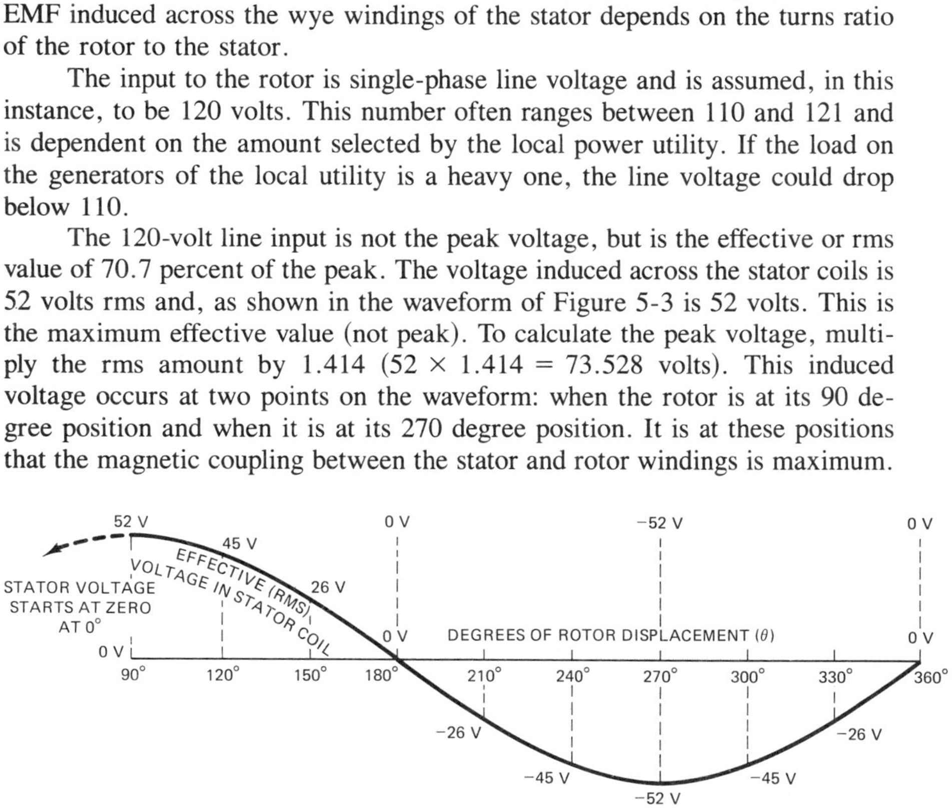

Figure 5-3 Synchro transmitter voltage values at various degrees of rotation.

Since the secondary consists of three wye arranged coils, the output of the synchro generator is three-phase AC and the separation or phase angle of these voltages is 120 degrees. The drawing of the waveform in Figure 5-4 shows these voltages. Except for the phase difference, these voltages are all alike and have a sine wave shape.

The maximum induced voltage from the primary (rotor) into the stator occurs when the axis of the rotor and stator are in alignment. But this is with reference to only one of the stator windings. The effective voltage induced into any one of the three wye windings is equal to the primary voltage (E_p) multi-

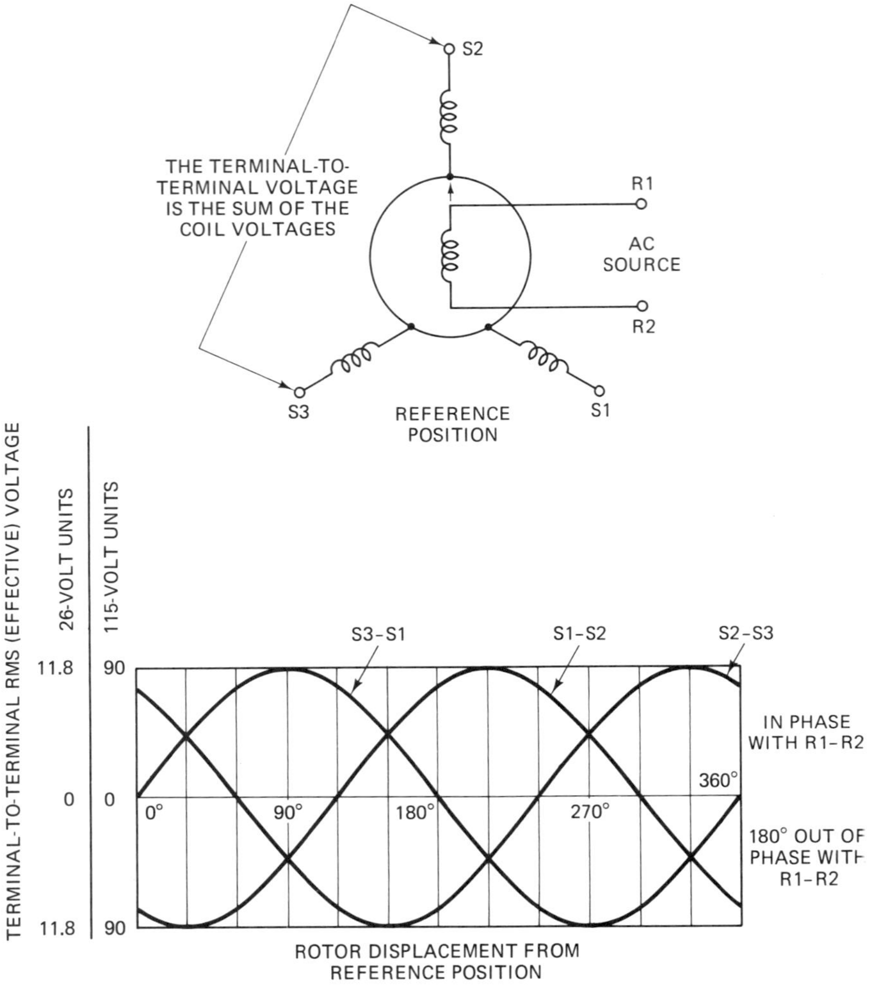

Figure 5-4 Synchro works on three-phase AC.

plied by the secondary to primary turns ratio (N) and the magnetic coupling between primary and secondary. But this magnetic coupling depends on the physical position of the rotor winding. This position is supplied in terms of the angle between them, identified by the Greek letter theta (θ). In terms of a formula this can be written as

$$E_s = E_p \times N \times \cos \theta$$

Cos in the formula is an abbreviation for cosine. The only variable is the angle θ, for the primary voltage and the turns ratio remain constant. The voltage induced across any of the secondary coils, then, depends directly on the angle.

Terminal-to-Terminal Wye Voltages

A characteristic of a wye winding is that the three coils have a common connection. This common connection, though, is inside the generator and is not available. However, the ends of the wye coils lead to terminals on the outside of the generator and so they are accessible for voltage measurements. The voltages induced across these coils are not directly additive since they are partially out of phase with each other. Further, the voltages will vary depending on the position of the rotor winding at any moment.

As shown in the graph of Figure 5-4, the terminal voltages are sine waves and can be either in phase with the input line voltage or out of phase with it. These voltages are led out of the generator to a terminal board on the rear. It is these voltages that are delivered, by wire, to corresponding terminals on the controlled motor.

Two events take place at the 0 degree to 360 degree line in this graph. The amplitude of the induced voltage in the stator coils reaches zero, and it is at this time that we get phase reversal. The frequency of each of the waveforms shown in this graph is 60 Hz and is the same as that of the input voltage. It takes each of the waves 1/60 of a second from the start of the wave to its completion.

Synchro Construction

The construction of the generator and motor of the synchro follows the same general procedure of the motors previously described. The stator consists of laminated steel slotted to accommodate the three wye coils. The laminations are placed directly inside the cylindrical metallic housing with the coils fitted into the slots. The slots are skewed to minimize cogging. The drawing in Figure 5-5 shows an end view of a single stator lamination. To minimize the flow of eddy currents, a thin coating of some insulating liquid is applied to both surfaces of the lamination. Since the laminations are a ferrous material, they contribute substantially to the inductance of each of the stator coils. The stator windings have no connections to an external voltage source, their only poten-

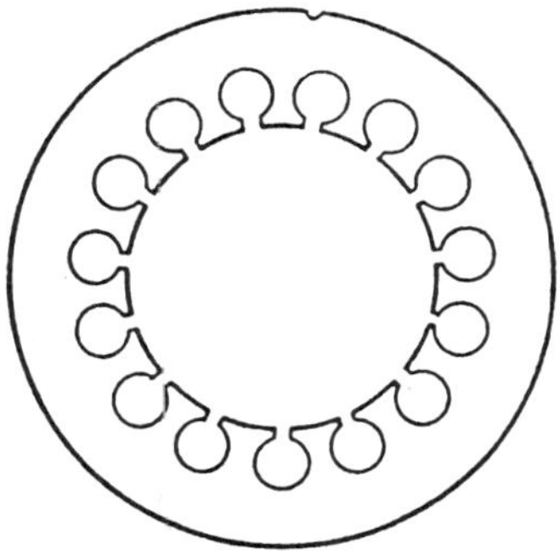

Figure 5-5 Stator lamination.

tial being an induced voltage resulting from the magnetic field of the rotating armature.

RECEIVERS

The construction of the controlled unit, the receiver, is identical to that of the transmitter. In some instances these two components are so designed that they can be interchanged. And while the line power is usually 60 Hz, the input power to the generator can be some other frequency, such as 400 Hz.

In the absence of any input power, the motor rotor is free to revolve and is restricted only by bearing, shaft, and friction between its slip rings and brushes.

When AC power is initially supplied to the generator, the rotor of that component will move. If, at this time, the receiver is switched to its on position, its rotor will also move, occupying an angular position identical to that of the rotor of the generator. This is not a one-way action however. Turning the rotor of the motor can cause the rotor of the generator to revolve in the same direction and angle.

It is also possible for the motor rotor to overshoot its correct angular position and, in making a correction, undershoot. In some instances it may oscillate, moving back and forth, a condition known as hunting. In extreme cases, the rotor will not only overshoot, but its momentum will carry it in the same direction, resulting in spinning.

The "following ability" of the motor rotor is influenced by the load being driven by the motor—the lighter that load the better the following ability.

Damping

Various methods are used to restrict the hunting or oscillatory tendencies of the motor rotor. In some small synchro units a shorted winding is used that is positioned on the quadrature axis at right angles to the direct axis. In other synchros, especially in larger types, mechanical damping is used. One method is to mount a heavy brass flywheel on the rotor shaft of the motor with the flywheel free to rotate around a bushing fastened to the motor shaft. A tension spring attached to the bushing rubs against the flywheel. If the rotor should accelerate or reverse its direction suddenly, the inertia of the damper opposes the change.

FUNCTIONING OF THE SYNCHRO

There are three steps in the functioning of a synchro as illustrated in Figure 5-6. Assume that at the start, as in drawing A, the two rotors, that of the generator and that of the motor, are in alignment. The two rotors receive simulta-

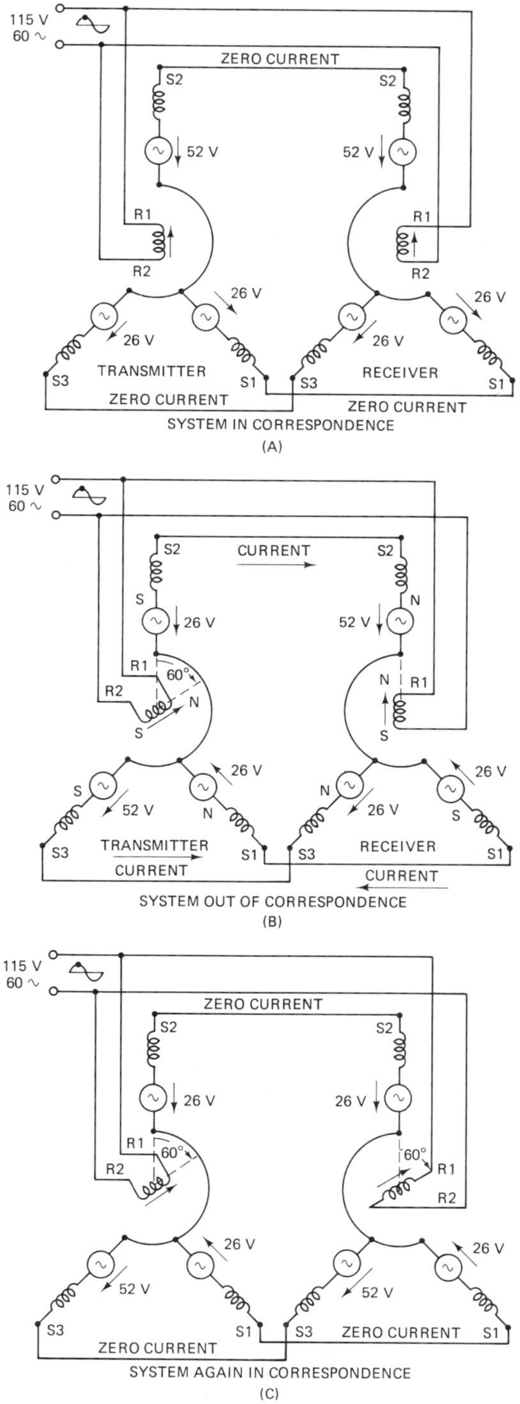

Figure 5-6 Steps in the functioning of a synchro: start (A), transmitter rotor advances a number of degrees (B), receiver rotor turns and makes angular movement corresponding to that of the transmitter (C).

neous AC excitement from the AC power line and actually the rotors are wired in parallel. The two units, the generator and the motor, are working under identical conditions. The voltages across the wye coils of the stator are the same for both. As a result, there is no current flow from the generator to the motor. There are induced voltages across the wye windings of both components, but since these potentials are equal there is no current flow between their wye windings. This situation is similar to that of two batteries wired in parallel. If each of the batteries has the same voltage there will be no current flow between them.

If, now, the rotor of the transmitter is turned through some angle, the two rotors will no longer occupy the same position. As shown in drawing B, the rotor of the generator has been turned 60 degrees in a clockwise manner. Note that the voltage across coil S3 has increased from 26 volts to 52 volts while that across S2 has had an exactly opposite reaction for its voltage has decreased from 52 to 26 volts. For this moment, though, none of the voltages across the stator coils of the motor have changed. Because of these voltage changes, currents will move between the two units as indicated by the arrows. And, as a result of these currents, a magnetic field will appear across the rotor coil, turning that coil through an angle of 60 degrees. With both coils, that of the generator and that of the motor, having the same angular displacement, the voltages across the motor coils will now be identical to those across the generator coils. This identity of voltages means there will be no further current flow between the two components and so there will be no further rotor movement. The shaft of the motor will have turned by an angle of 60 degrees (drawing C). This shaft action could turn some component by this amount. That component could be the arm of a variable resistor, a lever, or the control element of some other component.

Receiver Torque

As the rotor of the receiver motor turns, the angular difference between it and the rotor of the transmitter gradually becomes less. As a consequence, the voltage differences decrease and less current flows into the motor. This results in a decreasing motor torque.

TORQUE CAPABILITY

In a synchro system, the power source for the motor section can be regarded as dependent on the generator, which, in turn, receives its operating power from the AC line. The generator thus acts as an intermediary between the power line and the motor. If the generator operates a number of synchro motors simultaneously, it must be capable of meeting the power needs of those motors.

Whether the generator in a synchro system is capable of driving its following motor (or motors) depends on the load that is imposed. That load can be a light one, possibly consisting of nothing more than an indicator. Alternatively, the motor may be required to do some heavy work.

There are two ways of meeting the requirement for a heavy load. The first is to build the generator and its accompanying motor with a substantial power capability. If the generator is incapable of meeting the load demand, it can be followed by a power amplifier, shown in Figure 5-7. Under these circumstances the generator can be regarded as a voltage source, with the amplifier supplying the equivalent of a power source. The input to the amplifiers is voltage; their output is power.

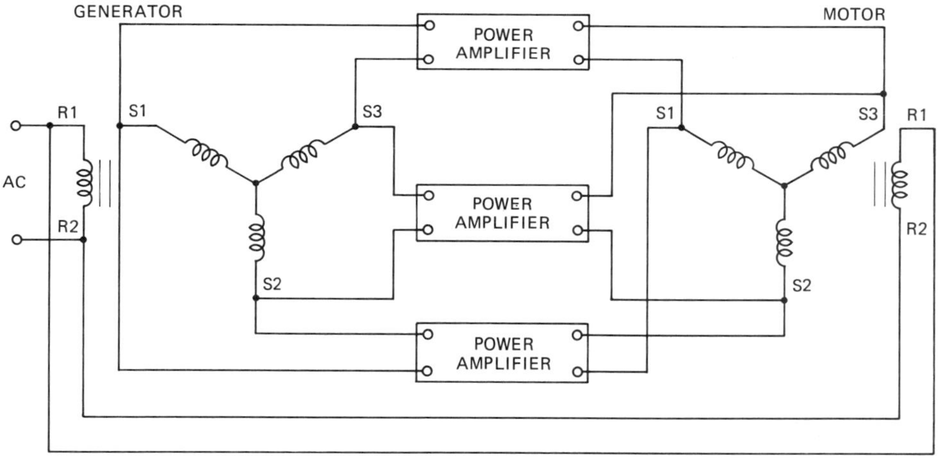

Figure 5-7 Power amplifiers for increasing the torque capability of the motor used in a synchro system.

6 ELECTRONICS FOR MOTORS

A variety of electronic components and circuits are used in conjunction with motors, as part of the motor structure, external to the motor, or as a combination of both. Electronics can be used to change the normal operating characteristics of a motor, improving its speed regulation, controlling its speed or output torque, or determining the amount of angular rotation of its shaft.

The electronics includes items such as diodes, transistors, thyrectors, thyristors, diacs, triacs, Schmitt modulators, silicon controlled rectifiers, relays, integrated circuits (ICs), logic, and Hall devices. It can also include transient pulse suppression networks such as snubbers and thyristors. These may be used individually or in various combinations.

A study of electronics and of semiconductor electronics and components is a prerequisite to an analysis of motor electronics. The material supplied here should not be regarded as an in-depth study but rather as a quick review.

DIODES

A semiconductor diode is a rectifier having AC input and pulsating DC output which can also be used for modifying sinusoidal AC into clipped AC waveform output. Figure 6-1A illustrates the symbol and identifies the anode and cathode. Diodes can be connected in various ways, as individual units, in series, in series back to back, in parallel, and in parallel back to back.

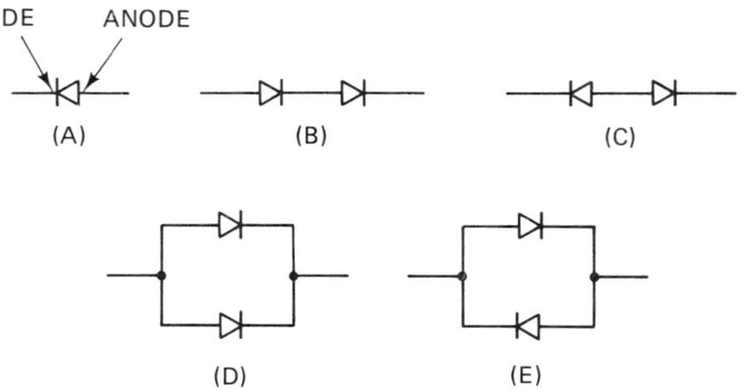

Figure 6-1 Semiconductor diode symbol (A); diodes in series (B); diodes back to back (C); diodes in parallel (D); diodes in parallel opposing (E).

Diodes can work as power line or signal voltage rectifiers and as electronic switches. If used without the application of a DC bias, the output of the diode will consist of a series of full half-wave negative or positive impulses depending on the relationship of the diode to the input signal. The diode can be biased by a DC voltage (Figure 6-2) so that only a small part of the input is effective in taking the diode out of the current cutoff resulting from the bias, thus producing an output. This output can be used as the trigger voltage to turn on or to turn off some other component.

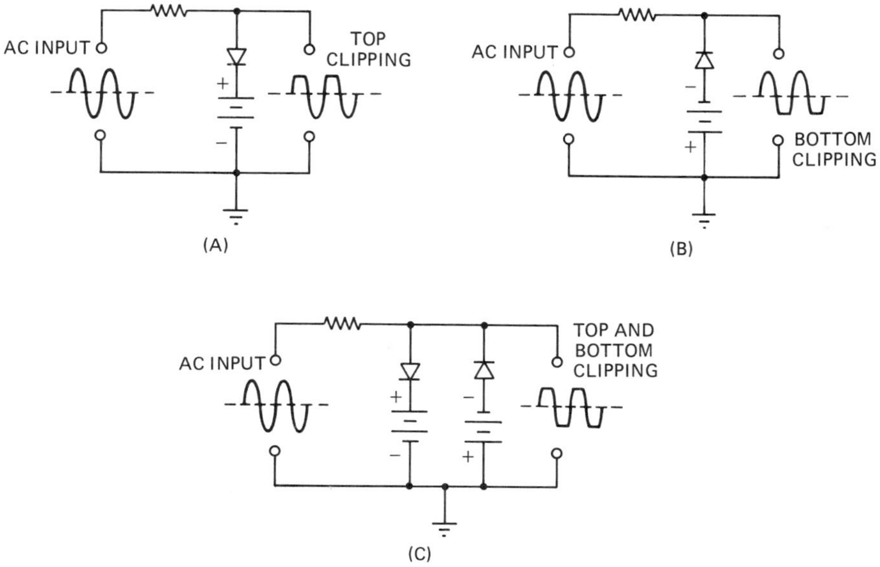

Figure 6-2 The diode can be biased to modify the output waveform. Positive output peak is limited (A); negative peak is limited (B); both output peaks are limited (C).

THE THYRECTOR

The thyrector is a semiconductor diode used to limit voltage transients to a limited value. They are sometimes placed across the AC input power line connected to circuits used to control the speed and regulation of motors. When so positioned, they protect the control circuitry from possible damage caused by transient voltages. Figure 6-3 shows the electronic symbol for a thyrector. Since the AC power line input voltage has both a positive and a negative half, it is possible for a transient to ride along with each. For that reason the thyrector is the equivalent of a pair of diodes placed back to back.

Figure 6-3 Thyrector symbol.

THYRISTORS

A thyristor is a semiconductor switch. These come in a number of different types, the most notable of which is the silicon controlled rectifier (SCR).

The Silicon Controlled Rectifier

The silicon controlled rectifier (SCR), whose symbol appears in Figure 6-4, is a three-terminal device consisting of a cathode (K), an anode (A), and a gate (G). In order to conduct, the SCR must be triggered by a voltage applied to its gate. In operation, the unit is normally in an open circuit condition but switches rapidly to conduction when an appropriate signal is applied to its gate. The SCR passes current in one direction only.

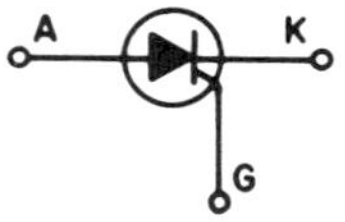

Figure 6-4 Symbol for silicon-controlled rectifier (SCR): A = anode; K = cathode; G = gate. (Courtesy Thompson Consumer Electronics, Inc.)

The anode of the SCR is made positive with respect to its cathode. This has the effect of forward biasing its P-N junction consisting of the anode and cathode. Consequently, a small current moves from the gate terminal, through an external load to the p-terminal anode.

The control circuit consists of the gate and the anode. SCRs can be low, medium, or high current types, with low current units capable of switching up to 1 ampere, medium current SCRs up to 10 amperes, and high current units having a switching capability of up to 2,500 amperes. Switching time is in the order of microseconds.

Unlike the usual semiconductor diode, the SCR does not conduct in the presence of forward bias until the gate is pulsed. The diode will then continue to conduct until the forward bias is removed or reduced below the conductivity level.

It is not desirable to apply a continuous DC gate signal because of the excessive gate power dissipation that will occur. In Figure 6-5, resistor R is used to limit the gate current. Diode D limits the amplitude of a negative gate signal.

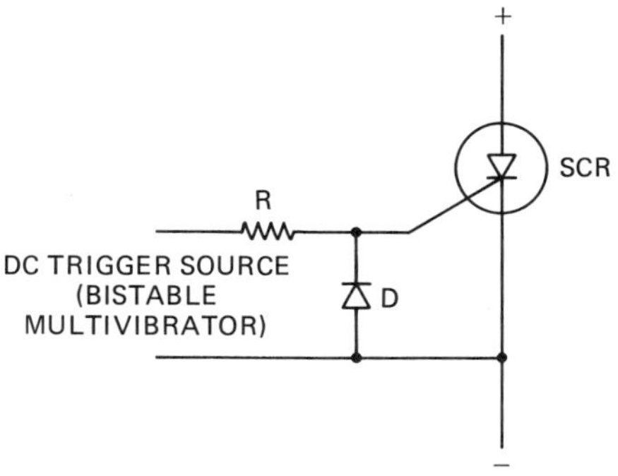

Figure 6-5 SCR circuit with gate current limited by R.

Triac

Like the SCR, this unit is a member of the thyristor group and is a silicon bidirectional thyristor. It consists of a pair of complementary semiconductors connected in parallel at the time they are manufactured. It is found in applications such as a light dimmer for incandescent lights and in electronics as a motor speed control. Its symbol is supplied in Figure 6-6.

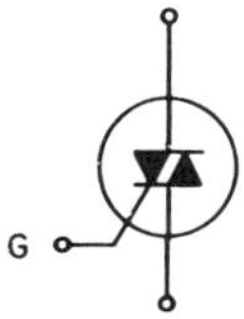

Figure 6-6 Triac symbol.

Diac

The diac (Figure 6-7) is a bidirectional current-limiting diode switch used as a variable resistance. Its input can be AC and the output can consist of a series of positive and negative trigger pulses. It is generally used as a trigger for a triac in motor control circuitry.

Figure 6-7 Diac symbol. The unit is a bilateral trigger diode.

TRANSISTORS

There are a number of different types of transistors, but the two basic units (Figure 6-8) are bipolar types. The N-P-N is illustrated in Figure 6-8A, the P-N-P in Figure 6-8B. P is an abbreviation for positive, N for negative. The two symbols are practically identical, the only difference being in the direction of the arrow for the emitter (E). In the N-P-N type this arrow points outward, in the P-N-P inward.

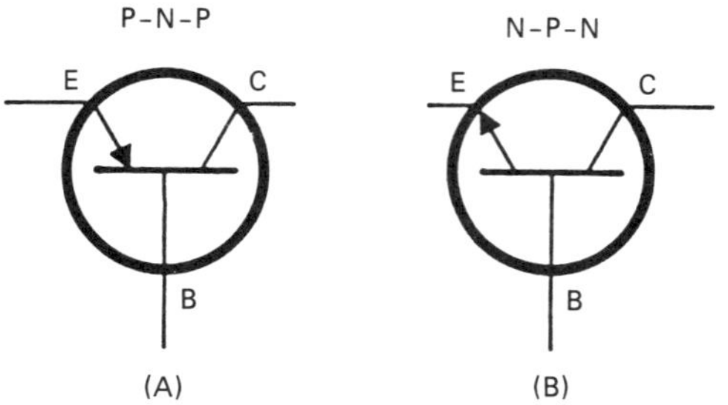

Figure 6-8 Transistor symbols: P-N-P (A) and N-P-N (B).

The bipolar transistor is a three element component consisting of an emitter (E), a base (B), and a collector (C). The transistor can be biased to establish its correct operating point. This can be done by two forms of bias: fixed bias and self bias, as indicated in Figure 6-9. The transistor Q1 is a P-N-P type with bias supplied by resistors R1, R2 and R4. R1 and R2 are in series shunted across the battery supply, E. The voltage produced by current flow through R2 supplies fixed bias for the transistor. The direction of current flow through R2 is such that the end of the resistor connected to the base is minus, the ground end is plus.

Self bias is supplied by resistor R4 and the amount of voltage it develops

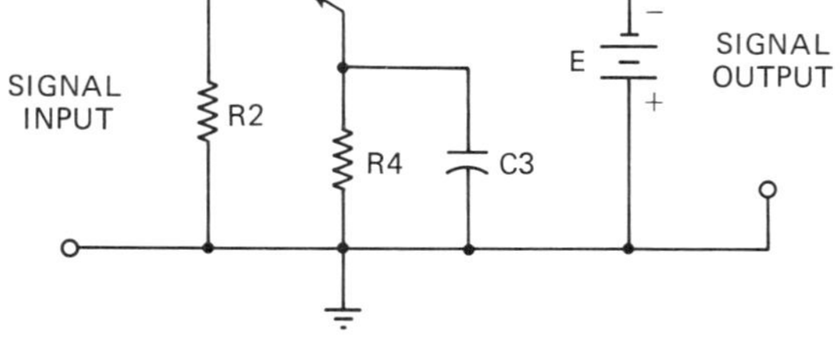

Figure 6-9 Transistor amplifier using fixed and self bias.

depends on the amount of current flow through the resistor at any moment. Capacitor C3, connected across R4, helps produce a smoother, nonvarying voltage across the resistor.

BASIC TRANSISTOR AMPLIFIERS

Transistors can be grouped into three amplifier types: grounded (common) base, grounded (common) emitter, and grounded (common) collector, as illustrated in Figure 6-10. These circuits all use N-P-N transistors, but P-N-P units could have been used instead.

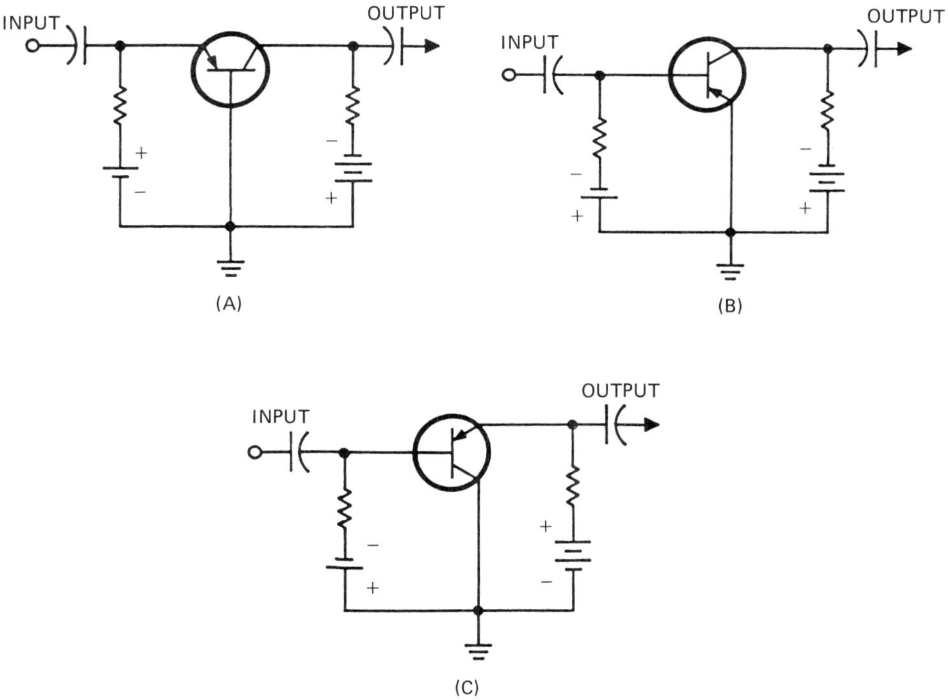

Figure 6-10 Basic transistor amplifier types: grounded base (A); grounded emitter (B); grounded collector (C). These are also known as common base, common emitter, and common collector types.

Grounded Base Amplifier

This amplifier (Figure 6-10A) does not invert the phase of the input signal. The signal is input between the emitter and collector, with the base as the common element.

Grounded Emitter Amplifier

This is the most widely used of the three amplifier types (Figure 6-10B). The unit uses self bias, supplied by the resistor connected to the emitter and bypassed by an electrolytic capacitor. Fixed bias is produced by the two series resistors R1 and R2. One end of R1 is grounded; one end of R2 is connected to the DC plus supply. The two resistors are shunted across the DC supply with the base connected to their junction point. This amplifier is characterized by the fact that its output signal is 180 degrees out of phase with the input signal; it is the only one of the three amplifier types to show this characteristic.

Grounded Collector Amplifier

The grounded collector amplifier (Figure 6-10C), also known as an emitter follower, is characterized by having a high input impedance and a low output impedance. The unit uses both fixed bias and self bias. Fixed bias is supplied by R1 and R2 shunted across the DC supply. Self bias is developed across emitter resistor R3. Electrolytic capacitor C1 in Figures 6-10 A and B is an emitter bypass capacitor that helps maintain self bias at a steady value.

HALL GENERATOR

A Hall generator, also known as a Hall effect device, consists of a section of semiconductor material such as indium arsenide or indium antimonide. A DC voltage is applied to opposite sides of the semiconductor wafer, as indicated in Figure 6-11. When the wafer is then positioned in a magnetic field so that the lines of flux are perpendicular to it, it generates a DC voltage. This voltage can be taken from the sides opposite those of the input DC voltage.

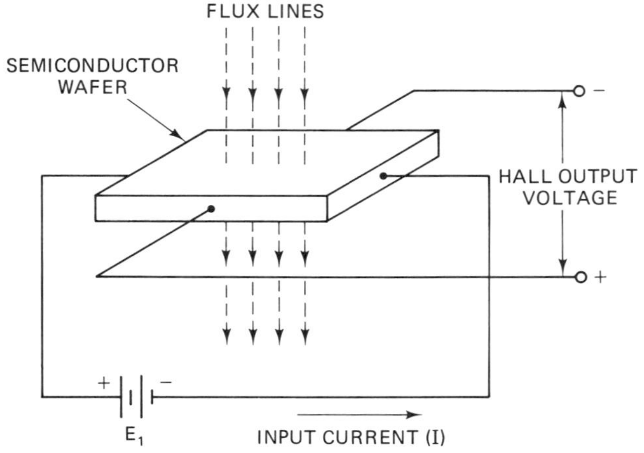

Figure 6-11 Hall generator.

The amount of output voltage depends on two factors, the strength of the magnetic field and the amount of input current supplied by the DC voltage source. The behavior of the semiconductor wafer is referred to as the Hall effect. These devices are used in brushless DC motors.

SERVOMECHANISMS

A servomechanism, more popularly known as a servo, is an automatic feedback control system. In the case of a compact disc player, for example, three servo systems are used to maintain the accurate flow of data. One servo controls the motor that supplies constant linear velocity for disc rotation. Another servo controls a motor that keeps the scanning laser beam precisely centered. The third servo adjusts laser beam focusing for discs that may be warped, however slightly.

There are many types of servos. In the compact disc player they are electromechanical with a feedback signal controlling the forward or reverse rotation of a motor shaft, or the lateral or vertical movement of a device. The servo-controlled motor (Figure 6-12) is sometimes referred to as an actuator.

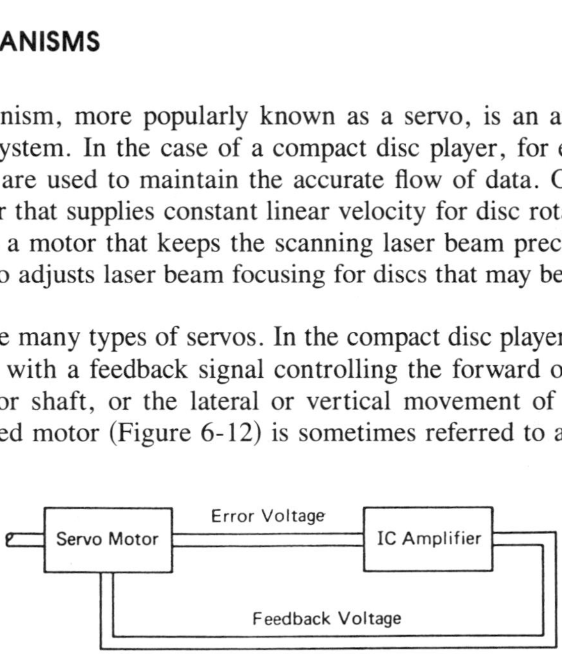

Figure 6-12 Servo motor controlled by error feedback voltage.

Servos have several distinct advantages. They can be made to be completely automatic and they can control mechanical movements in very small increments. Servos can be analog or digital, depending on the kind of control signal used.

In the servo system an error voltage is developed if the device controlled by the servo motor is not precisely positioned. This error voltage is fed into an integrated circuit (IC) in which the voltage is amplified. The error voltage, referred to as a correction voltage or feedback voltage, following amplification, is supplied to the servo motor to operate it in a direction that decreases the error voltage. When that error voltage drops to zero, there is no further feedback; consequently, there is no further correction movement of the shaft of the servo motor. The feedback error voltage can be DC and can be positive or negative. A negative voltage causes the motor shaft of the servo to revolve in one direction; a positive voltage results in rotation of the servo motor shaft in the opposite direction.

SENSORS

A sensor is a device that detects an operational change in some component or a circuit consisting of a number of components and is often found associated with servomechanisms. The "sense of change" may be expressed in the form of an AC voltage or as a positive or negative DC voltage pulse. The stimulus can be a change in heat, light, sound, a current or magnetic field strength, or some motion such as the revolution of a motor shaft.

Figure 6-13 shows the basic setup of a sensor arrangement. The sensor is connected to an operating circuit or, in this case, a motor which supplies an output in the sense that it varies from a preset speed. That speed is the reference, and any increase or decrease in that speed is the deviation. This deviation is changed by the sensor into an equivalent voltage which is then used to modify the behavior of the source or operating voltage, increasing or decreasing it accordingly. The feedback signal may increase the speed of the motor or decrease it.

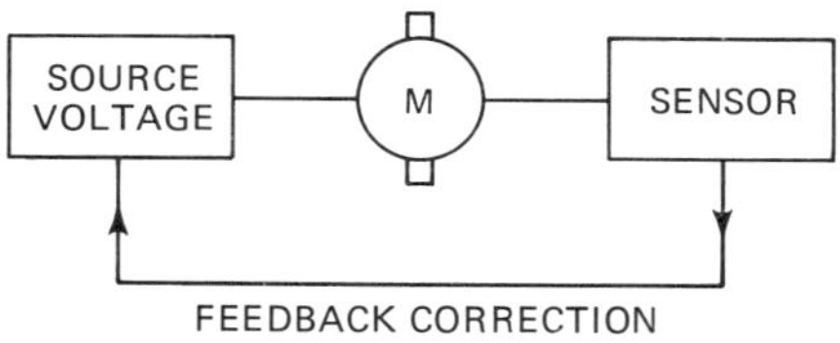

Figure 6-13 Basic sensor arrangement.

Sensors can be mechanical, electromechanical, or electronic. The type being used depends on the motor and its function. Sensors, regardless of type, are automatic. Their accuracy and working speed depend on their design.

RELAY MOTOR CONTROL

The system shown in Figure 6-14A uses a double-pole double-throw (dpdt) relay to control the direction of rotation of a miniature brushless DC motor. The relay is remotely controlled and is connected to the output of a radio receiver. A single, short-term pulse sent from a remote transmitter is picked up by the receiver whose output actuates the relay in one direction. The absence of a transmitted pulse means that the relay contacts are returned to their original position since the relay is no longer energized.

The purpose of the relay is to connect the motor's voltage input terminals to the battery. The action of the relay is such that the polarity of the DC voltage is transposed. As a result, the motor can be made to turn in one direction or the other.

A technique such as this can be used for control of model boats, autos, and planes. The advantage of this technique is that just a single battery is required. Another method, shown in Figure 6-14B uses a much simpler type of

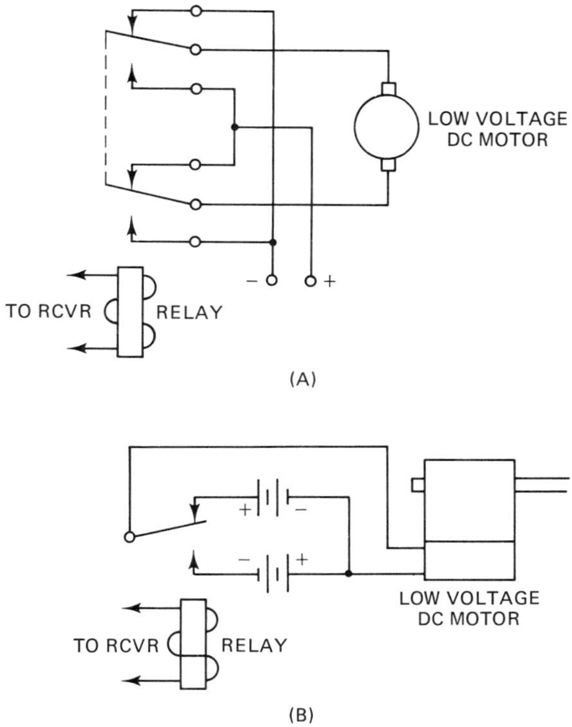

Figure 6-14 A dpdt relay controls the direction of motor rotation (A); a simplified circuit requires use of two batteries (B).

relay, a single-pole, double-throw type (spdt) but this requires the use of two batteries.

In this circuit, the relay is also wired in the output of a radio receiver with a single transmitted pulse (or the absence of a pulse) causing the low voltage DC motor's armature to turn in one direction or its reverse.

LOGIC CIRCUITS FOR MOTORS

Logic circuits, also known as gates or decision making circuits, long used in computers, are also applicable for the regulation and speed control of motors. These circuits use diodes or transistors, or some combination of the two.

One of the simplest of all logic circuits is a single biased diode (Figure 6-15), an arrangement that permits the passage of current through the diode when an input signal voltage pulse exceeds the amount of bias applied to the diode.

The diode in this illustration is reverse biased since a negative voltage, supplied by the battery, is applied to the anode of the diode, and a positive

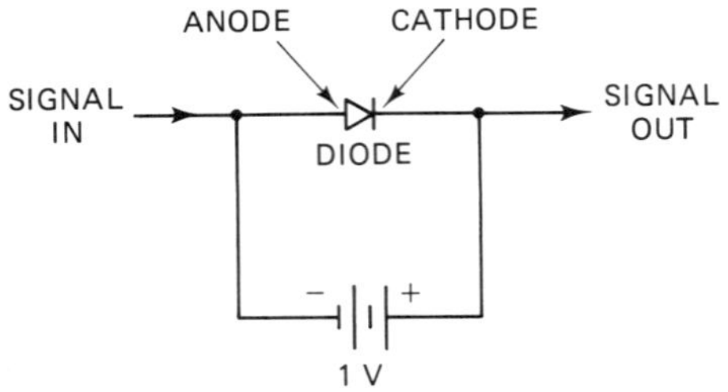

Figure 6-15 A biased semiconductor diode does not conduct until the positive amplitude of the input pulse is greater than the battery voltage.

voltage to the cathode. The battery voltage selected in this example is 1 volt, but any other voltage could have been used. As a result, the diode will not conduct for any positive input pulses until they reach a strength of 1 volt. Thus the diode works as an automatic switch.

The AND gate. A pair of diodes can be connected in series and, although electronic, they are represented in Figure 6-16 as a pair of mechanical switches. These switches have three possible operating conditions. Switch A can be open and switch B closed; switch A can be closed and switch B open; or both switches, A and B, can be open or closed. Current can only move from the input to the output when switch A and switch B are closed; hence the circuit is referred to as an AND circuit. The two mechanical switches, wired in series, can be replaced by a pair of biased diodes or, as shown in Figure 6-17, by two transistors connected in series.

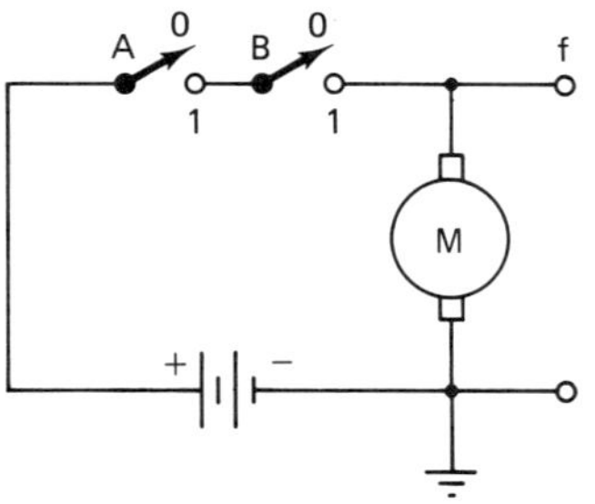

Figure 6-16 AND circuit. Switches A AND B must be closed for a current to reach the motor.

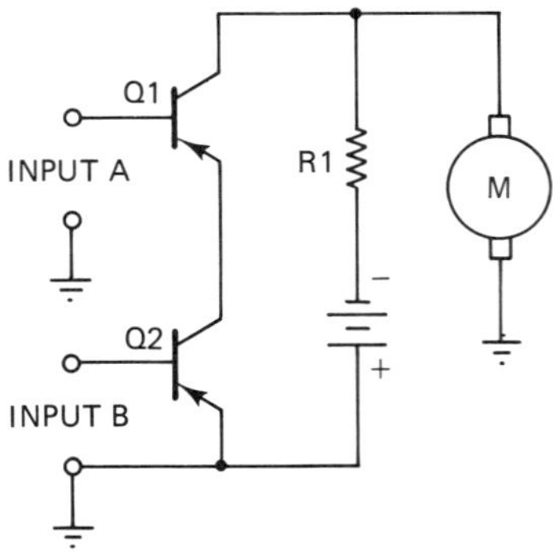

Figure 6-17 Transistorized AND circuit.

In the absence of signals for input A AND input B, transistors Q1 and Q2 are cut off and no signal appears across the output resistor R1. The same situation applies if there is a signal at input A but none at B, or if there is a signal at B but none at A. However, a negative signal at both inputs A and B will cause both transistors to conduct, thus supplying an output across resistor R1, the common load for both transistors.

OR gate. Figure 6-18 shows a pair of mechanical switches wired in parallel, representing either a pair of diodes or transistors. If both switches are open, no signal can move from the input to the output. However, if either switch A OR B is closed (or if both switches are closed), a signal can proceed from the input to the output.

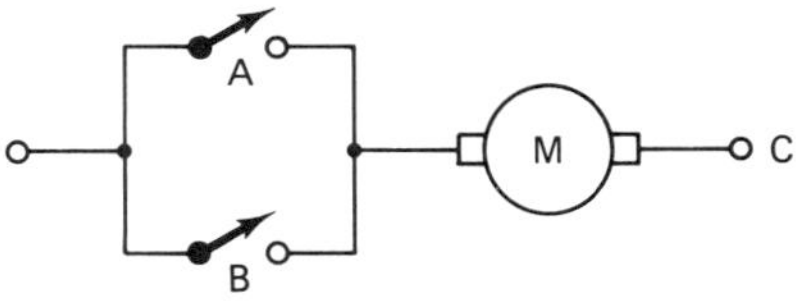

Figure 6-18 OR circuit. Either switch A OR switch B must be closed for a current to reach the motor.

Figure 6-19 is the transistor version of the OR gate. The two transistors are wired in parallel with R1 as the common load resistor. There will be signal output if either transistor Q1 OR transistor Q2 or both receive a signal input. However, if both transistors do not receive a signal input they will be nonconductive and no signal voltage will develop across the common load resistor R1.

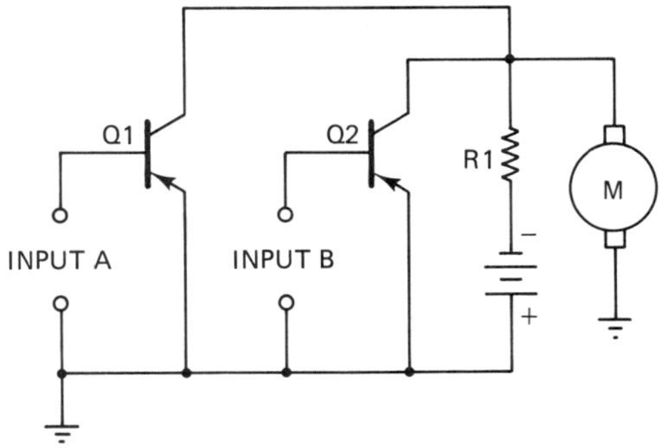

Figure 6-19 Transistorized OR circuit.

Symbols for gates. Symbols are available for individual components such as resistors, capacitors, diodes, and transistors. Instead of showing these as separate parts, they can all be combined into a single, special symbol. Figure 6-20 illustrates the symbols used for AND gates and OR gates. The gates may include diodes or transistors or both.

AND-OR gate. By combining an AND circuit with an OR it is possi-

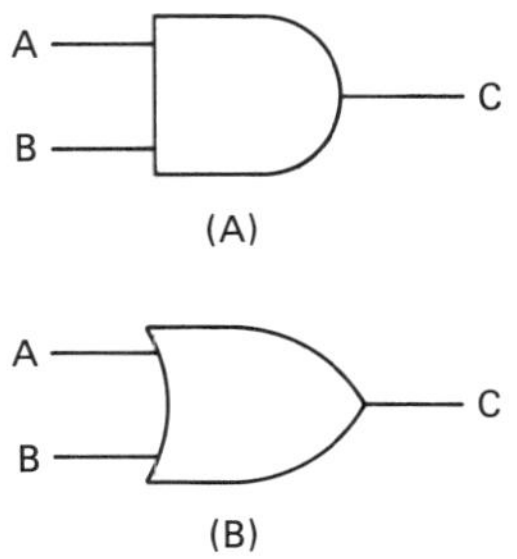

Figure 6-20 Symbol for AND gate (A); OR gate (B).

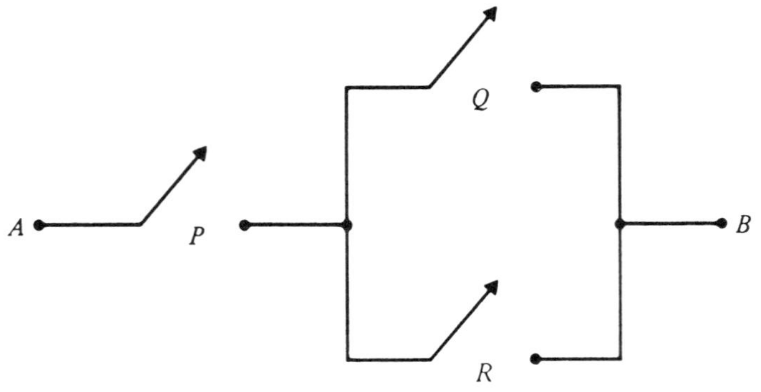

Figure 6-21 AND-OR arrangement. P is the AND switch. Switches Q and R are the OR switches. For a signal to travel from A to B switch P must be closed AND switch Q OR R.

ble to obtain an AND-OR arrangement, as indicated in Figure 6-21. In such a circuit, the AND switch and one or the other of the OR switches must be closed for a signal to travel from the input to the output.

NAND gate. Logic circuits can be operated by positive input pulses, but negative pulses can also be used. It is possible to develop an AND circuit that is responsive to negative pulse input instead of positive; such a circuit is called a negative AND or NAND. The NAND gate can be represented by its own symbol (Figure 6-22).

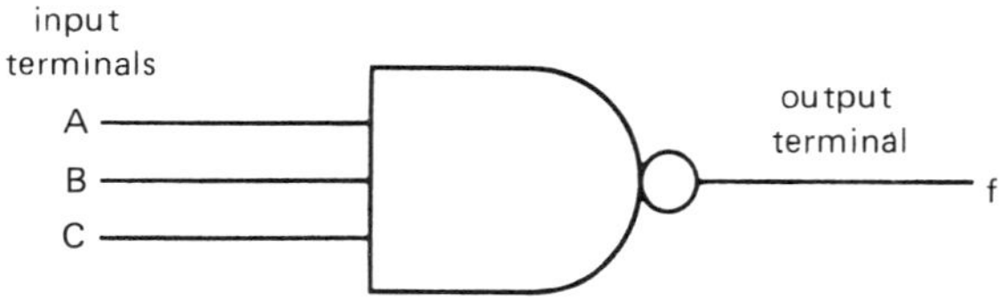

Figure 6-22 Symbol for a negative AND or NAND gate. It may have a number of inputs, with three of them, A, B and C, shown here. The output can be indicated by any letter, but in this drawing is represented by f.

NOR gate. It is also possible to have negative OR circuits, called NOR. The NOR symbol is shown in Figure 6-23. And, just as AND and OR circuits were put together to form an AND-OR, so too is it possible to combine a NAND and NOR to obtain a NAND-NOR.

Both positive and negative types of gates are used since electronic pulses are available in three classifications: positive pulses, negative pulses, and combined positive/negative. Positive pulses and negative pulses are DC. Combined positive/negative pulses are AC.

Gating circuits can be combined with electromechanical components such as motors and relays. The gating circuits can operate these for specific

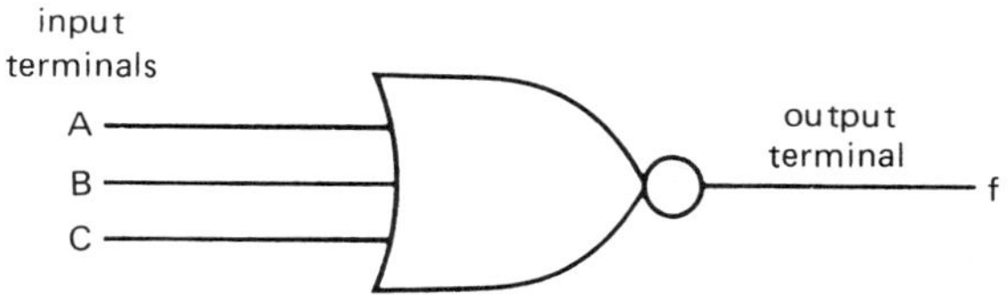

Figure 6-23 Symbol for NOR gate, also with 3 inputs.

amounts of time, turning them on or off as required. The rotor shaft can then be operated continuously or for predetermined amounts of time. With stepper motors (see Chapter 8), the shaft can be made to move through a desired angle during a specific time period.

In some instances, the motor may be connected to a defeat switch, giving the user the option of overriding the action of the gating circuit. Gates can be single-pulse input types or may have multiple input. With the multiple input gate, pulses from various circuits can trigger the gate, initiating some desired action by the motor.

INTEGRATED CIRCUITS

An integrated circuit (IC) consists of various components such as diodes, transistors, resistors, and coils mounted on a semiconductor base known as a substrate. The substrate not only supports the parts but, in the case of diodes and transistors, is actually a part of them.

The substrate, made of silicon, and the parts engraved on it or mounted on it form what is known as a chip. The chip, when mounted in its protective package, is referred to as an integrated circuit. Sometimes, though, an IC and a chip are terms that are used synonymously. The IC has terminals which may be located on one side, two sides, or on all sides, as in Figure 6-24. The circuits contained in the IC aren't always detailed in its circuit drawing, and so all that is available is its outline.

Electronic circuits used in connections with motors may be detailed in the form of ICs. Often enough, the IC symbol will be shown, with a brief indication of its function, plus the part number of the IC but with no further explanatory material.

OPERATIONAL AMPLIFIER

This component, sometimes simply referred to as an op amp, is an IC characterized by high stability and good linearity. Its symbol is a triangle, usually shown with its apex pointed to the right (Figure 6-25). In some instances just the triangle is used (A), or it may also have polarity symbols (B). In some cases further information such as inverting (inv) and noninverting (non-inv)

Figure 6-24 This IC performs such varied functions as motor speed control, frame sync detection, error detection and correction, and data interpolation. It is used in a compact disc player.

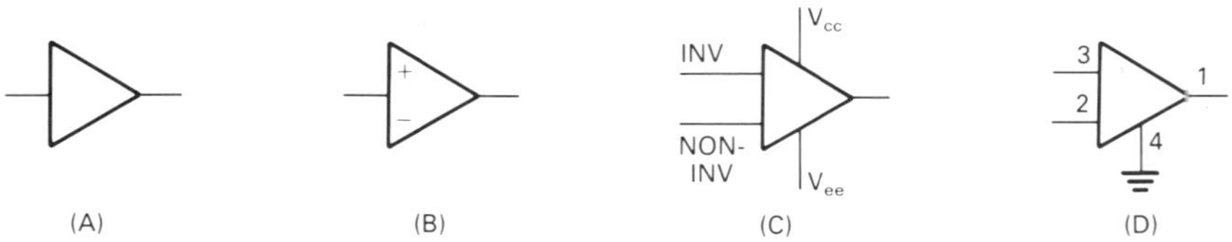

Figure 6-25 Op amps: basic symbol (A); with polarity symbols included (B); inverting-noninverting type (C); symbol with numbering included (D).

are used (C), plus voltage indications such as V_{cc} and V_{ee}, representing collector voltage and emitter voltage respectively. Each op amp has its own numbering system (D).

An op amp is a high-gain, direct-coupled differential amplifier. Op amps have a very high input impedance and as a result impose a very light load on the circuit delivering a signal. The output impedance is low. Op amps also have a wide bandwidth.

Motors are electrical devices that, no matter how detailed or elaborate their structure, are basically dependent on the attraction and repulsion of magnets. Electronics for motors consists of various techniques for control. With the help of electronics and electronically-controlled motors, mechanical devices can be made to be self-correcting, that is, if they produce an error they can use that error to correct their movement. Thus, a laser beam in a compact disc player must follow a path that is a fraction of a hair in width. Should it deviate, it will produce an error voltage which will operate a motor to eliminate that deviation.

While the motor for a compact disc player or an audio or video cassette deck is a miniature type, electronic control can also be used for motors having a substantial power rating.

CONTROL TECHNIQUES FOR SERIES DC MOTORS

The speed of a series DC motor can be controlled by adjustment of its source voltage. There are various methods for doing so.

Resistance method. A variable resistor such as a wire-wound rheostat can be inserted in series with the source voltage leading to the motor. Since the current flowing through the field coils and armature winding will also pass through the rheostat, there will be a considerable power loss across this resistive element. The method does have the advantage of being simple and inexpensive and supplying fairly smooth control.

Battery switching method. With this technique, two or more batteries, wired in series, can be switched in or out of the power source circuit. This method is crude and not very practical, but it does eliminate the power loss of the resistor method.

Chopper technique. A chopper is an electronic switch that can be put in series with the motor. Choppers have several advantages. They are electronic, can be made to have various switching times, are automatic rather than manual, and are dependable. Choppers are circuits that produce pulsed waveforms that resemble square waves (Figure 6-26).

Figure 6-26A is that of a pulsed waveform that has a short ratio of on-time to off-time. Since the on-time is so low, the average input source voltage

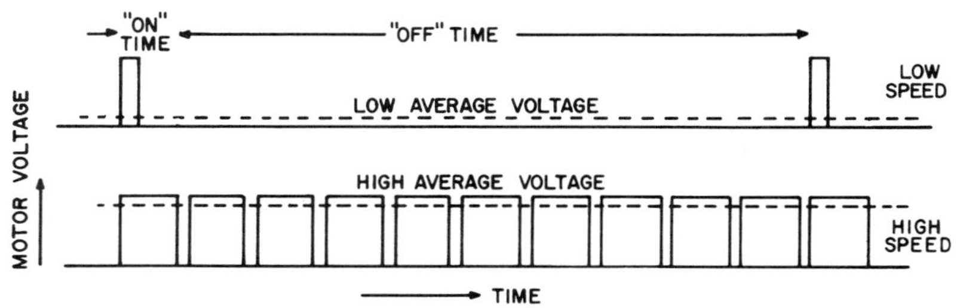

Figure 6-26 Chopper voltage for motor speed control. (Courtesy Thompson Consumer Electronics, Inc.)

to the motor is also low, and so the speed of the controlled motor is correspondingly low.

Figure 6-26B shows that the pulses are now in rapid succession. As a result, the average source voltage input to the motor is much higher, producing a high motor speed. The pulse repetition rate can be varied and in this way any motor speed, up to maximum, can be obtained.

THE CHOPPER

The circuit of the chopper (sometimes called a commutation circuit) is shown in Figure 6-27. It consists of a pair of silicon controlled rectifiers SCR1 and SCR2, controlled semiconductors which work as switches. The two switches take turns in controlling the flow of current through the motor's series field and armature windings. The current flow is unlike that of a sine wave and is in the form of unidirectional pulses.

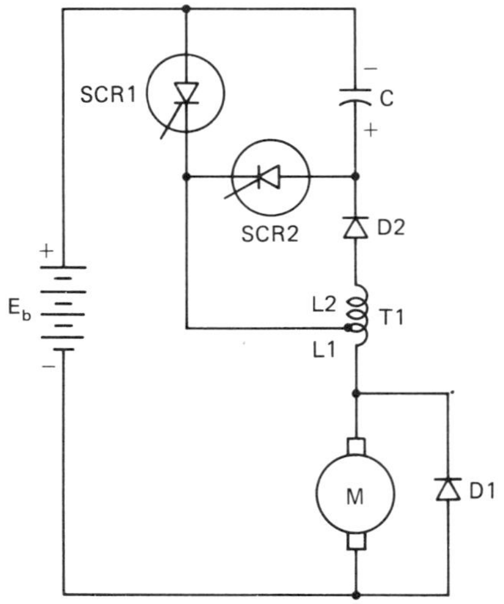

Figure 6-27 Chopper circuit. (Courtesy Thompson Consumer Electronics, Inc.)

LIGHT ACTIVATED SILICON CONTROLLED RECTIFIER

Another variation of the SCR is the light activated silicon controlled rectifier (LASCR). The LASCR is a silicon controlled rectifier housed in a container having a transparent window to permit the entry of light. The LASCR can control large amounts of current and can be used when light control of a motor is wanted. Its symbol appears in Figure 6-28 and is almost identical with that of the SCR.

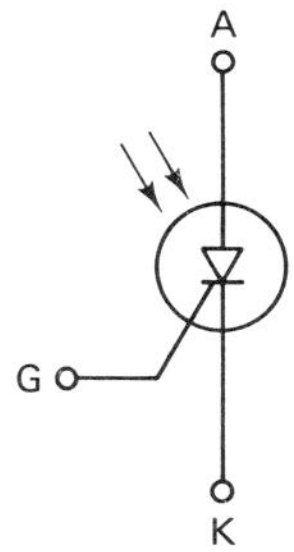

Figure 6-28 LASCR symbol. (Courtesy Thompson Consumer Electronics, Inc.)

AUTOTRANSFORMER CONTROLLER

An autotransformer (Figure 6-29A) used as a motor speed controller can be a step-down type. It reduces the source voltage to the motor and can be designed to increase current availability. These controllers have an off, run, and at least one starting point. More than one starting position may be included, thus permitting motor acceleration in steps.

The output of the autotransformer is AC and its frequency is that of the power line to which it is connected. As in the case of power transformers having a primary and secondary winding not physically connected, the autotransformer output can be fed into a rectifier, and may be followed by a filter, to supply a DC source. The amount of DC voltage output can be adjusted through selection of one or more taps on the autotransformer winding.

DC MOTOR SPEED CONTROL

DC motors lend themselves to external motor control since the DC power required can be supplied by a power supply operated from the AC power line. It is true that DC power can be supplied by a battery but this method requires a series connected rheostat. Not only is there a power loss in the rheostat but the battery itself requires attention since it must be recharged. Further, the battery has a limited working life.

The alternative electronic method, a simple approach, is shown in the combined block-circuit diagram in Figure 6-29B. The electronic circuitry may include a power transformer followed by rectifier circuitry. The rectifier changes the input AC to a unidirectional voltage and current. This can be followed by a filter to supply a smoother DC.

Speed control using this arrangement can be handled in a number of different ways. The primary of the power transformer can have a tapped primary connected to the AC input. Selection of any particular tap means the DC output voltage can be changed. Instead of taps, the transformer can be a type that has a variable primary. Depending on the setting of the transformer, the DC

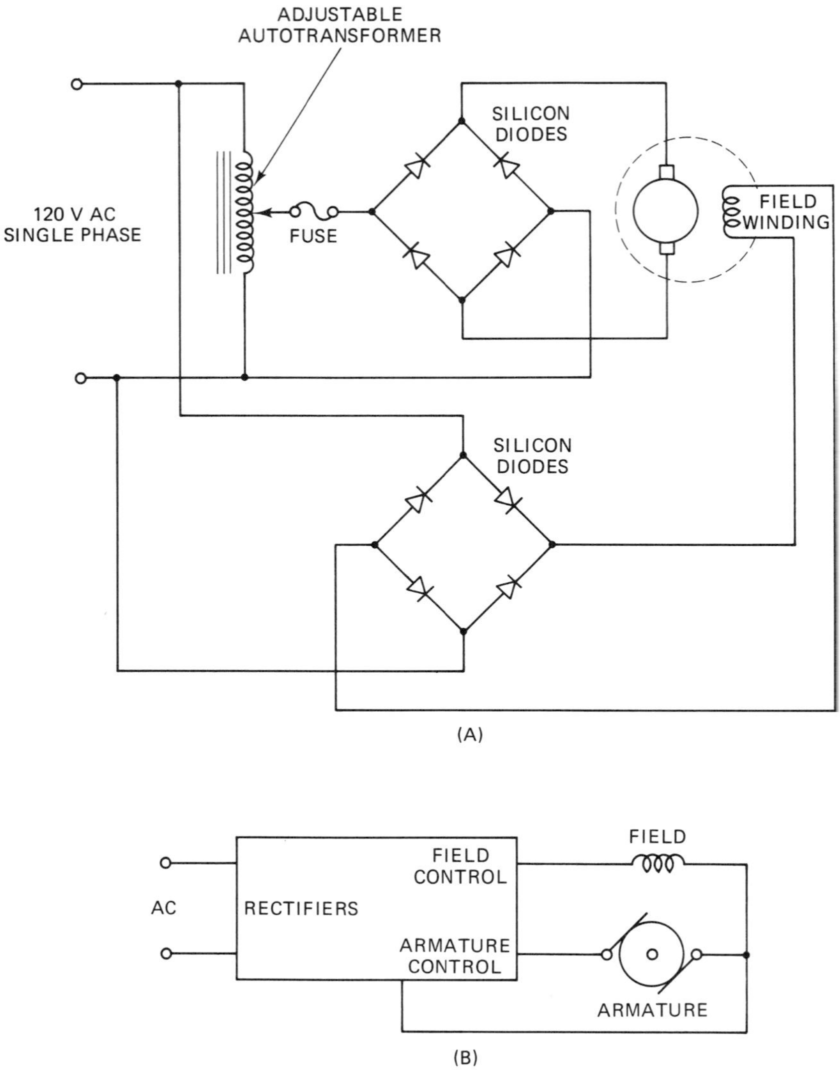

Figure 6-29 Autotransformer used as motor speed controller. Control is obtained by connecting motor to different taps on the transformer (A); overview of electronic circuitry for speed control of DC motors (B).

output of the power supply can be made to vary smoothly and widely over a wide range. Changing the voltage delivered to the armature varies its rotational speed. Thus, if the voltage is reduced by 50 percent, so is the speed, since the speed is proportional to the applied voltage.

The chief disadvantage of this speed control setup is the need for an AC power source.

ELECTRONIC POWER SUPPLIES FOR DC MOTORS

Power line DC voltages are not generally available for DC motors. For such motors the DC supply can be furnished by wet-cell type batteries or DC generators. A convenient method is to use AC power, stepping the line voltage down to a suitable amount through the use of a power transformer followed by a rectifier system and a filter. The voltage output of the power supply can then be used to drive a DC motor.

Various types of power supplies are used. One of these is the full-wave center-tap type shown in Figure 6-30A. The primary of the power transformer, T, is connected to the AC line via a single-pole single-throw switch, S. In some supplies the primary winding carries a fuse. The transformer is a step-down type and is connected to a pair of semiconductor rectifiers followed by a filter consisting of capacitors C1 and C2, plus an iron core choke. The capacitors are electrolytic units having a high amount of capacitance, often several thousand microfarads. The filter choke, L, is in series with the line connected to the motor and so must be able to carry the entire amount of motor current.

Another type of motor power supply is illustrated in Figure 6-30B. This is a full-wave bridge rectifier and has the advantage of not requiring a transformer with a center tapped winding. Unlike the full-wave center tap type, the bridge rectifier takes advantage of the full voltage of the secondary winding. The filter section is the same as that used by the power supply in Figure 6-30A. Both circuits in A and B carry fuses in series with the plus DC line connected to the motor.

Both circuits, A and B, use single phase AC obtained from the AC power line. The power supply in C, however, has its input connected to three-phase AC. A three-phase variable autotransformer is used to control the amount of AC input to transformer T2. This power transformer has a delta wound primary and a wye wound secondary. The variable autotransformer determines the amount of AC voltage delivered to the primary of the power transformer and in this way controls the DC output of the power supply.

In all three circuits the secondary winding of the power transformer must be able to carry the starting current of the motor. This is also true of the filter choke, L.

Electronic control systems can be inserted in series between the power source and the motor windings. For DC motors the electronic control may also include a rectifier system for changing the line voltage from AC to DC. The rectifier system can be looked on as a partial power supply only since it does not include a filter system.

One such arrangement is shown in Figure 6-31. The AC line voltage is input to a full-wave diode bridge consisting of four diode rectifiers D1, D2, D3, and D4 with the DC output shunted by series resistors R1 and R2. Resis-

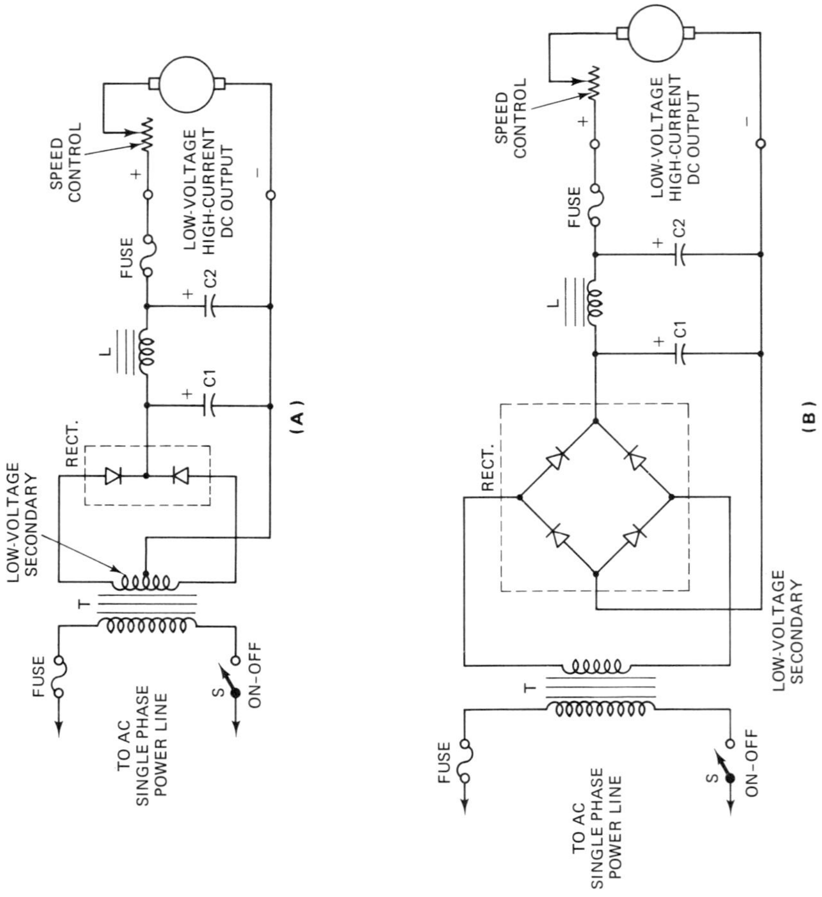
SPEED CONTROL
LOW-VOLTAGE SECONDARY
RECT.
L
FUSE
C1
C2
+
–
LOW-VOLTAGE HIGH-CURRENT DC OUTPUT
T
FUSE
S
ON–OFF
TO AC SINGLE PHASE POWER LINE
(A)
(B)

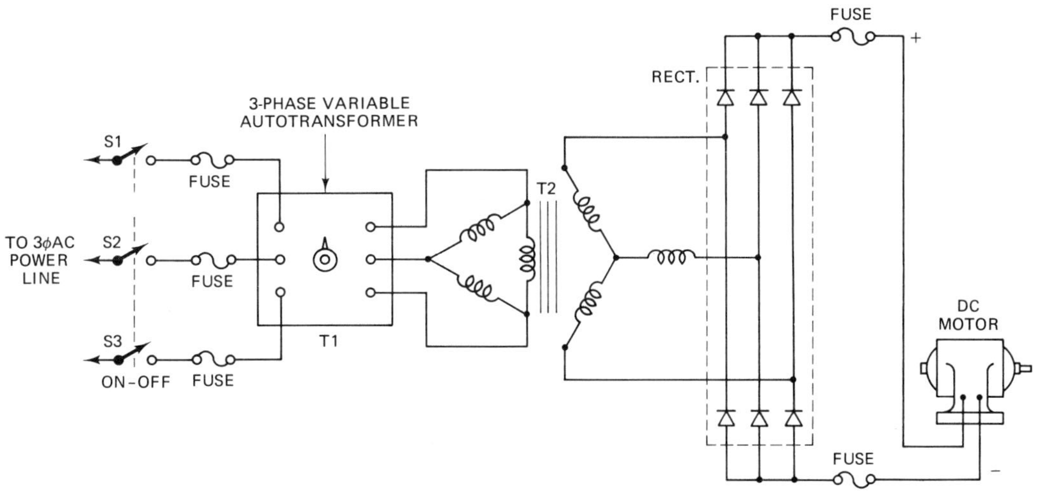

(C)

Figure 6-30 Full-wave power supply (A); bridge rectifier (B); three-phase supply (C).

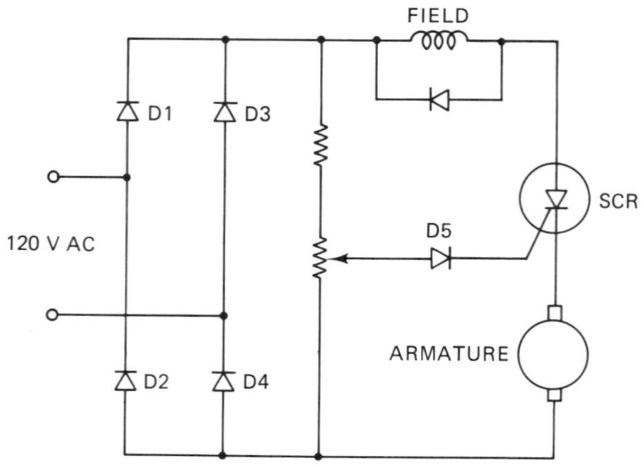

Figure 6-31 Electronic speed control of a series-wound DC motor. (Courtesy Thompson Consumer Electronics, Inc.)

tor R2 is a two-watt unit and controls the DC input, via diode D5, to the control element of the SCR. The output of the bridge circuit also supplies the necessary DC source voltage for the field and armature windings of the motor.

This circuit is somewhat unusual in that the current controlling SCR is inserted in series between the field winding and the armature. Since this is a series network, the same amount of current will flow through all three. Speed control is obtained by the amount of DC voltage put on the gate electrode of the SCR. R2 is a low wattage resistor since the field and armature current does not flow through it. The SCR, in this application, can be looked on as a variable resistor but unlike a variable resistor does not suffer a substantial I^2R loss.

With this arrangement, the speed of the series wound DC motor is controlled by the setting of R2. The entire control unit can be mounted on its own chassis.

The advantage of this circuit is its low cost and simplicity, but it does have disadvantages as well. It is possible for the motor to hunt. Another disadvantage is that the connection from the field winding to the armature must be broken open to allow for the insertion of the connections to the SCR.

SILICON UNILATERAL SWITCH

The silicon unilateral switch (SUS) whose symbol is shown in Figure 6-32A, behaves like an avalanche diode shunted by a typical semiconductor diode as in Figure 6-32B. Basically it is a miniature SCR, but it is equipped with an anode gate, instead of the usual cathode gate. It finds application in the speed control circuit for 120-volt operated shunt-wound motors.

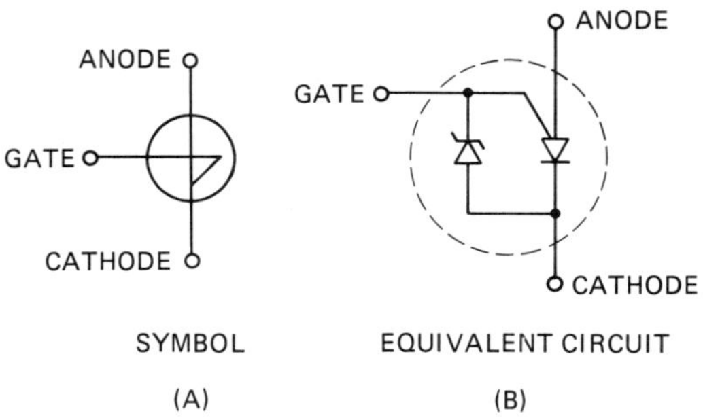

Figure 6-32 Symbol for silicon unilateral switch (A); equivalent circuit (B). (Courtesy Thompson Consumer Electronics, Inc.)

ELECTRONICALLY SPEED CONTROLLED SHUNT-WOUND DC MOTOR

This motor and its electronic control using a silicon unilateral switch are shown in Figure 6-33. The control system is supplied with its own full-wave bridge rectifier system whose input is connected to the 120-volt AC power line. The DC output of the rectifier is applied directly to the shunt field winding of the motor which is also in parallel with series resistors R1 and R2. The silicon

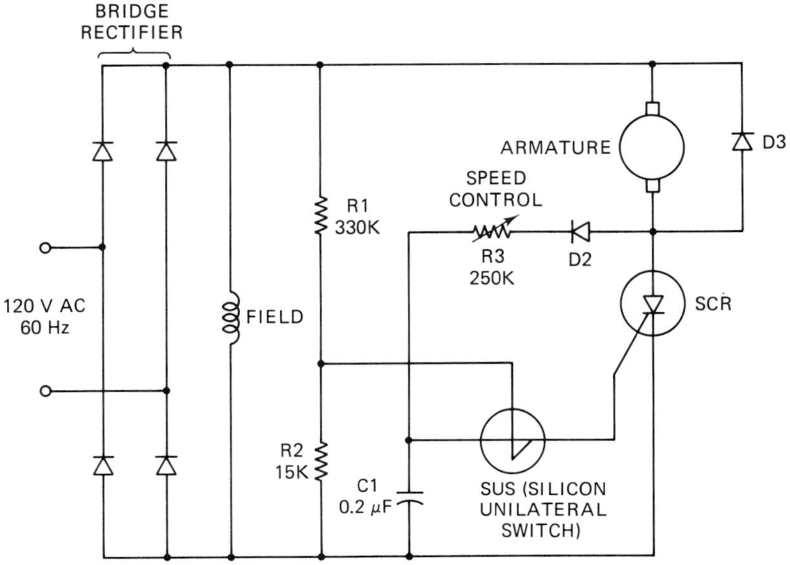

Figure 6-33 Speed controlled shunt-wound DC motor. (Courtesy Thompson Consumer Electronics, Inc.)

controlled rectifier is in series with the armature winding and controls the amount of current passing through it, thereby governing its speed.

A semiconductor rectifier, D3, is wired in shunt with the armature coil. When the SCR turns off, the magnetic field around the armature collapses which induces a voltage across that coil. The diode provides a current path for the resulting current that is produced. Without this diode the current would pass through the SCR, keeping it from turning off.

When the SCR is off, capacitor C1 starts to charge by current flow through the armature, rectifier diode D2, and speed control R3. When the voltage across C1 becomes large enough it will allow a voltage pulse to be applied to the SCR gate. This permits power to be applied to the armature's winding, producing rotation. After C1 discharges, there is no further trigger voltage for the silicon unilateral switch and it opens. There is also no further trigger pulse on the SCR and so current flow through the armature ceases. The speed control resistor, R3, governs the rate of charge of capacitor C1, governing the firing time of the SCR.

This circuit has the advantage of supplying a wide range of speed control.

MOTOR BLOWER CONTROL

A circuit arrangement using a pair of diacs and a single triac appears in Figure 6-34. It is intended for controlling a blower motor used for a furnace. The AC input to this circuit is 120 volts AC, 60 Hz, single-phase. The motor that is used is an induction type.

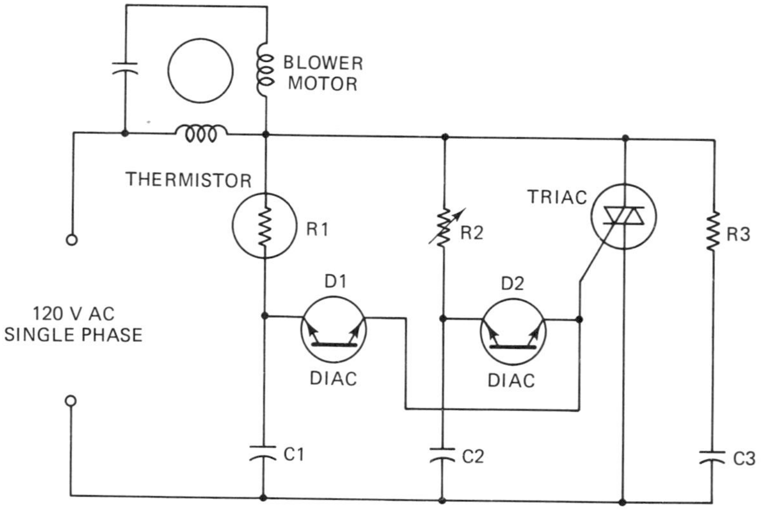

Figure 6-34 Motor blower control.

Because of its application, that of a blower used in conjunction with a furnace, there is no requirement that the speed of the motor be precise. It is the temperature of the air that is of prime importance.

One of the key elements in this diagram is the thermistor R1. Resistors are generally temperature sensitive, but the thermistor is much more so. A thermistor is a semiconductor and can have a negative temperature coefficient with its resistance decreasing with an increase in temperature, and with this resistance varying substantially. The resistance may decrease by as much as 5 percent for each degree Centigrade rise in temperature.

The initial control setting is handled by variable resistor R2 in series with C2, supplying a phase shifted voltage to operate diac D2. This voltage works as a trigger for the triac with R2 adjusted so that the blower motor is off for a specific ambient temperature. If the temperature should rise above a certain amount, there will be a current increase through the thermistor. This will develop a trigger voltage that will be applied to the input of the triac. This will overcome the existing trigger maintained by diac D2. The triac, though, is in series with the motor coil, and the effect will be to increase the current flowing through it. This will cause the motor to start and to reach running speed, thus operating the blower fan.

R2 can be set so that the blower fan is always operative but at minimum speed. The action of the thermistor with a rise in temperature will then increase current flow through the motor winding, increasing the speed of the blower motor.

ELECTRONIC CONTROL FOR THE UNIVERSAL MOTOR

As explained previously in the chapter on AC motors, the universal motor can operate using either AC or DC as a power source. This motor belongs to the series-winding motor family and as a fractional horsepower motor is widely used in small appliances such as vacuum cleaners, fans, driers, and electric drills. With this motor, speed increases as the load decreases.

The simplest speed control consists of a variable resistor in series with the power line connected to the motor. The disadvantage, of course, is the considerable power loss in the resistor since the full motor current flows through it. The resistor can also get quite hot.

There are two other possible speed control methods. One is to change the frequency of the AC power supply, but in household use this is impractical since the power line frequency is remarkably constant. The other is to change the amount of applied voltage.

One method of electronic speed control for a universal motor is shown in Figure 6-35. The line input is 120 volts AC supplied by a power outlet. This voltage is rectified by a half-wave power supply; so the power output is pulsating DC. The amplitude of this half-wave voltage is represented by the vertical arrow identified as V1. V1 is used as a reference voltage.

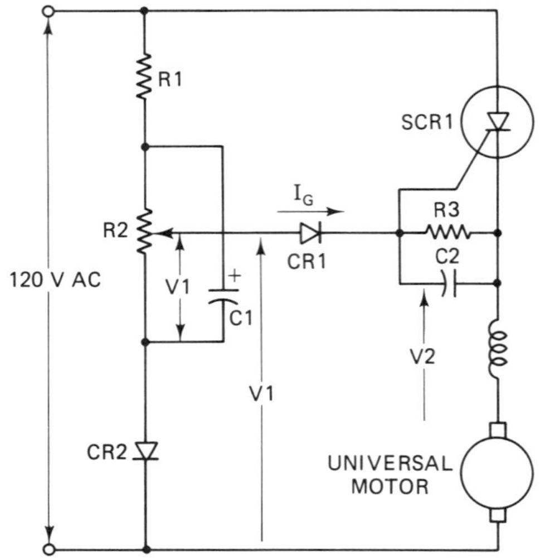

	LOW UP TO 1 AMP NAMEPLATE	MEDIUM UP TO 3 AMP NAMEPLATE	HIGH UP TO 15 AMP NAMEPLATE
R2	10K 1 W	1K 2 W	1K 2 W
R1	47K 1/2 W	3.3K 2 W	3.3K 2 W
R3	1K 1/2 W	150K 1/2 W OPTIONAL	150K 1/2 W OPTIONAL
C1	0.5 μF 50 V	10 μF 50 V	10 μF 50 V
C2	0.1 μF 10 V	0.1 μF 10 V OPTIONAL	0.1 μF 10 V OPTIONAL
SCR1	GE C106B	GE C22BX70	GE C33B

Note for 220 V, 50/60 Hz Operation: Double value of R1 and use at least 400 V semiconductors (SCR and diodes)

Figure 6-35 Electronic speed control for a universal motor. (Courtesy Thompson Consumer Electronics, Inc.)

When running, the motor develops a counter EMF, and in the diagram this is identified as V2. The greater the speed of rotation, the higher the amount of this voltage.

Since diode D1 is a half-wave rectifier, the current flowing through R1 and R2 will consist of half-wave sine pulses. Current will flow only during the positive half of the AC input. If the back EMF of the motor is greater than voltage V1, diode D2 will be reverse-biased, and as a result, the SCR will not be triggered with current flow through it reduced. This will cause the armature to rotate more slowly. But as it does so, its back EMF will decrease, and so voltage V2 will become a little less than voltage V1. The effect of this will put diode D2 into conduction and the SCR will be triggered, permitting more current to flow to the motor whose rotation will then increase.

This comparison of voltages V2 and V1 will result in pulses of current passing to the motor tending to keep it operating at a moderately constant speed. The actual speed in rmp can be set by adjusting resistor R2.

With variable light loads the effect on the motor may be a rapid increase and decrease of motor current. This can cause a condition of hunting or cogging plus an increase in the motor's mechanical noise, a condition that is alleviated by the inclusion of capacitor C1 in the circuit.

SPEED AND REGULATION CONTROL

The regulation of a motor is a reference to its variation in speed with changes in load. A motor whose speed remains essentially constant despite increasing or decreasing amounts of load is said to have good regulation while a motor whose speed is sensitive to load changes has poor regulation. These designations, poor and good regulation, are generalized descriptions, and there is no formula for the more precise determination of regulation.

Generally, the regulation of a motor is determined by its basic design and the load it is required to handle. If, for example, a load is substantially lighter than the torque output of a motor and if that light load remains fairly constant, the operating speed of the motor could remain equally constant, depending on the original design of the motor.

An electronic speed control, such as the one illustrated in Figure 6-36 has a double function. Primarily, it is intended for governing speed, but at the same time it works to maintain a fairly constant speed level.

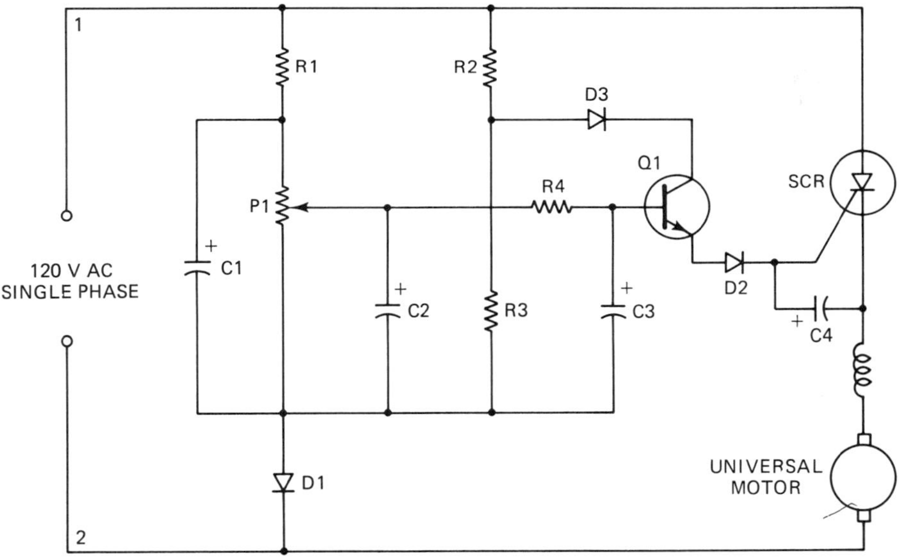

Figure 6-36 Speed and regulation control for a universal motor.

The circuit resembles that of Figure 6-35, but with modifications. Start at point 1 and trace the series path through the SCR and the motor winding back to point 2 to see the resemblance.

Shunted across the AC line is a series network consisting of R1, potentiometer P1, and a diode rectifier, D1. Because of the presence of the diode, the current flowing through it is pulsating DC, and so the bias voltage applied to the base of transistor Q1 is also DC and determines the operating point of this transistor. Q1 works as an emitter follower and supplies the gating voltage for the SCR. The setting of potentiometer P1 determines the operating speed of the motor, but regardless of the load, the speed will remain fairly constant.

The SCR will trigger under the following conditions: its gate drive input must be equal to the CEMF (counter EMF) of the motor plus the triggering voltage it receives from P1 via Q1. Of these two voltage inputs, that supplied by Q1 to the gate input of the SCR can be regarded as a reference voltage.

If the motor speed should happen to decrease, so will the counter EMF. However, this action will let the SCR conduct with a lower drive voltage from N-P-N transistor Q1. Since the SCR conducts, the average voltage supplied to the motor increases, and so the motor speed also increases. If an opposite condition takes place, that is, if the motor speed should happen to increase during normal motor operation, a condition exactly opposite that just described takes place and the average voltage supplied to the motor will decrease.

In effect, then, the universal motor is forced to acquire the constant speed characteristics of the shunt-wound DC motor. The circuit in Figure 6-36 is sometimes referred to as feedback type speed control. Diode D2, connected to the emitter of Q1, is used to make sure that pulses of current from Q1 are of the correct polarity for the gate of the SCR, while the electrolytic filter capacitor, C3, works as a reservoir to help maintain a fairly constant voltage level supplied by Q1 on the gate of the SCR.

SHUNT MOTOR SPEED CONTROL

The 120-volt supply in Figure 6-37 will supply a sinusoidal voltage which may be accompanied by voltage transients. These transients can have two unwanted effects. Their amplitude may be such that they can trigger the diac and SCR into conduction at unwanted times, resulting in unwanted motor speed. Or, the transient voltage may be so high that damage will result to various component circuits. The thyrector eliminates or minimizes line voltage transients. The single phase AC input can be supplied directly from an AC power outlet or from the secondary of a 1-to-1 turns ratio of a power transformer. The advantage of

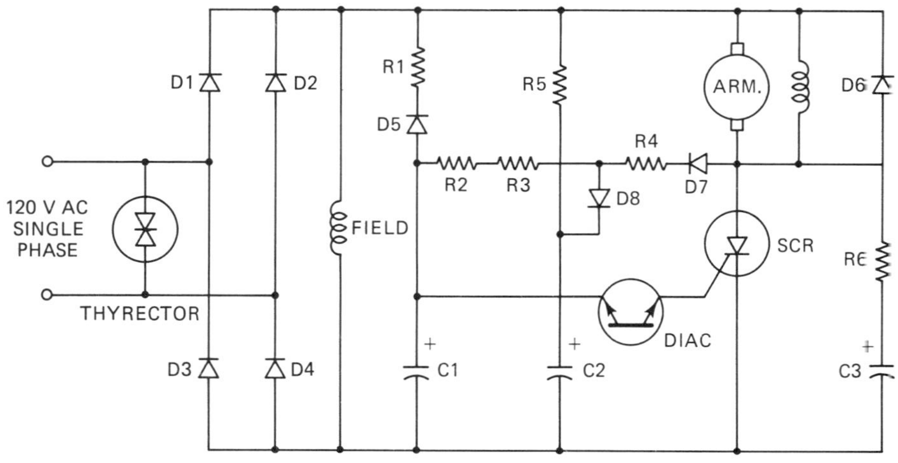

Figure 6-37 Shunt-wound DC motor speed control.

the transformer, not shown in Figure 6-37, is that it supplies isolation from any other components connected to that branch of the power line.

Following the thyrector is a full-wave rectifier system supplying DC to the field winding of the shunt-wound DC motor and to the armature connected in series with an SCR. The advantage of the full-wave bridge rectifier is that it supplies a higher voltage output than the two diode full-wave type.

The triggering time of the SCR is determined by two factors: the time it takes capacitor C1 to charge to the breakdown voltage of the diac and the breakdown level characteristic of the diac. Potentiometer R2 is the speed control. And, as in the case of other speed controls, such as the one described earlier in Figure 6-36, this system uses a feedback arrangement to improve speed regulation. If, for example, the motor should run slower, the amount of counter EMF would be reduced. As a consequence, capacitor C1 would charge more rapidly through diode D5 and through series resistors R2, R3, and R4. The effect of this would be to advance the charging time of C1. This would trigger the diac and the SCR sooner; and so current flow through the armature winding would increase. This increased current flow would counteract the decrease in counter EMF, and as a result the armature would rotate at a constant speed.

If, however, the rotation of the armature tended to increase, an opposite effect would take place. The counter EMF would increase, but this would result in a decrease in charge across C1. The firing time of the diac and the SCR would be such that the flow of current through the armature winding of the motor would decrease, lowering the rotational speed.

REGULATION AND SPEED CONTROL FOR PERMANENT MAGNET MOTORS

An improvement in motor speed regulation as well as speed control can also be applied to permanent magnet motors as well as to wound winding motors, such as the universal and shunt connected types.

The circuit diagram is illustrated in Figure 6-38. As indicated, the circuit can be considered as consisting of three sections: a Schmitt modulator, a driver and output, and current feedback sensing circuitry. The DC motor is part of the collector circuit of transistor Q4. The arrangement of the circuitry is such that the current through the armature is made to vary to keep the speed-torque constant.

As an example, assume that the motor load increases, resulting in a decrease of motor speed. In turn, this action will yield a higher base-emitter voltage in transistor Q5, a power type. This voltage change will be sensed by current feedback transistor Q6, effectively changing the voltage across electrolytic capacitor C4, connected to the collector of this transistor via a 1K ohm resistor, R17. A feedback line is connected between the junction of this resis-

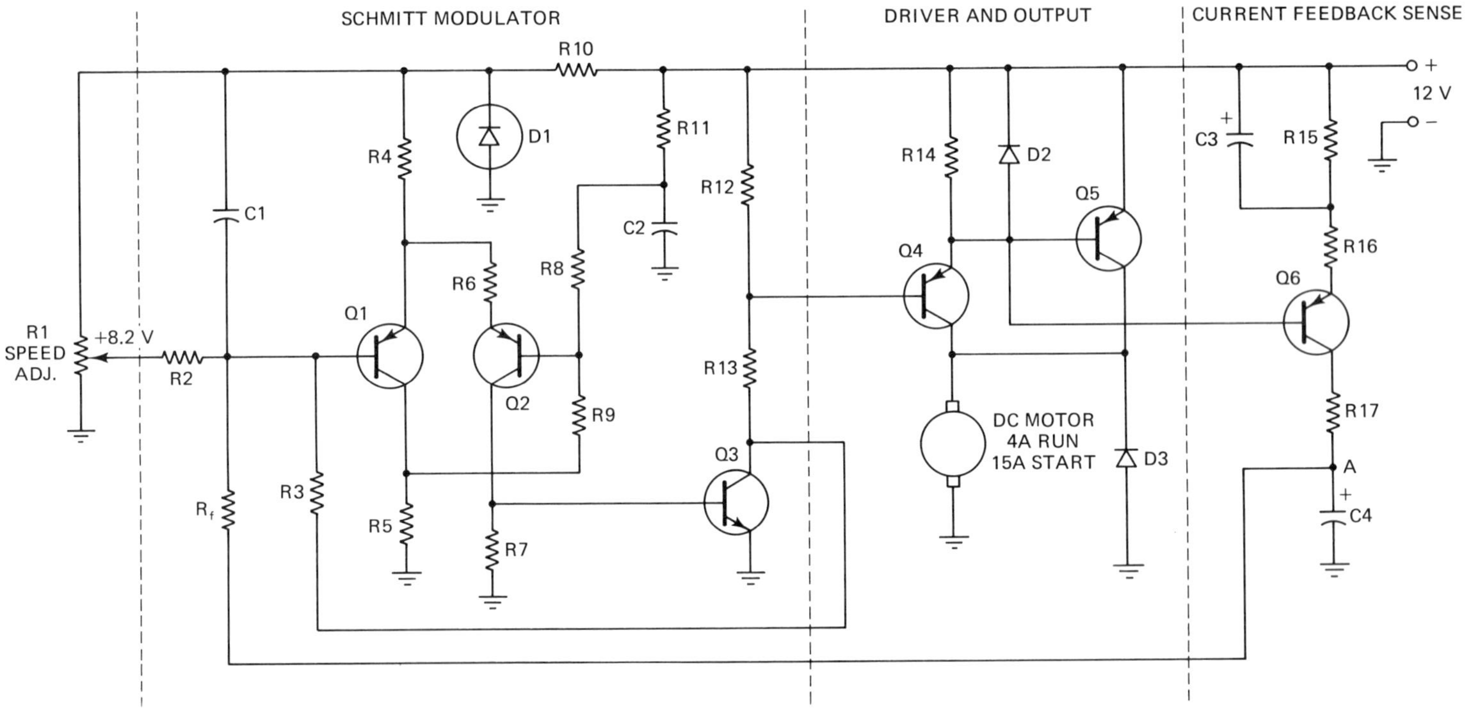

Figure 6-38 Control system for a DC motor.

tor (point A) and capacitor C4, via feedback resistor Rf, to the base input of transistor Q1. The effect of the feedback voltage is to reduce the conduction time of this transistor. The effect on the following stages will be such that current flow through the motor armature will increase and the motor speed will be revised upward. Conversely, if the motor speed increases, an opposite feedback action will take place, ultimately resulting in a decrease in armature current with a consequent lowering of motor speed.

The speed adjustment of the motor is handled by potentiometer R1. While the source voltage is 12 volts, the presence of resistor R10 drops the source voltage by about one-third, and so there is approximately 8.2 volts developed across R1. This DC voltage is applied as bias between the base and emitter of Q1 and determines its initial conduction time. Fixed bias is also supplied by Q1 by resistor R5, and so at no time does the bias on this transistor drop to zero. The maximum resistance value of R5 and R1 is the same and is 1K ohms. Resistors R11, R8, R9, and R5 are in series and are in shunt with the 12-volt DC supply. Transistor Q2 obtains its bias from this network.

D1 is a zener diode. Its purpose is that of a voltage regulator, keeping the DC input to the Schmitt modulator at a constant voltage level.

TRIAC SPEED CONTROL

The circuit in Figure 6-39 uses a diac and a triac for speed control of a universal motor. The AC drive voltage can be used for either 120 volts or 240 volts single-phase, 60 Hz input. For 120 volts, R1 is rated at 100K ohms, 1/2 watt, and for 240 volts it is 200K, 1 watt. C1 is 0.1 μF, 200 working volts DC

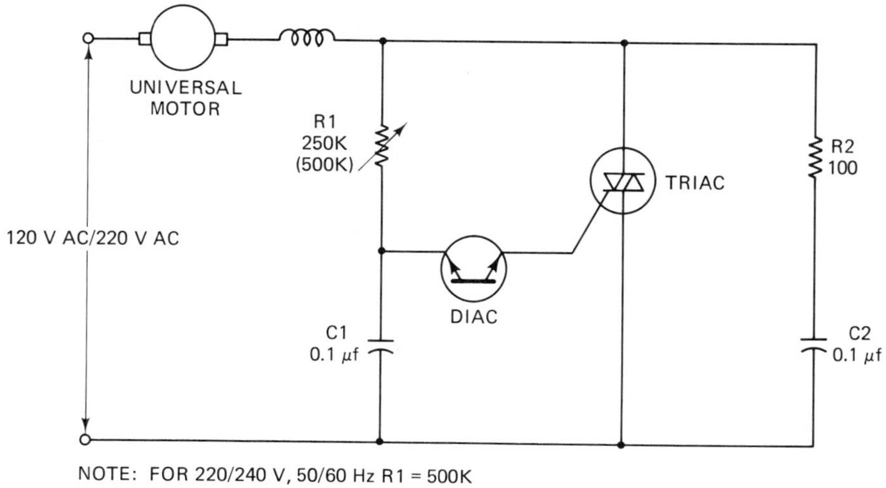

Figure 6-39 Diac-triac speed control. (Courtesy Thompson Consumer Electronics, Inc.)

(WVDC), and while its capacitance remains the same for 240-volt input, its working voltage is doubled to 400 WVDC. The same approach applies to C3. For 120 volts AC it is rated at 0.22 μF, 200 WVDC. Its capacitance remains the same for 240 volts AC input, but its WVDC is increased to 400 volts.

R1-C1, a phase shifting network, is shunted across the AC input via the universal motor windings, with R1 acting as the speed control. R2 and C2 represent another phase shifting network with these two components shunted across C1. R2 and C2 act as a voltage divider with the diac connected to their meeting point. The triac is equivalent to a pair of SCRs connected back to back. R3 and C3, wired in series and shunted across the triac, work as a "snubber." This network is actually in shunt with the motor and is used to absorb the inductive kickback pulse that can develop when the motor is turned off.

The triac, a bidirectional gated semiconductor, is in shunt with the universal motor and the AC source voltage and works as an electronic switch. The triac can be regarded as a variable resistor, controlling the amount of current supplied to the windings of the universal motor. The conductivity of the triac is governed by potentiometer R1. The triac can be controlled by either either AC or DC or by a transistor.

TORQUE CONTROL

The electronic control of a motor extends to more than just speed and regulation. Other control factors include starting and stopping the motor, reversing rotation, angular positioning of the shaft, torque and horsepower.

The drawing in Figure 6-40 shows an electronic technique for controlling the torque of a DC motor, either a shunt-wound type or one that uses perma-

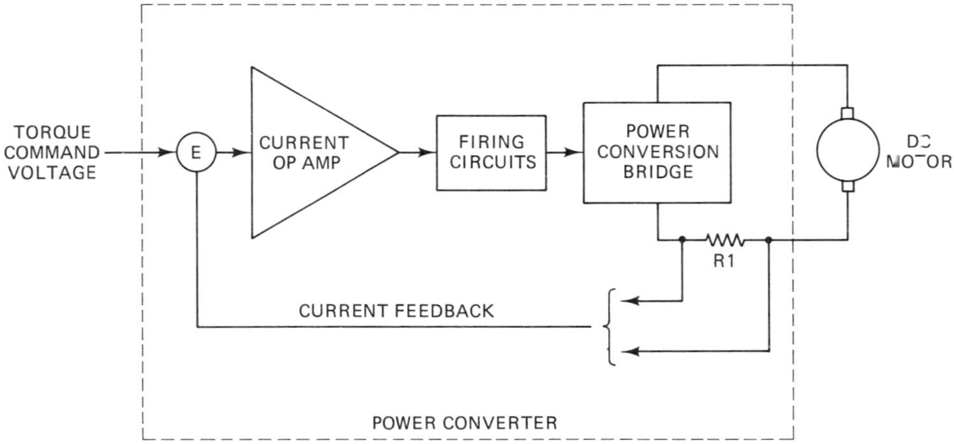

Figure 6-40 Circuit for torque control.

nent magnets. The circuitry, shown enclosed in dashed lines, is a power converter that makes use of current feedback, and is a technique that controls torque, but it does so without relying on an increase of rotational speed; that is, the average armature current remains constant.

The torque of a motor is dependent on the power developed by the rotating armature. Power is commonly considered as proportional to current, but this assumes current is the only factor involved. Power, however, is also the product of voltage and current, that is, $P = E \times I$. If more torque is required and if the current is to remain constant, the alternative is to increase the voltage supplied to the armature.

Resistor R1 in the diagram is wired in series with the armature and so the armature current flows through it. The voltage developed across R1 as a result of this current flow is fed back to an IC which triggers firing circuits to supply more voltage to the armature, when armature current decreases. The reverse takes place for an increase in armature current.

An increase in the load would normally result in a corresponding rise in armature current. However, the feedback circuit works to raise the armature voltage; consequently the voltage increases and the current quickly moves back to its original level.

HIGH-CURRENT MOTOR CONTROL

The electronic control of a motor depends on the motor, its voltage and current requirements, its speed (whether variable or fixed), and whether or not the rotation is to be reversed. As a consequence of these varying needs, the electronic control used for one motor is not necessarily suited for another. Depending on design, the electronic control may either be in series or in parallel with the motor's windings.

There are three basic types of electronic control: control of speed only, control of direction of rotation only, and control of speed and direction of rotation. These are basic listings for a subheading including low-current and high-current controls. Sometimes the only difference between the two is simply in the ratings of the components in the control circuitry.

HIGH-CURRENT MOTOR SPEED CONTROL

The combined block and circuit diagram in Figure 6-41 is intended for high-current motor speed control. The details of the rectangular block are those of a commercially available positive adjustable 5-ampere voltage regulator (Figure 6-41B). Thus, with the use of this component and the few additional parts shown in Figure 6-41A, it is possible to easily set up a high-current speed control. The block consists of an adjustable 4-terminal positive voltage regulator

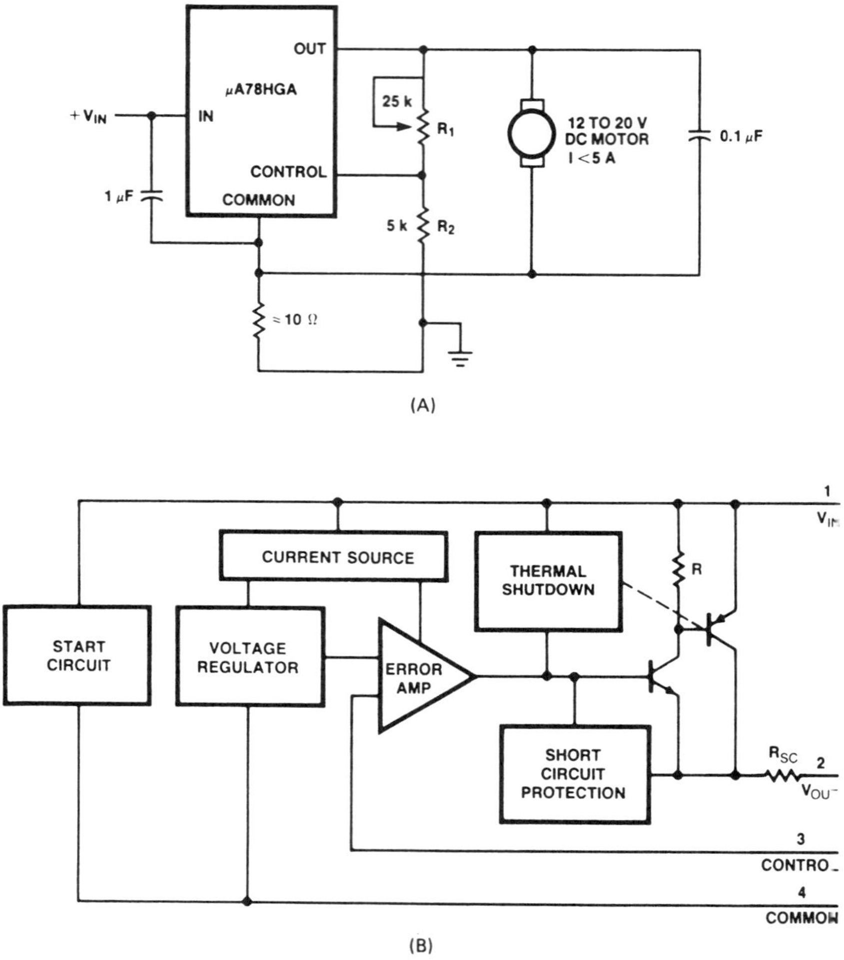

Figure 6-41 High current motor control circuit (A); details of a control block (B). (Courtesy Sanyo Fisher (USA) Corporation)

capable of supplying in excess of 5 amperes over a 5-volt to a 24-volt DC output range. Only two external resistors, R1 and R2 in Figure 6-41A are required to set the output voltage. The regulator in B is a hermetically sealed monolithic chip.

TACHOMETER SPEED CONTROL SYSTEM

A tachometer is a device for indicating the speed of a motor and this can be done by connecting the tachometer to the motor shaft. The tachometer, actually a DC voltage generator, has a DC output based directly on rotational

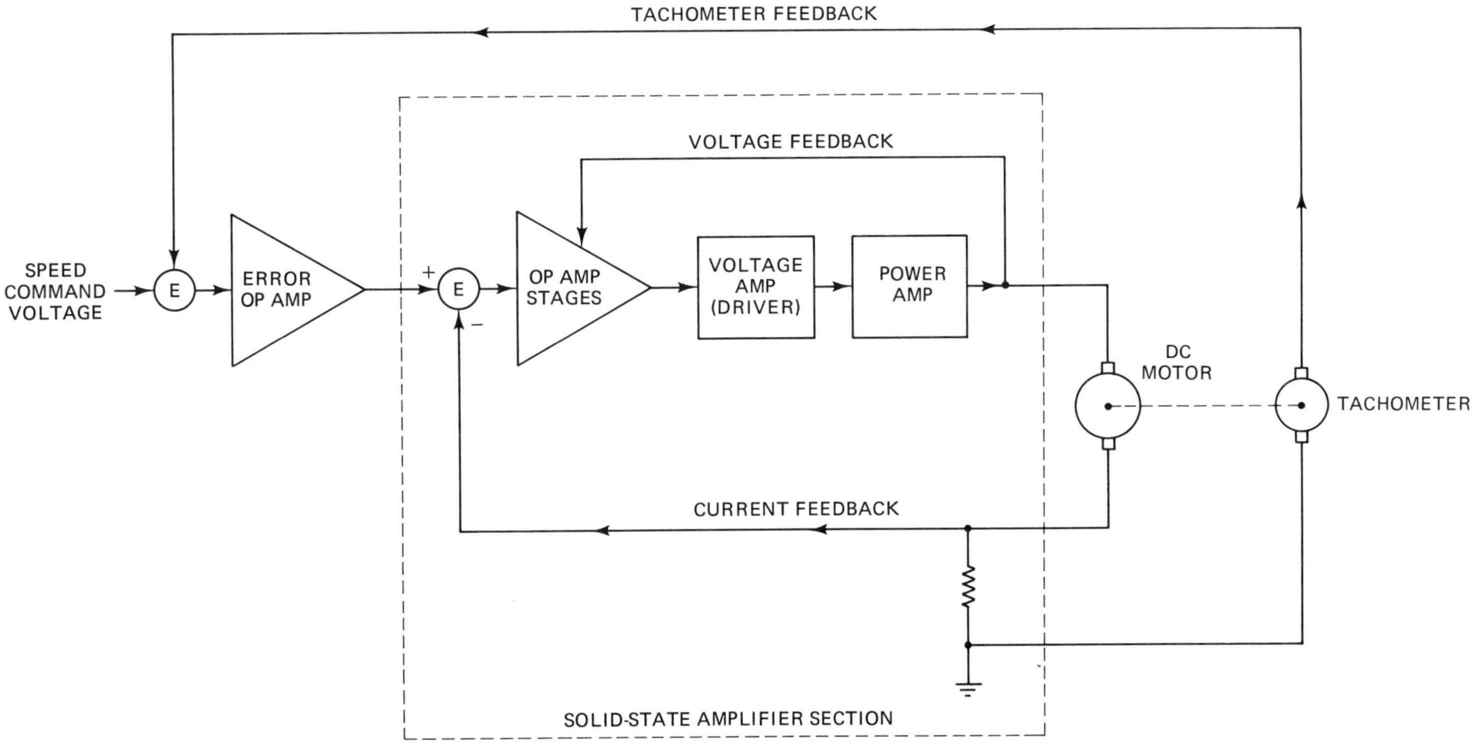

Figure 6-42 Tachometer feedback motor control system.

speed. The polarity of the tachometer's output voltage can also be used to determine the direction of rotation of the motor, since that voltage will be positive with one direction, negative with the other. As indicated in Figure 6-42 the voltage supplied by the tachometer is fed back to the input of the speed control system where it is compared with the speed command voltage. The difference between these two voltages is an error voltage. This error voltage may be regarded as a signal which is then fed into a transistor amplifier. This amplifier consists of a voltage amplifier (driver) followed by a power amplifier. The feedback arrangement counteracts any change in rotational speed, increasing the voltage across the motor when the armature speed decreases; lowering the voltage when the armature speed increases.

To maintain linearity within the transistor amplifier, two types of feedback are used: voltage feedback and current feedback. The voltage feedback line is from the output of the power stage to the input of the driver. Current feedback is from the motor to the input of the operational amplifier (op amp).

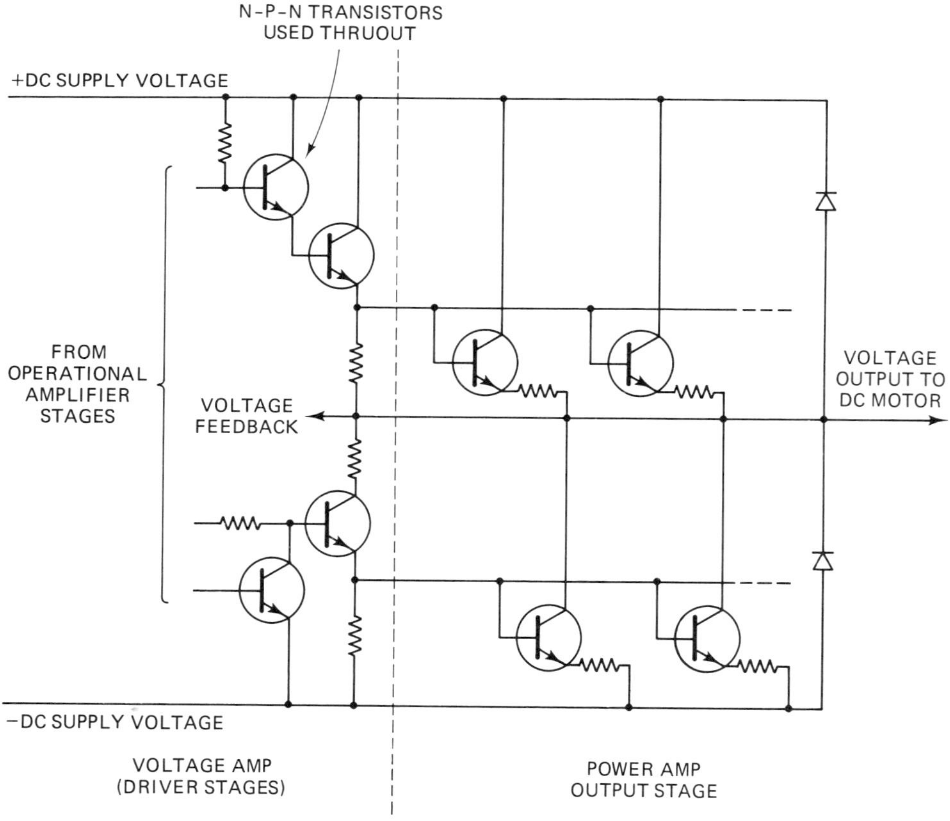

Figure 6-43 Voltage amplifier driver and power amplifier circuitry for tachometer feedback motor control system in Figure 6-42.

The drawing in Figure 6-43 shows the transistor circuitry for the amplifier. The driver and the power amplifier output stages all use N-P-N transistors. The lower half is a mirror arrangement of the upper half. Either separate power supplies are used to supply the positive and negative DC voltages required, or else a single bipolar power supply can be used. This arrangement is needed so as to accommodate motor reversal, if required.

MEASURING SHAFT ROTATIONAL SPEED

The tachometer, physically coupled to the motor shaft, is a voltage generator with the voltage developed proportional to shaft speed.

The speed can be shown on a meter that is part of the tachometer system and is indicated in rpm. For motors that are unidirectional, the meter pointer begins at zero at the extreme left. For bidirectional motors, the meter is a zero center type, with a movement of the meter pointer to the left indicating rotation in one direction, movement of the pointer to the right indicating movement in the opposite direction.

The meter display can be analog or digital. In the analog method, the meter is a d'Arsonval, similar to the type used in a volt-ohm-milliammeter (VOM). The tachometer, a permanent-magnet unit, requires no outside voltage source. The meter scale is linear and can be calibrated for rotational speeds of 0-10, 100, 250, 500, 1,000, 2,000, 3,000, 4,000, 6,000, 8,000, 10,000 and 12,000 rpm.

The scale is normally marked rpm, but some are made to read "0 to 100 Percent of Full Speed" or "0 to 100 Percent of Capacity." Dual and multirange indicators are also available.

In the digital speed indicating tachometer, the instrument has a panel display using light emitting diodes (LEDs). Binary coded decimal (BCD) output is also available for direct connection to microprocessor-based control systems or printer terminals.

While normally providing a continuous reading, the digital unit also allows any reading to be displayed indefinitely, regardless of speed changes, through logic or switch actuation of a hold signal. Additionally, the digital display is accurate, nonambiguous, and can be read at distances up to 18 feet.

Three bidirectional ranges are standard: 0-199.9 rpm, 0-1999 rpm, and 0-19,999 rpm. The digital tachometer, unlike the analog type, can read accurately to one decimal place. The analog unit reads whole numbers only, and numbers between scale markings must be estimated. The analog speed indicator does not require operating power input; the digital speed indicator does need operating power for its electronics, and this can be either AC supplied by a power outlet, or by a DC voltage source.

WHEATSTONE BRIDGE CONTROL

The diagram in Figure 6-44A shows four identical resistors connected in a series-parallel arrangement known as a Wheatstone bridge. Resistors R1 and R2 are in series, as are R3 and R4. If identical values of current flow through

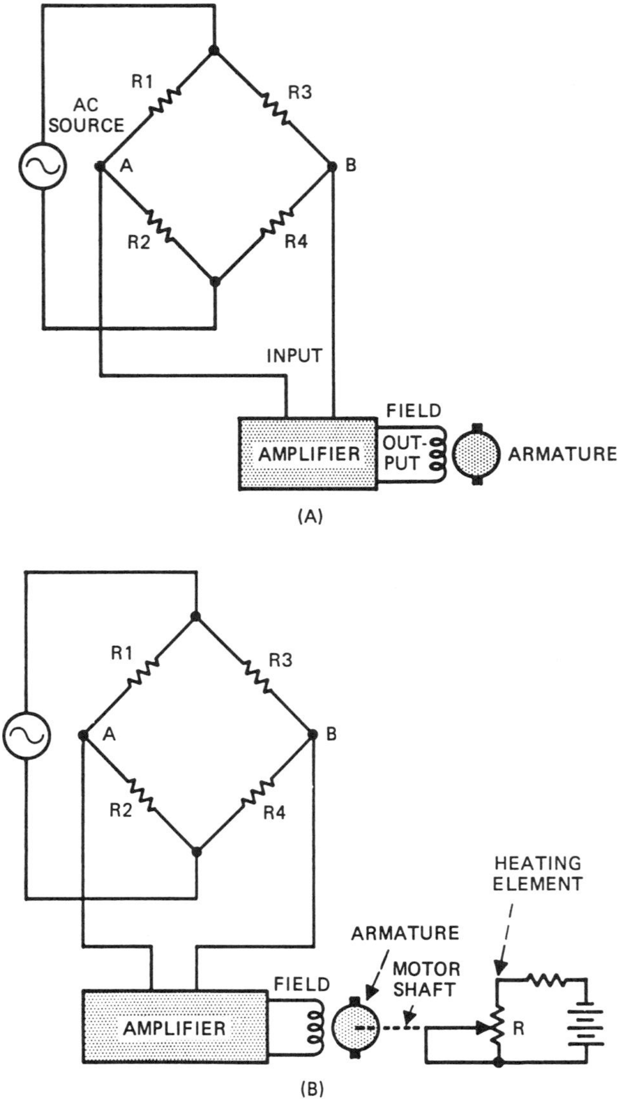

Figure 6-44 Wheatstone bridge motor control (A); motor operated temperature control (B).

these resistors they will have equal voltage drops. When this happens a measurement across points A and B will show no voltage difference.

Connected across A and B is an amplifier whose output supplies the field current for a shunt-wound motor. In the absence of a voltage input to the amplifier, the field receives no current and so the motor is inoperative.

Figure 6-44B illustrates a motor control circuit using a Wheatstone bridge. The shaft of the motor is connected to the wiper arm of a wirewound variable resistor. This resistor is designed to give off heat with the amount of heat depending on the amount of current flowing through it.

Assume that resistor R4 is a temperature-sensitive element and that it is in or near a liquid or gas kept at a constant temperature by the heating element. If the temperature should change, R4 will change its resistance. This produces a voltage across points A and B. This voltage is amplified and results in a current through the field winding, causing the motor's armature to turn. But as it turns it moves the wiper arm of the heating element R, either increasing or decreasing the heat of the element as needed. Although shown in the diagram as a separate heating element it is actually R4.

When the temperature of the liquid or gas is correct once again, R4 (the heating element) will have the same resistance as all the other resistors of the Wheatstone bridge, but at this moment, the voltage across points A and B will be zero. When this happens the amplifier has no input, but with no input there is no longer current available for the field winding and so the motor stops.

Figure 6-44B is a specific example of how to control temperature, but this is just one of many possible industrial system operating factors. It might be necessary for a liquid in a tank to always be at a certain level or for an object always to have a predetermined weight.

The general method of control for these systems is shown in Figure 6-45. The error measuring device could be a Wheatstone bridge or some other electronic device. Whatever it may be, a change produces an "error voltage" which is amplified. The amplifier feeds current into the field winding and the motor starts to turn. The motor operates a correction device. Information in the form of a voltage or current is sent back to the error-measuring device to bring it back to a state of balance or "no error."

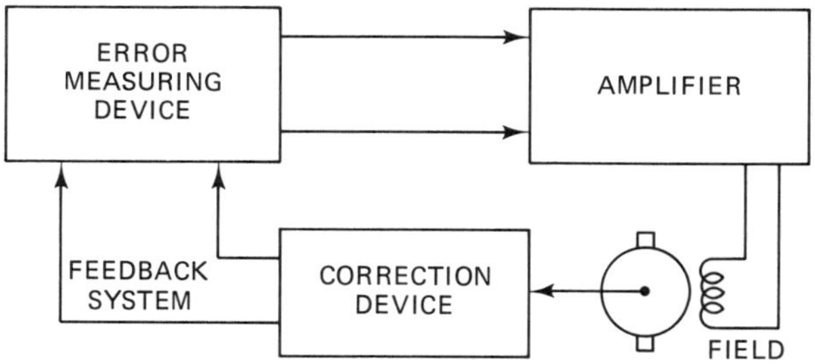

Figure 6-45 Feedback error correction system.

MICROCOMPUTER MOTOR CONTROL

In a cassette tape deck, the motor for operating the tape reels can be controlled by a microcomputer, an integrated circuit (IC) shown as IC 500 in Figure 6-46. The IC is responsible for controlling the mechanism of the tape player and for outputting the drive signals for the reel motor. These are input to the reel motor drive circuitry which, in turn, causes the reel motor to rotate in the necessary direction, either forward or reverse.

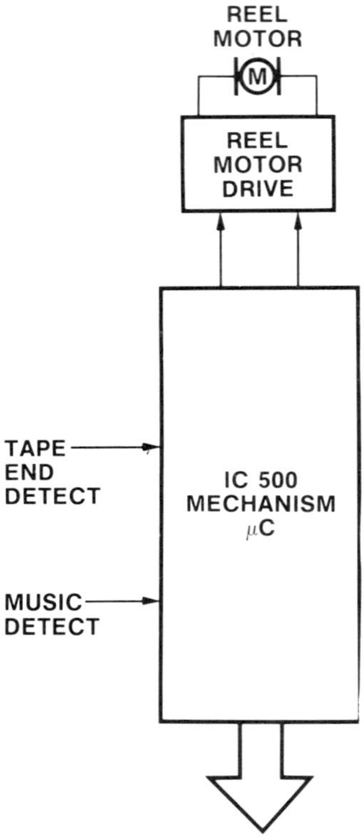

Figure 6-46 Motor control by a microcomputer. (Courtesy Thompson Consumer Electronics, Inc.)

The "tape end detect" signal is input directly to the IC and allows the microcomputer to detect the end of the tape and automatically reverse direction. The automatic reversal may be instituted in either the record or playback mode of operation.

Figure 6-47 is a view of the entire tape transport mechanism. This drawing shows that two motors are used, a capstan motor and a reel motor. The reel motor provides drive to the takeup and supply reels via a gear assembly. The tape is moved by a pinch roller which holds the tape against a capstan.

Figure 6-48 is a rear view of the tape transport mechanism. This shows that the capstan flywheels are belt driven via the capstan motor. Because of the

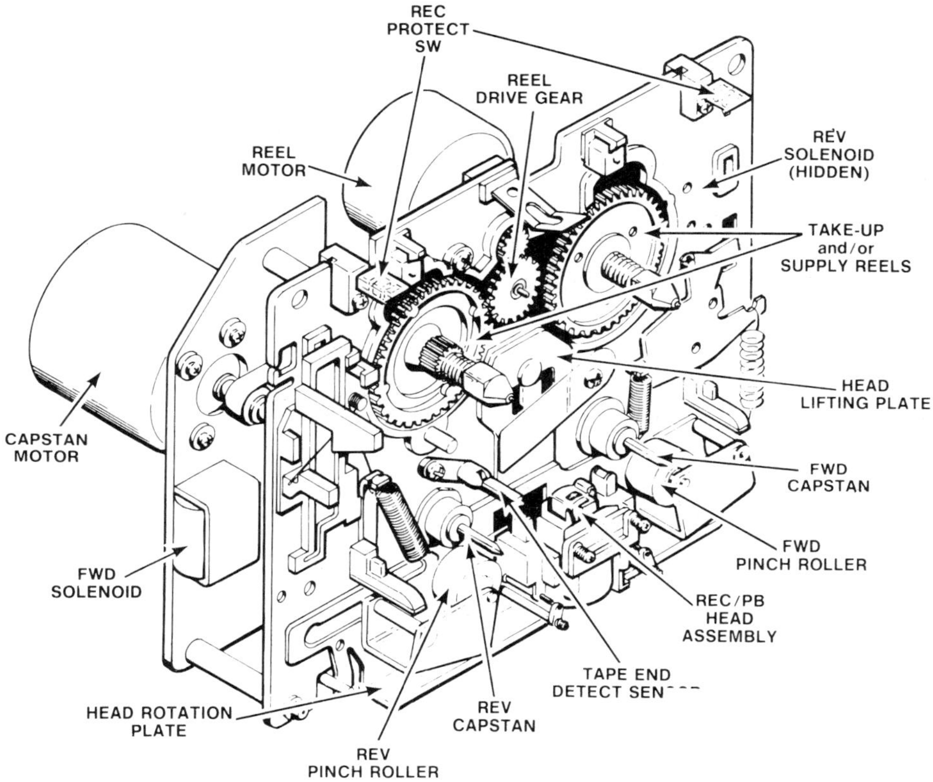

Figure 6-47 Tape transport mechanism. This unit uses a capstan motor and a reel motor. (Courtesy Thompson Consumer Electronics, Inc.)

belt arrangement, the capstan motor is required to turn in only one direction but has the ability to drive the forward and reverse capstans in opposite directions.

MOTOR CONTROL IN A VCR

The movement of tape in a video cassette recorder (VCR) involves two actions: the tape is pulled along by a motor-operated tape transport system and at the same time, each head, in turn, wipes against the tape.

Two separate motors are required in the VCR for this system, one for each of the tape reels (the supply reel and the takeup reel), both of which are housed inside the video cassette. When the video cassette is inserted into the VCR it becomes a part of the transport system.

Other motors are required in the VCR but they are not part of the transport. The video heads are mounted on a drum which is made to revolve by a

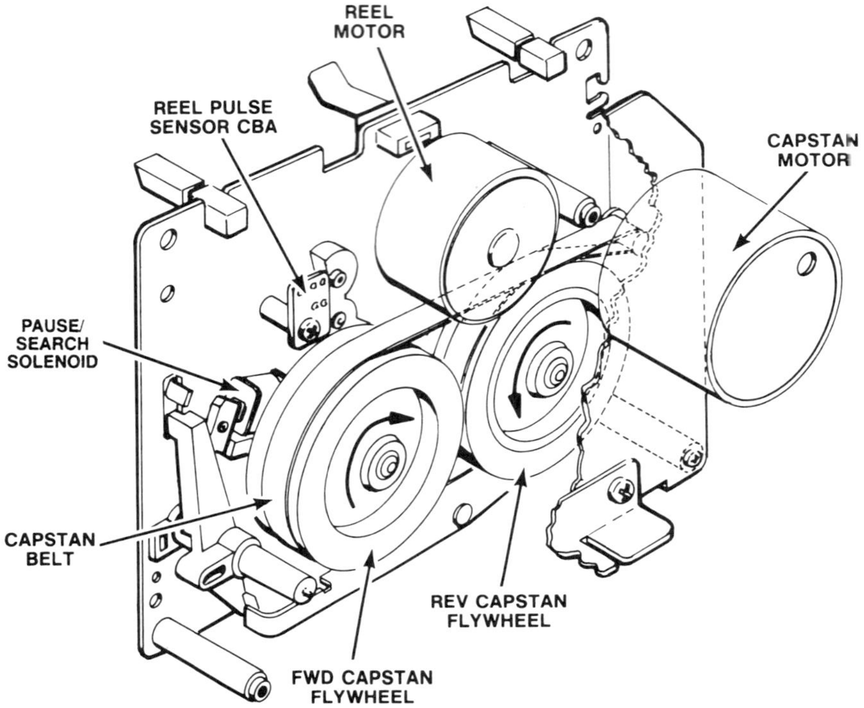

Figure 6-48 Rear view of tape transport mechanism. (Courtesy Thompson Consumer Electronics, Inc.)

brushless DC motor. The motor action is such that the two tape reels, the supply and takeup reel, are made to rotate clockwise, with the head drum turning counterclockwise.

Servo Control

It is essential for the head motor to rotate at as constant a speed as possible; and so the speed of this motor is controlled by a servo system. Any increase or decrease in motor speed is sensed by the servo, with any speed change interpreted in the form of a signal voltage. This signal is sent to a motor drive amplifier connected to the motor input. As a result the motor is kept on its assigned speed.

Number of Motors

There are more motors in a VCR than in an audio cassette recorder-player. The usual VCR will have a minimum of three, but may have as many as seven. One motor is needed to pull the tape out of the video cassette and to wrap the tape around the head assembly, an action that takes place as soon as the start button is depressed. A motor is needed to rotate the capstan, and still another to rotate the drum that contains the head assembly. The tape is positioned between the motor-operated capstan and an associated pressure roller. It is the rotation of the capstan that moves the tape.

There could also be another motor to keep the tape at optimum tension. And a pair of motors are needed for the tape transport.

Digital Auto Tracking

VCR's are not only used to supply pictures in regular motion, but in slow motion as well as in still frame. The problem with these two latter operating conditions is that they are usually accompanied by electrical noise. A technique for overcoming this is to use digital auto tracking involving the capstan motor; this is illustrated in Figure 6-49.

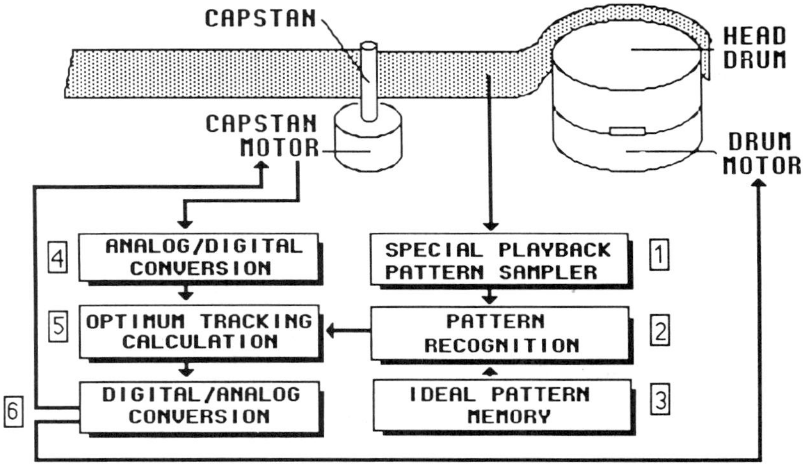

Figure 6-49 Method for removing electrical noise from slow motion and still frame pictures supplied by a VCR. (Courtesy Mitsubishi Electric Sales America, Inc.)

Digital auto tracking uses digital signal processing for the automatic adjustment of slow tracking. The circuit that is used is in the form of a chip.

During slow motion or still frame operation of the VCR, the playback signal is monitored by (1) the special playback pattern sampler. This circuit puts the signal into a form that can be recognized by (2) the pattern recognition circuit. This signal is then compared to an ideal pattern memory in (3). This memory is programmed at the factory of the VCR manufacturer with the patterns generated by a perfect still or slow-motion image. The difference is then sent to the optimum tracking calculation computer (5).

Meanwhile, the signal from the capstan motor has been converted from its analog form to digital and is fed into the optimum tracking calculation circuit (5).

The computer analyzes the two inputs from the recognition pattern (2) and the analog-to-digital conversion circuitry (4) and sends the appropriate control signals to adjust the noise out of the frame to the digital-to-analog conversion circuitry (6). Once it is converted, the output signal is supplied to the capstan and head drum motors which respond by changing the position of the tape and the head relative to each other. The result is noise-free slow-motion and still-frame pictures.

MOTORS FOR TURNTABLES

Typically, two basic types of motors are used for turntables, the synchronous and the induction motor. Both types can give good results. The number of poles contained in these motors is significant; less expensive units use only a two-pole motor. Such a motor has poor speed regulation and is a direct cause of wow and flutter. A four-pole motor is a definite improvement.

A hysteresis synchronous motor usually has a large number of poles and its speed consistency is dependent on the frequency of the power source. This is accurately controlled by the electric power company and results in good speed consistency, but since hysteresis synchronous motors are high speed units, such motors tend to increase rumble because of higher rotation and increased vibration.

The induction motor rotates at lower speeds, effectively reducing rumble. An induction motor, though, is subject to voltage variations, and therefore the drive system must include some flywheel effect to smooth rotation, thereby controlling wow and flutter.

One of the problems in improving direct drive turntable performance is reducing wow and flutter due to the cogging action of the motor. With its limited number of poles and its relatively slow operating speed, most conventional motors used in direct drive systems also have an inherent problem of low starting torque.

A method for overcoming these problems is shown in Figure 6-50. In effect, the platter becomes the turntable's motor. This motor, with its 120-pole design, practically eliminates cogging action and lowers wow and flutter to a totally inaudible 0.03 percent. The linear motor direct drive system further reduces turntable rumble to an extremely low −70 dB, far below hearing level.

How It Works

A continuous band of ferrite material, containing 120 magnetic poles, is attached to the inside bottom rim of the platter. To start platter rotation, each pole opposes one of the electromagnetic drivers on the base top. Each of the 120 poles is attracted or repelled as it passes the driving electromagnets. An electronic sensing device monitors the platter's speed and acts as a servo-feed-

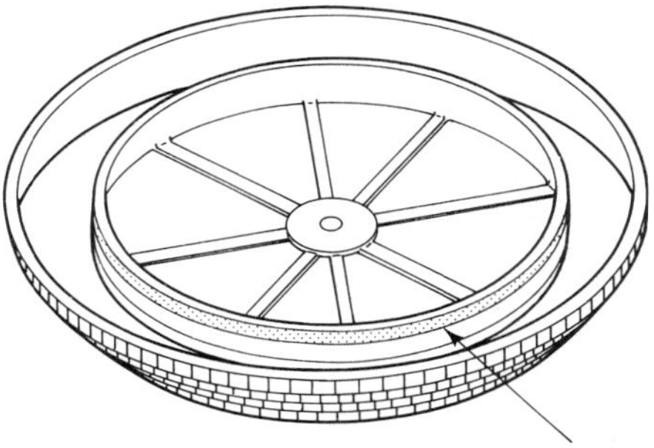

Figure 6-50 Record platter functioning as turntable motor. (Courtesy Sanyo Fisher (USA) Corporation)

back control to maintain the correct speed accuracy. The speed accuracy of this system is independent of line voltage fluctuations.

PLATTER MOTOR CONTROL

There are various arrangements for platter motor control for a turntable and one of the techniques is illustrated in Figure 6-51. In this circuit the platter motor circuit is controlled by a single integrated circuit, identified in the drawing as IC 1, the platter motor phase locked loop (PLL). This IC receives turntable start/stop commands and turntable speed information from the system control microcomputer.

One of the features of IC 1 is that it has an internal crystal controlled oscillator whose natural operating frequency is divided by 64. This oscillator divider supplies a reference signal for both speed and phase control portions of the PLL.

The motor speed detector, an integral part of the motor, generates the frequency generator (FG) signal. This signal is applied to a high-impedance high-gain amplifier which drives the speed and phase control circuits. The speed control circuit within IC 1 compares the reference and FG signals, thus supplying a frequency difference (FD) signal at pin 9.

The phase control circuit compares the phase of the reference and the FG signals, supplying a phase difference (PD) signal at pin 8. The frequency difference signals and the phase difference signals are added and are then applied to the inverting input of an operational amplifier. The noninverting input of the op amp is connected to R10, the speed adjust control. The op amp output is input to transistor Q1 which drives the two switching Hall amplifiers.

The switching Hall amplifiers are built into the platter motor. Each of these amplifiers supplies two inputs to the differential amplifiers of IC 4, in turn supplying drive to the drive amplifiers that power the platter motor.

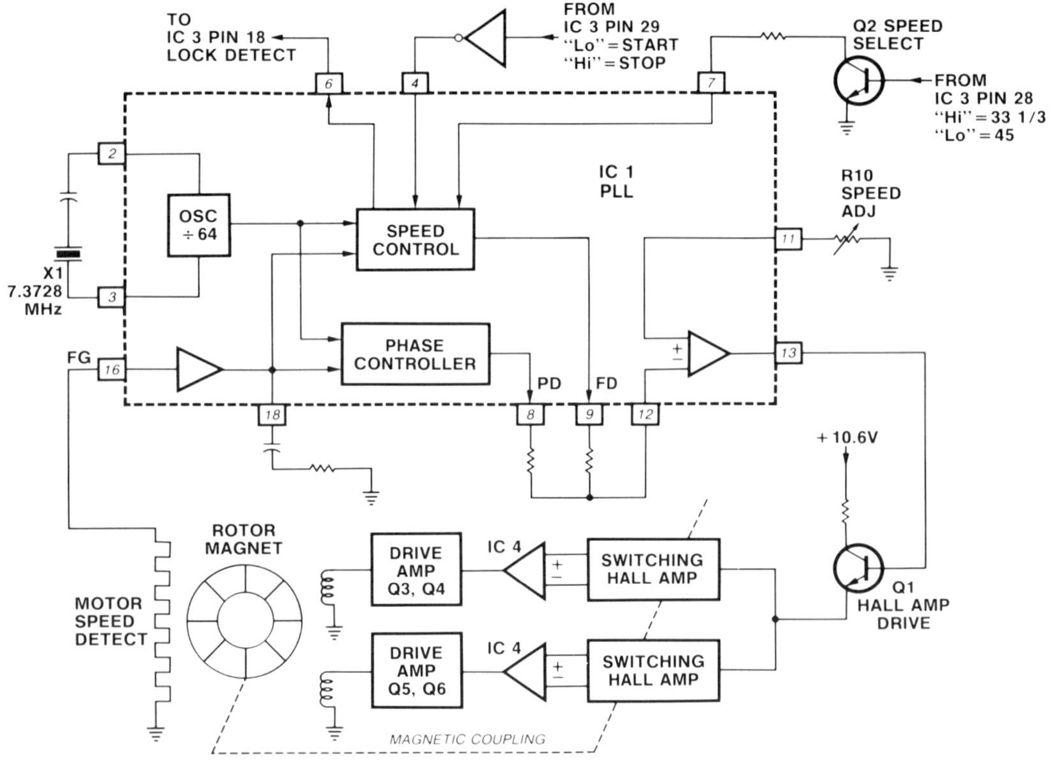

Figure 6-51 Simplified platter motor control for linear tracking turntable. (Courtesy Thompson Consumer Electronics, Inc.)

PLATTER MOTOR DRIVE

The platter motor drive (Figure 6-52) receives its input drive bias from the platter motor control PLL IC 1 (refer to the preceding illustration). The signal is delivered to the base of Q1, the Hall amplifier driver. This driver (Q1) supplies drive to both switching Hall amplifiers.

The switching Hall amplifiers are magnetic sensitive devices which are coupled to the platter motor rotor, an eight-pole magnet. Each of the Hall amplifiers has two outputs of opposite polarity. As the magnetic fields of the rotor magnet pass over the switching Hall amplifiers, the outputs change polarity. The outputs of the switching amplifiers are applied to the inputs of the two differential amplifiers of IC 4. A positive-going signal is supplied to pin 3 of IC 4. At the same time a negative-going signal is input to pin 2 of IC 4. The resulting output at pin 1 is positive-going and is applied to the base of Q5 and the base of Q6 in Figure 6-52. A positive signal on the base of Q6 turns it off. With a positive signal on the base of Q5 this transistor turns on, thus conducting through the motor winding. As the motor rotates, the output polarities of the switching Hall amplifiers reverse and are applied to pins 2 and 3 of IC 4. The result is a negative output signal at pin 1 of IC 4, which is delivered to the base of Q5 and the base of Q6, turning Q5 off and Q6 on. This reverses the current flow in the motor winding, resulting in an AC motor drive signal with a frequency of 2.2 Hz at 33-1/3 rpm and 3 Hz at 45 rpm. The two windings of the platter motor are positioned such that their respective drive signals are approximately 90 degrees out of phase.

MOTORS FOR COMPACT DISC PLAYERS

Compact disc players use several motors (Figure 6-53). One of these is used to operate the disc table and has its speed controlled by a motor servo. There is also another motor for loading and unloading the disc. A third motor is used for driving the laser beam pickup. The pickup drive and motor are both servo controlled.

There are various kinds of motors used in compact disc players, but these all have similar requirements. They must have sufficient torque to drive a mechanism directly or via a gear arrangement. They must be responsive to small signal levels, start and stop rapidly, and not do any hunting. Hunting is the motion of a motor shaft on both sides of the zero point of the feedback signal.

Some motors are given special names by compact disc player manufacturers but they are usually linear DC motors, that is, a motor whose rotation is directly and linearly proportional to the amount of drive voltage.

Another type of unit is the Hall motor. This motor makes use of the Hall effect to give rotation proportional to field strength. The Hall effect is the de-

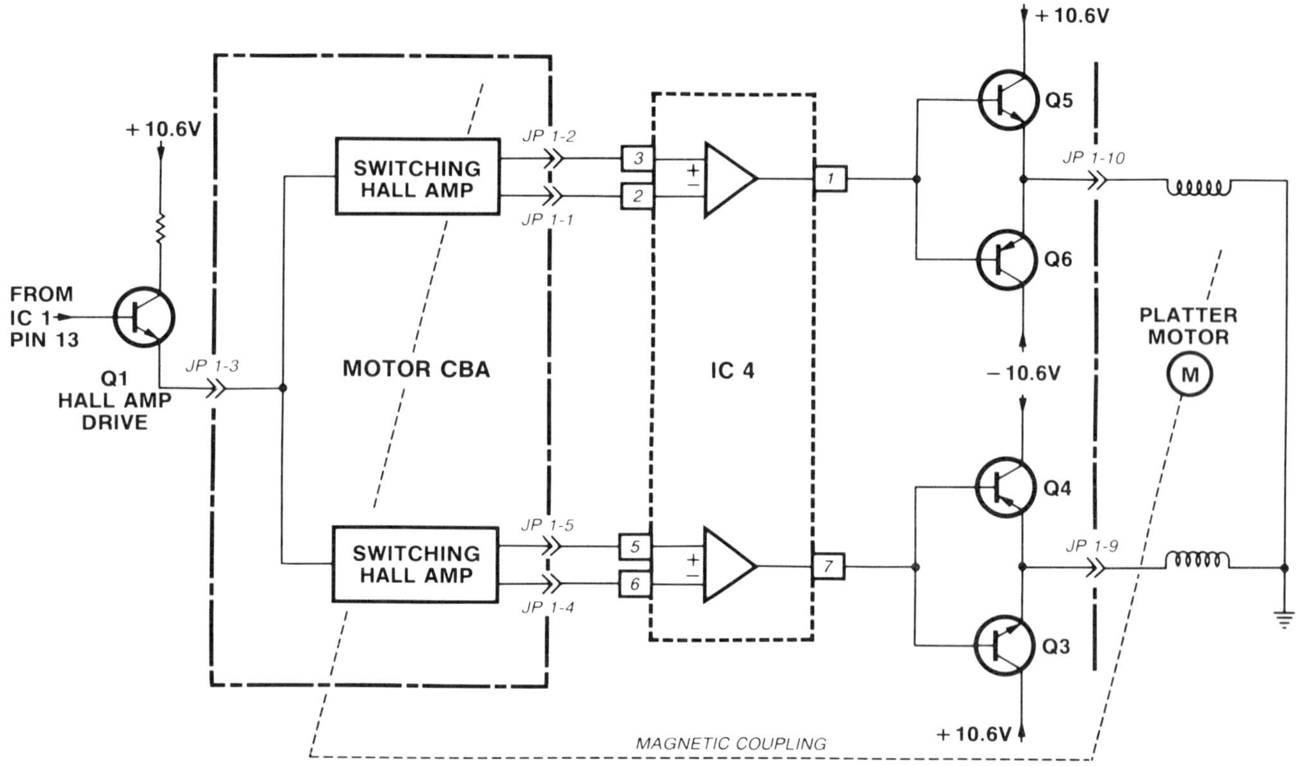

Figure 6-52 Platter motor drive for linear tracking turntable. (Courtesy Thompson Consumer Electronics, Inc.)

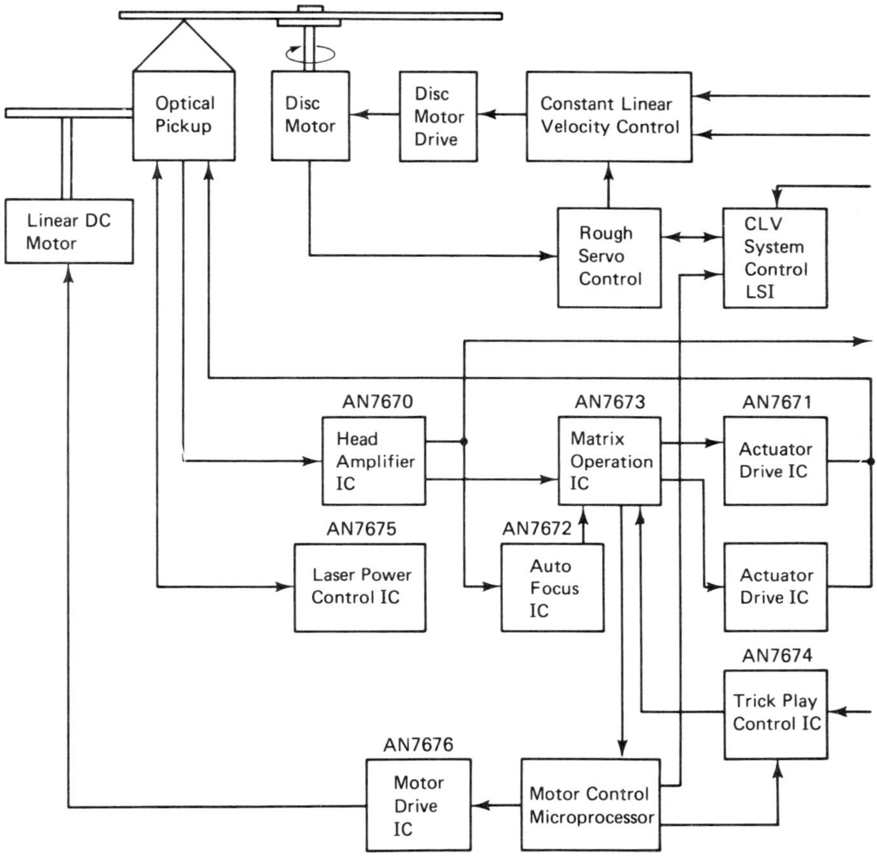

Figure 6-53 Motor system for compact disc player.

velopment of a transverse electric field in a current carrying conductor placed in a magnetic field.

Motor Tracking Mechanism

The optical pickup system of a compact disc player must be able to read the disc's encoded signals quickly and accurately, as well as to provide aid in error correction. For these reasons the laser beam must be servo controlled. Among the various servo control systems, the stability of the tracking servo is important in order for the optical pickup to realize its full potential.

The purpose of the tracking servo is to assure accurate tracking of the compact disc's laser beam along the microscopic 1.6 micrometer track even under the most severe disc conditions. The tracking servo mechanism must balance the need to move at a high speed with an accuracy of less than a micron, as well as to move slowly from the inner to the outer track across the entire disc.

In a conventional compact disc player, the torque of a DC motor is applied to the laser pickup through a complicated worm and reduction gear assembly. Such systems are difficult to design properly because the mechanism's "play" (backlash and friction) does not allow an optimum smooth, high-speed movement. Smooth movement depends on having a low gear ratio, while high speed access demands a high gear ratio design.

Disc Motor Control Loop

The data from the compact disc must be supplied as steadily as possible to a following digital signal processing circuit. Without motor control of the disc, however, the speed with which data bits would arrive would depend on the position of the laser pickup. If the disc is allowed to rotate at a constant speed, the data bits will be delivered to the signal processing circuitry at a variable speed. For example, if the laser pickup reads in the middle of the compact disc, the reading speed is much higher than near the edge. For this reason the disc motor is operated at a variable speed with that speed determined by the location of the laser beam assembly with respect to the disc. The disc motor speed is constantly adjusted so that the laser pickup reads the encoded data on the disc at a steady rate of 1.2 to 1.4 meters per second.

HALL DC MOTOR WITH PHASE LOCK LOOP

The Hall device can be combined with phase lock loop (PLL) circuitry for motor speed control.

The Phase Lock Loop

The basic concept for PLL functioning is detailed in the block diagram of Figure 6-54. The unit uses a frequency and phase comparator, having a pair of signals for its input. One of these signals is supplied by a voltage controlled oscillator (VCO). The other signal is a reference voltage furnished by a precisely controlled oscillator. The two signals are compared in the phase comparator, and if they are identical in phase and frequency, the result is zero output from the comparator. If, however, there is some difference between the two signals, the comparator supplies a DC output voltage. Depending on the signal relationships, this output voltage can be either positive or negative.

The output voltage is first brought into a low-pass filter which eliminates any remnants of the two signals which were originally brought into the comparator, and so the output of the filter is solely a DC voltage whose amplitude depends on the difference between the two original input signals. This signal is then input to a DC amplifier. The output of this amplifier is supplied to the VCO, changing the signal it supplies to the comparator. The effect is to bring

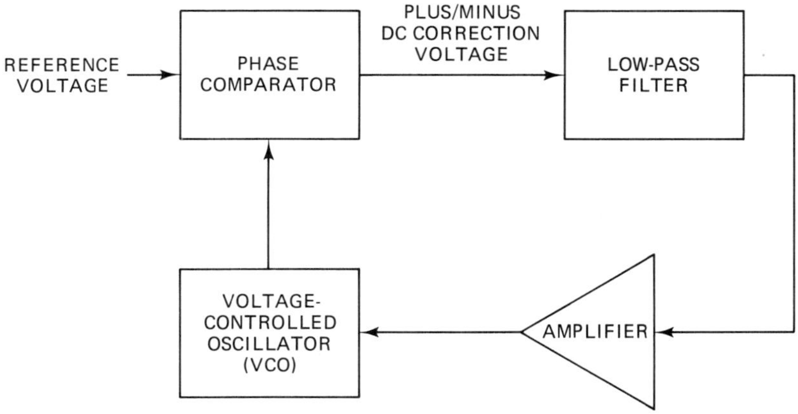

Figure 6-54 Phase lock loop system.

the two signals into agreement so that they are mutually cancelling. The result is zero signal voltage output from the comparator.

Hall-PLL Combination

The PLL circuitry can be combined with a Hall device to control the speed of a motor so that it remains constant. The block diagram appears in Figure 6-55. The phase comparator receives pulsed signals from the Hall-device controlled motor with these signals compared with those supplied by a

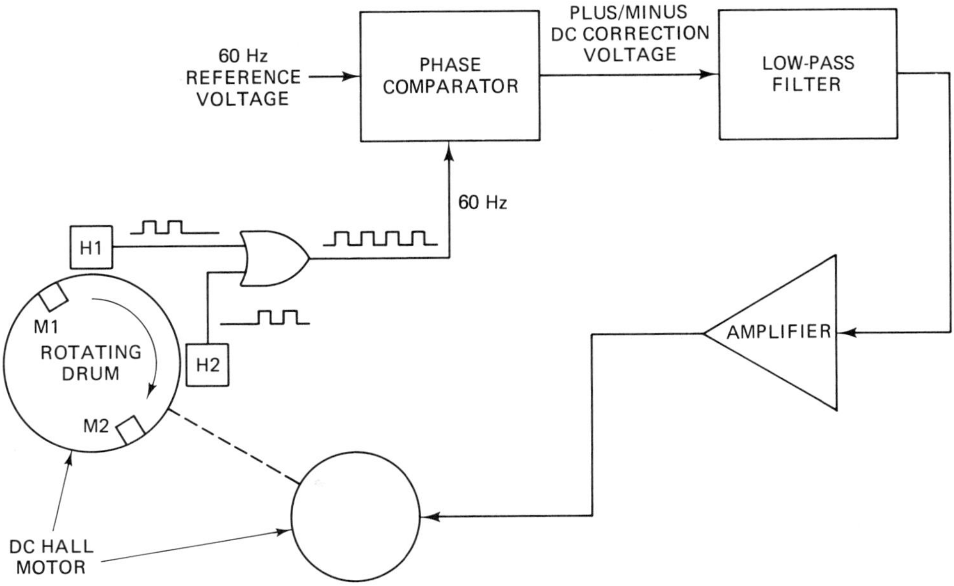

Figure 6-55 Speed control of Hall motor using phase lock loop.

60 Hz precision signal generator. The output of the phase comparator is supplied to a low-pass filter, amplified, and then used as the DC voltage supply to the Hall motor. If this voltage is zero there is no change in motor speed. A voltage increase or decrease will either raise or lower the motor speed.

The motor speed can be increased, and then kept constant at the increased speed, or can be decreased, and also kept constant at the decreased speed by feeding the pulsed voltage output of the Hall device into a frequency changing network before being supplied to the phase comparator. The frequency changing network can have a single or multiple frequency capability. With a multiple frequency capability it is possible to have a Hall-controlled motor having a wide range of speeds, with each speed that is selected kept constant.

COMBINED MOTOR SPEED AND DIRECTION CONTROL

For certain motor applications it may be necessary to control both motor speed and the direction of armature rotation. A way of doing this is illustrated in the circuit diagram in Figure 6-56.

The motor used is a permanent magnet, brush type DC, with the DC supplied via full-wave bridges and also an external DC source. Since the motor is the working load, the maximum current available to it is the amount that can be handled by the external circuitry. Upgrading these current-wise will permit the use of a larger motor.

Operating power is supplied by a stepdown transformer, T1, whose primary is connected to the 117-volt AC power line. The secondary winding supplies 24-volts AC. The circuit uses a pair of full-wave bridge rectifiers, with the first consisting of diodes D1 to D4 and the second consisting of diodes D5 to D8. D1 and D2 supply minus 12 volts DC to the cathode (C) of silicon controlled rectifier 2 (SCR 2), while diodes D3 and D4 furnish 12 volts plus to the anode (A) of SCR 1.

The circuit has a pair of opto-isolator couplers, U2 and U3. These provide input-to-output circuit isolation and are also a way of supplying gate current to the two power SCRs. By using the opto-isolator couplers the input control voltage can be driven by almost any logic-output source, or interfaced to a computer, without the motor or SCR switching noise being coupled back into the control circuitry.

Directional Control

Point A is a DC voltage input point and is intended for a plus potential of 5 to 12 volts. When this is done, opto-isolator U2 will deliver a gate current to SCR1, turning it on. The effect will be to supply a positive voltage drive to the motor. The motor will now turn in one direction. If the control voltage is removed from point A and connected to point B the motor will reverse its direction.

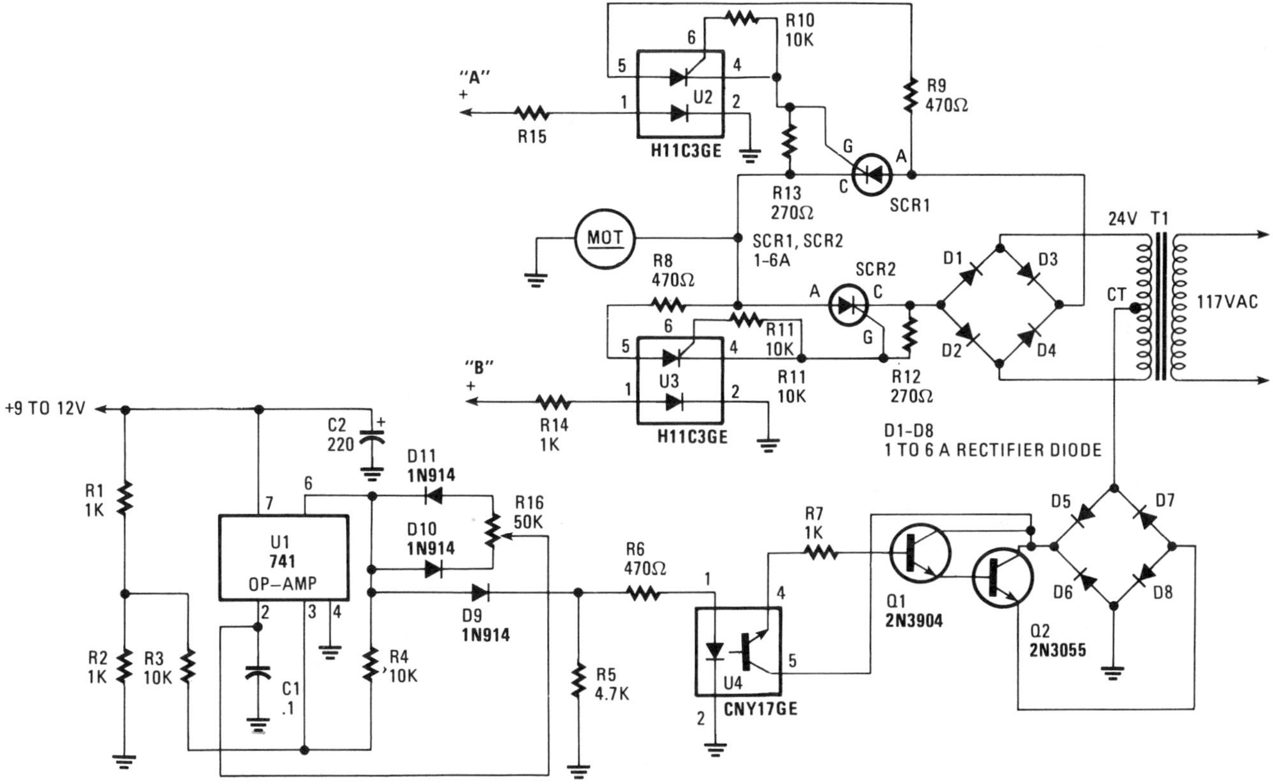

Figure 6-56 Circuit for motor speed and motor direction control. (Courtesy Hands-On Electronics, a Gernsback Publication © June, 1988)

Speed Control

Diodes D5 and D6 are part of a full-wave arrangement between the center tap of T1 and ground. Therefore, no matter what polarity is being fed to the motor, the voltage between the emitter and collector of N-P-N transistor Q2 is always correct. Since neither the positive nor the negative output of the bridge circuit is at ground level, another opto-isolator coupler, U4, is used to supply DC isolation between the speed control and the power switching circuit. That allows the speed-control circuit to work with a common ground.

By putting a jumper between pins 4 and 5 of U4 transistors, Q1 and Q2 would be biased on and with a positive voltage on points A or B; the motor would run at full speed. By making and breaking the circuit between pins 4 and 5, the motor's speed could be lowered. A much more practical method than manual switching would be to use a pulse-width modulator to supply finely controlled switching.

Table 6-1 supplies a list of parts for this motor speed/direction circuit.

TABLE 6-1. Parts List of the Motor Speed/Direction Control Circuit

Semiconductors
D1–D8—1-to-6 ampere silicon diode
D9–D11—1N914 general-purpose small signal diode SCR1, SCR2—1–6 ampere silicon-controlled rectifier
U1—LM741 general-purpose, op-amp, integrated circuit
U2, U3—H11C3GE, or similar, SCR-output opto-isolator/coupler, integrated circuit (Digi-Key)
U4—CNY171GE, or similar, transistor-output optoisolator/coupler, integrated circuit (Digi-Key)
Q1—2N3904 general-purpose N-P-N silicon transistor
Q2—2N3055 N-P-N silicon power transistor
(All resistors are 1/4-watt, 5% units unless otherwise noted.)
R1, R2, R7, R14, R15—1000-ohms
R3, R4, R10, R11—10,000-ohms
R5—4700-ohms
R6, R8, R9—470-ohms
R12, R13—270-ohms
R16—50,000 ohms, 100-turn potentiometer
Additional Parts and Materials
C1—.1-μF, 100-WVDC Mylar capacitor
C2—220-μF, 25-WVDC electrolytic capacitor
T1—24-volt center-tapped 1-to-6A transformer
Printed circuit or perfboard materials, small DC motor (9 to 12-volts), 9 to 12-volts DC power supply, enclosure, IC sockets, wire, solder, hardware, etc.

MOTORS IN THE FUTURE

For many years motors were strictly electrical. Early motors were controlled by a varying voltage input either by using fewer or more batteries or by using resistors in series with those batteries. The development of motors for AC voltage input was an important step forward, but even more significant was the union of electronics with the electrical design required by motors.

The most significant step forward came with the realization that electronics could stop and start motors with a precision that manual control could not even begin to match. Electronics can not only start and stop motors, but it can make the motor shaft turn forward or in reverse, make a motor shaft rotate a previously determined number of degrees or fractions of a degree, and do so repetitively and accurately. With electronics, motors can control other motors. With electronics, we were able to enter the world of size reduction, followed by miniaturization, and finally microminiaturization.

Microminiature Motors

These are incredible, and sizewise can be made the width of a hair. The usual motor is driven by dynamic electricity, which is a current produced by an AC or DC generator or by a battery. Microminiature motors, practically invisible to the naked eye, are driven by electrostatic charges. This is the kind of voltage generated when you comb your hair or walk across a rug. Rub a dry comb vigorously and it can pick up 1/4-inch squares of newspaper. The end result is motion and that applies to any motor: large, small, and micro.

Using technology developed in the manufacture of integrated circuits and working with knowledge of the silicon chip, the microminiature motor has its parts etched onto silicon. The motor consists of two parts: a fixed hub and a rotating arm.

The concept of attraction and repulsion is applicable to electrostatically operated microminiature motors as it is to the much larger electromagnetically operated types. The problem with electrostatic charges is not so much in developing them as it is in avoiding them. The voltages encountered electrostatically can be in the thousands or even millions of volts. Unlike motors that are current operated, electrostatic voltages can be induced at a distance, and so the need for a physical connection is eliminated. Despite the high voltages involved, however, electrostatic induction requires a short spatial separation.

7 BRUSHLESS DC MOTORS

There are numerous ways of categorizing motors: by torque, horsepower, amount of power input, whether the motor is DC, AC, or universal, by the type of motor frame, by whether the motor is single-phase or polyphase, by whether the motor is a synchronous type, split-phase, or shaded-pole, by size, whether it is a series wound, shunt, or compound. It is difficult to believe that some motors do not come within the headings of such listings.

And yet there are some motors that require a classification of their own: permanent magnet motors, brushless motors, pulse controlled motors, motors for heavy loads, for light loads, for low-speed operation, and those intended for high speeds. There are also geared and nongeared motors, stepper motors, miniature motors, and motors that are microwatt types (millionth of a watt). There are some motors, such as those designed to fit into a watch case, so small that several can be supported on a fingernail. There is such a variety of motor types that some simply cannot be classified easily. In some instances, motors are made to special order or else an existing motor type will be modified to meet specific working requirements.

PERMANENT MAGNET MOTORS

Brushless DC motors are permanent-magnet DC motors using electronic switches to send power to the windings directly instead of brushes and a commutator. The electronic switches are solid state, and with these it is no longer necessary to rotate the windings and a commutator to energize the appropriate coil at the proper time. The windings are therefore conveniently placed in the stator rather than the rotor and the permanent magnets are attached to the rotor.

Speed and Voltage Range

The speed range of brushless DC motors (BLDC) is wide and in some instances is in excess of 50,000 rpm. Operating speed is dependent on the load with the variation between full-load rpm and no-load rpm based on motor size (Table 7-1). Designs with low speed and high torque are practical and often gears and/or belt speed reducers can be eliminated.

TABLE 7-1. Operating Characteristics of Some Selected Brushless DC Motors

Drive* Circuit	Full Load Amps	Full Load Torque	Full Load rpm	No Load rpm	Volts	Starting Torque (oz-in)
B	3.0	8	6000	8500	24	48
B	4.5	30	3000	5000	24	80
B	4.0	10	10000	12500	24	60
U	0.5	4	2200	3000	24	14
U	1.2	4	4400	6000	24	8
B	8.0	38	5000	5600	24	112
B	5.0	28	4000	4800	24	125
B	10.0	56	4000	5300	24	200

*U - Unipolar
B - Bipolar

The source voltage is low and is typically less than 42 volts DC, but there are lower and higher voltages. 24 volts is commonly used.

Centrifugal blowers and axial flow fans can be thermistor controlled so that only the volume of cooling air required is supplied. The blower or fan speed can be made to vary directly with the ambient or hot spot temperature resulting in minimum rotor speed and acoustic noise. Consequently, the power consumed by blowers and fans is minimized. High peak torques, accurate speed control, high efficiency, and long life make the BLDC motor the optimum choice for many applications.

These motors can supply quiet, high-speed operation and, because they are DC powered, are immune to AC power supply variations. They can be used on portable DC power and have safe, low-voltage operation. They are easily adapted to programmable controls in appliance applications where various speeds, duty-cycles, reversibility, and braking are required. The safe, low-voltage, arcless operation of brushless DC motors also makes them suitable for hazardous environmental installations.

Applications

Electronically-commutated brushless DC motors are used in a wide range of applications: in office equipment, copying machines, fans or blowers, water, air, and chemical pumps, centrifuges, scanners, printers, tape drives, and gas analyzers. They are also used in business machines, commercial ventilating systems, optical scanners, medical equipment, and power tools.

THE WORKING BLDC (BRUSHLESS DC)

Figure 7-1 shows the basic working arrangement of a BLDC. The transistors form a bipolar circuit and consist of N-P-N and P-N-P transistors. Hall-effect devices located near the rotor sense its position and feed information to a drive chip. The chip turns on the power transistors in the proper sequence to supply power to the windings, creating a true rotating magnetic field. The rotor magnets lock into this field, converting electrical into mechanical power. In effect

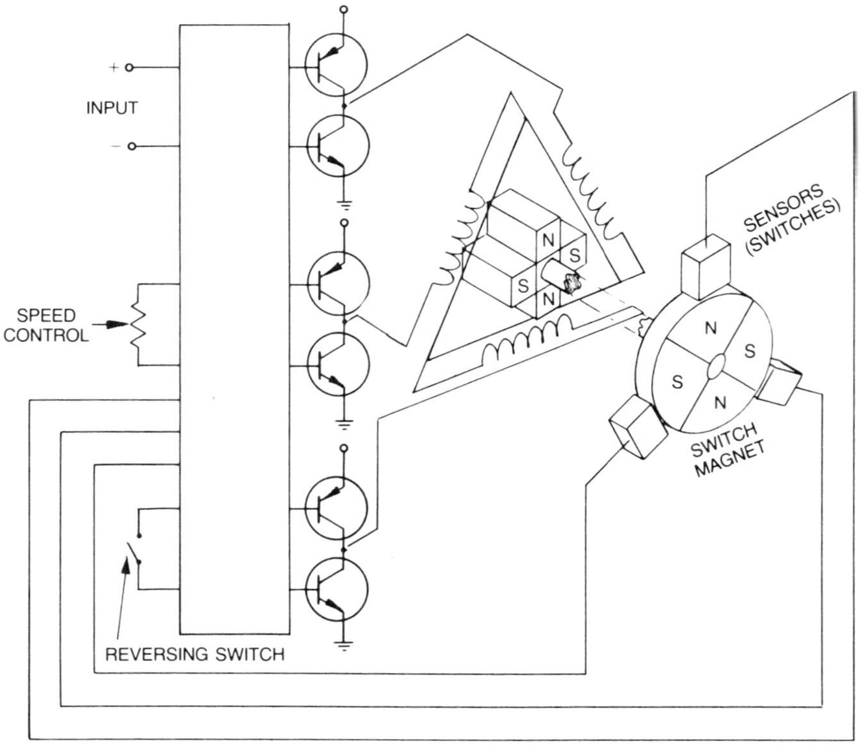

Figure 7-1 Operational setup of brushless DC motor. (Courtesy Eastern Air Devices, Inc.)

the electronic commutator switches the stator windings at the appropriate rotor positions to generate continuous torque and rotation.

Electronic Commutation

Electronic commutation consists of a rotor position sensor and a switching amplifier. Position sensors can be either magnetic or optical. If magnetic, a Hall switching device is used; if optical, a light-emitting diode (LED) phototransistor arrangement is employed. The stator windings can be designed for two-phase or three-phase wye or delta.

The electronic commutating switches are controlled by a multipole switch that is attached to the shaft. The switch (a permanent magnet for Hall switches or a stamped disc for optical switches) and the sensors are radially positioned to sense the angular position of the magnetic poles on the rotor.

The three Hall-effect devices are spaced at intervals near the rotor. They sense magnetic flux and provide logic outputs. These signals are used by the logic circuitry to provide three-phase waveforms (Figure 7-2).

When the switch is activated, it biases transistors or logic circuitry, causing a direct current to flow into specific stator windings. These stator windings generate a flux field which is in quadrature with the rotor poles. The stator magnetic poles are of such polarity and position that the rotor poles are attracted, causing rotation in the preferred direction.

A representative arrangement is a four-pole three-phase wye connected star with a ceramic magnet rotor. The rotor position is sensed by the Hall effect switches mounted on a printed circuit board located at the lead end of the motor. These switches generate square wave pulses which are used by the control to develop the necessary coil switching and speed regulation.

The sensors, working like switches, are turned on and off in correct sequence and timing to assure continuous rotation and optimum torque output. Figure 7-3A shows the phasing diagram for unipolar drive, Figure 7-3B, the phasing diagram for bipolar drive. The bipolar type, using N-P-N and P-N-P transistors, is more efficient and permits higher outputs than the unipolar drive but it is more expensive.

Construction

Brushless DC motors use conventional outer stator laminations with rotating permanent magnets (Figure 7-4). This construction yields maximum heat dissipation, resulting in improved bearing life, and minimizes mechanical problems at high speeds. The rotor magnets are ferrite, rare-earth, or Alnico (aluminum, nickel, cobalt). Hall effect devices, optics, or vane coupled coils are used for commutation. Motors can be supplied with or without electronic controls. In some instances the control circuit can be positioned within the motor.

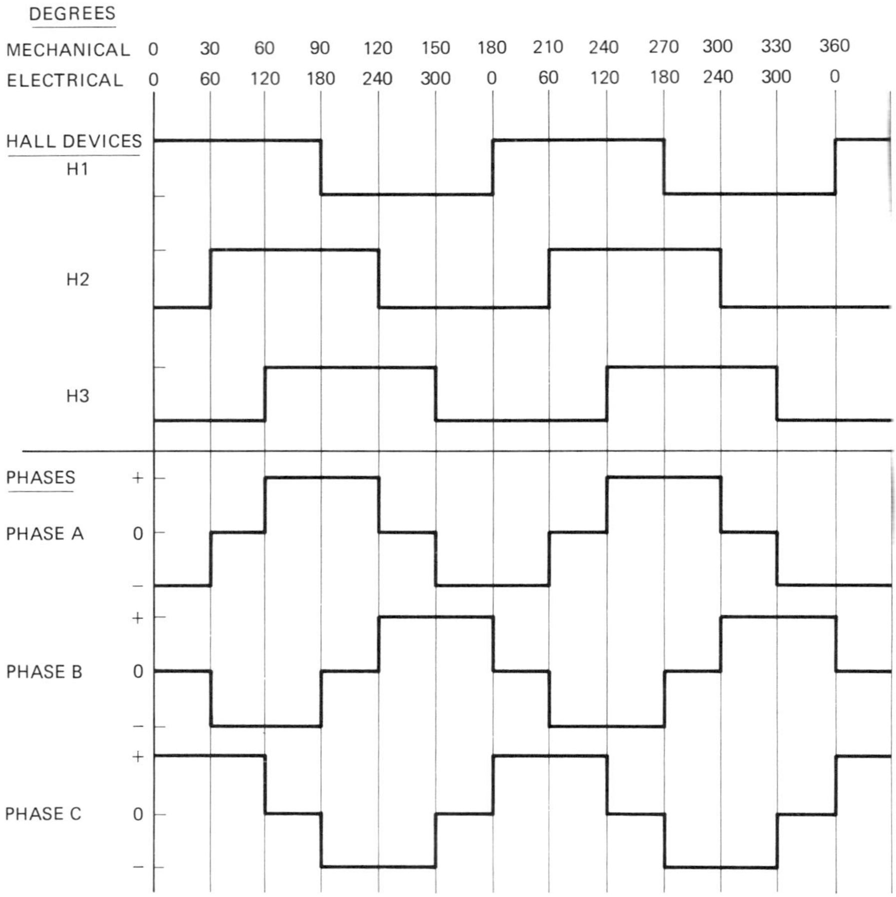

Figure 7-2 Commutation timing diagram, clockwise rotation. (Courtesy Fasco Industries, Inc., Motor Division)

Brushless DC motors can be built with the rotor inside (standard construction) or outside (inverted construction).

Speed Control

Speed control is offered for most brushless DC motors. The speed can be made independent of the load or supply voltage, or it can be made dependent upon a tachometer output signal or can use some other sensor.

There are two types of speed controls, velocity feedback and phase-locked-loop. In the velocity feedback circuit the speed correcting torque is proportional to the amount that the actual speed differs from the command speed. In the phase-locked-loop technique the motor speed is proportional to the frequency of a clock oscillator. The speed correcting torque is a function of the

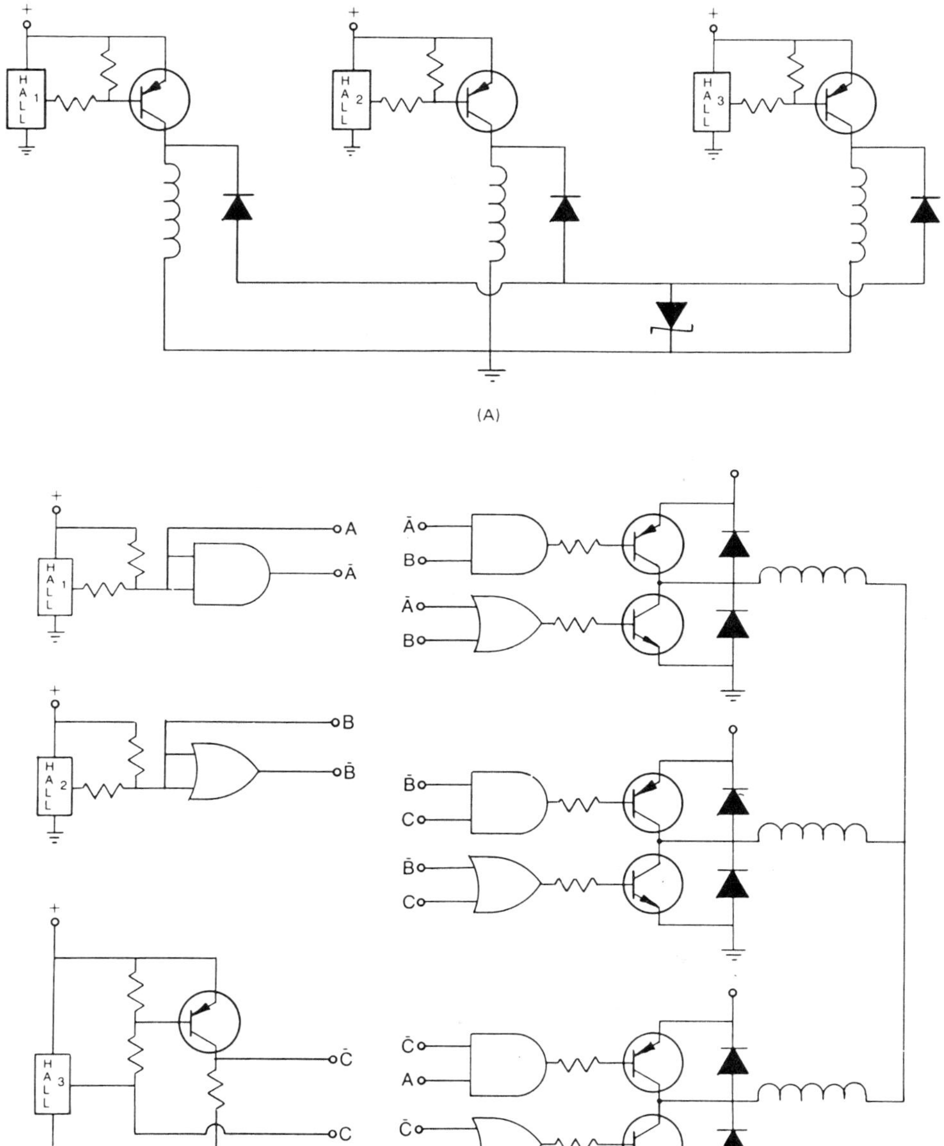

Figure 7-3 Unipolar (A) and bipolar (B) drive circuits. (Courtesy Eastern Air Devices, Inc.)

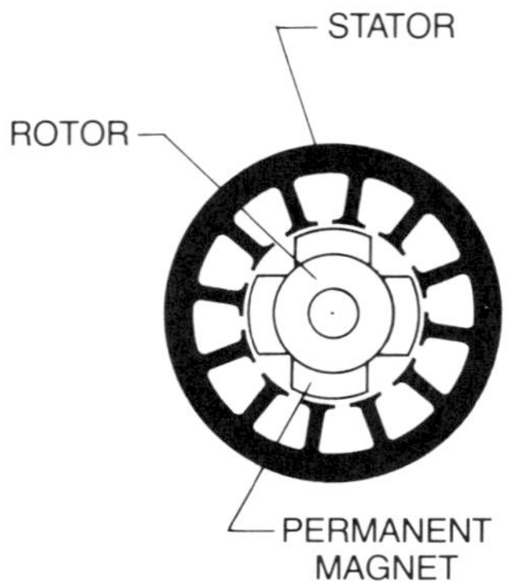

Figure 7-4 Motor construction. (Courtesy Eastern Air Devices, Inc.)

phase lag between the motor and the clock. This phase lock control is very precise because motor speed accuracy is equal to that of the clock frequency. Feedback signals can be obtained from motor sensors (Hall devices, optics, etc.), tachometer, optical encoders, and resolvers.

Slotless Stator

It is possible, especially with the newer high energy magnet materials to make a surface wound armature without slots and eliminate the problem. With a brush type motor, the armature (conductor carrying member) rotates. With a brushless DC motor, the field generally rotates and the armature is stationary. The rotor position must be sensed electrically, and this signal must be processed and amplified to cause the windings to be switched as previously described. An economical method of field position sensing is by means of Hall Effect devices (Figure 7-5).

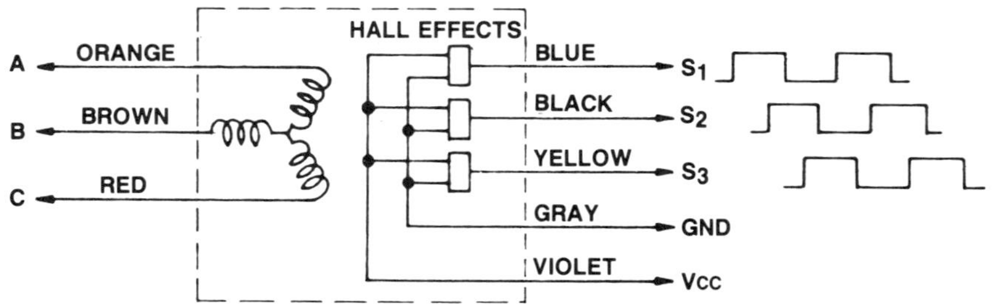

Figure 7-5 External connections for counterclockwise rotation viewing lead end of motor. (Courtesy Motronics Corp.)

These solid-state devices are sensitive to magnetic flux density and put out a voltage of approximately 5 volts when they see a north pole and will be at ground potential when they see a south pole. The Hall device itself supplies an analog signal which is roughly proportional to flux density. The Hall devices used for commutation usually have a Schmitt trigger built into the package which delivers a constant voltage square wave and is TTL (transistor-to-transistor logic) compatible.

Optical Devices

These are sometimes used for position sensing, and if an encoder is to be used for other functions, it can easily be modified to add a track (or tracks for three phase) sensors properly spaced. The track will have one opaque segment and an equal transparent segment for each pair of poles (two opaque and two transparent segments) for a four-pole motor, etc.

There are other methods of sensing but the Hall device or optical methods are the predominant techniques. In the case of Hall devices, the main field magnet is usually used to trigger them, but it is sometimes desirable to use a separate commutation magnet which is usually stamped from flexible magnetic sheets.

As with any motor, the field can either be the inside or the outside member. Each type, as indicated in Table 7-2 has its advantages and disadvantages.

TABLE 7-2. Advantages and Disadvantages of Outer and Inner Rotors

	Outer Rotor	Inner Rotor
Inertia	Higher—sometimes an advantage	Lower—sometimes an advantage
Torque/Power	Higher	Lower
Cost	Lower	Higher
Balance	More critical	Less likely to need balancing
High speed	Magnets on inside diameter; Bonding not critical	Magnets on outside diameter
Printed Circuit Board	Usually easier to package Hall devices	Hall device packaging can be a problem on small motors

Braking

Braking is easily accomplished with brushless DC motors. The permanent magnet rotor makes it possible to achieve reasonable braking by simply shorting the stator windings. Improved braking characteristics are obtained by applying reversed rotation logic to the driver.

Reversibility

Reversibility is readily obtained by a simple on/off switch for the bipolar and unipolar drives with minor changes to the control circuitry. Unipolar drives without logic require that all stator coils be reversed. This is done with six-pole double-throw switch for three-phase designs.

Current Limiting

Current limiting is supplied in some brushless DC motors. Without current limiting, starting currents can be more than ten times the full load current, requiring expensive high current motor drive transistors and additional cooling. By limiting starting current to approximately 20 percent above full load, it is possible to design economical motor and controller systems. Starting torque and acceleration characteristics are adversely affected because of the reduction in starting current and torque.

Brushless DC motors offer an alternative to conventional AC induction motors in that they are immune to AC power variations and eliminate mechanical speed control elements such as belts, pulleys, and gear transmissions. They offer a much higher controllable speed and have rapid acceleration and deceleration and instant reversibility. They also operate at safe, low voltages with very high efficiency and reliability.

Brushless DC motors also provide solutions to the inherent disadvantages associated with brush-type DC motors; solutions to noise, brush wear, electromagnetic interference, the generation of heat, erosion of commutator surfaces.

BRUSHLESS DC MOTOR ANALYSIS

Figure 7-6 is a simplified illustration of how torque is developed in a permanent magnet DC motor. Simply stated, torque is produced if a current is made to flow in the conductors which comprise the armature. The force can be expressed as

$$f = KBLIZ$$

where f is force, K a constant, B is the air gap flux density, L is the inductance, I is the current measured in amperes, and Z is the number of conductors in series. Since, in a motor, the conductors rotate about a central shaft, the

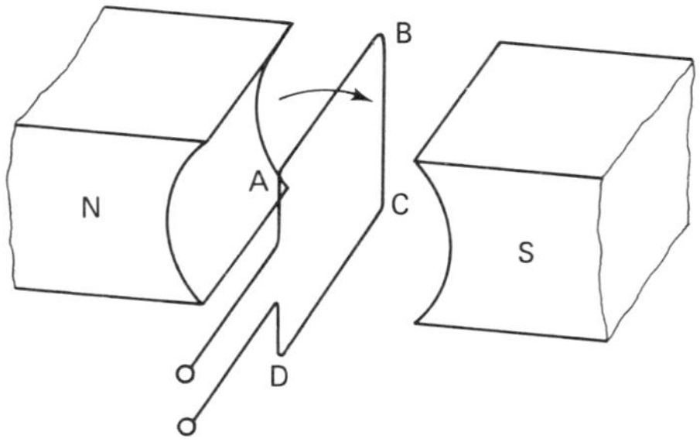

Figure 7-6 Basic concept of torque in a permanent magnet motor.

torque can be written as

$$T = f \times r$$

where $T = KrBLIZ$ and r is the radius at the air gap.

The illustration in Figure 7-7 shows the armature coil in the zero torque position. The maximum torque is 90 degrees from the position shown. As the conductors rotate from the maximum torque position, torque drops off in sinusoidal fashion and becomes zero when the coil has moved 90 mechanical degrees for a two-pole motor, 45 mechanical degrees for a four-pole motor, and so on.

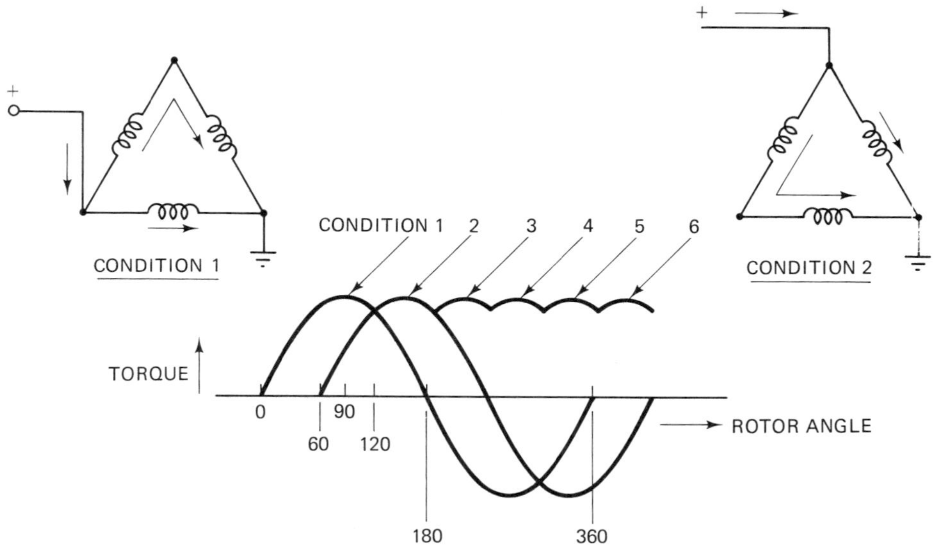

Figure 7-7 Development of torque.

In an actual motor there is more than one coil, with each angularly displaced from another, so that when the torque from one coil has dropped off, current is switched to another coil which is properly located to produce maximum torque. In the motors described in the chapter on DC motors this switching is done mechanically with brushes and a commutator. In a brushless motor the position of the coils with respect to the field is sensed electronically and the current is switched to the appropriate coils by means of transistor switches.

It would be possible to convert a brush type motor to one that is brushless by bringing out all of the leads that are attached to the commutator and providing switches for each lead, but this would involve a large number of switches. Instead, a polyphase winding is used that is similar to that used in AC motors; it switches the phases as a function of shaft position. Two, three, and four phases are used by various manufacturers, and with a little gimmickry it can be done single phase. Three phase provides optimum performance although it does require more electronic components. Three-phase windings include various types: delta bipolar, wye bipolar, and wye unipolar.

To see how torque is generated in a brushless motor, consider the delta bipolar as an example. Apply a steady DC voltage by closing switches S1 and S6 in drawing A of Figure 7-8. This can be regarded as condition 1.

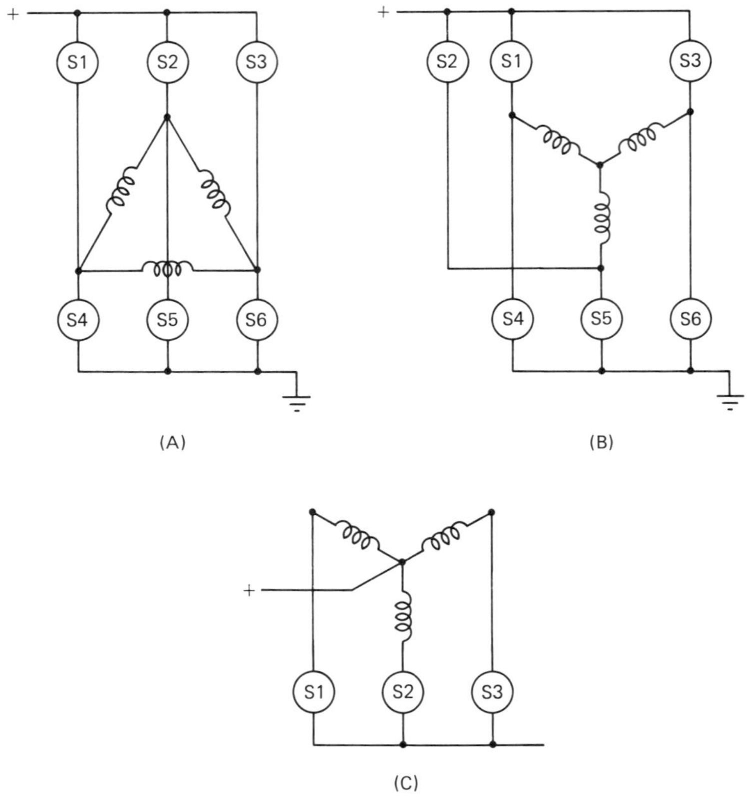

Figure 7-8 Transistor switching: delta bipolar (A); wye bipolar (B); wye unipolar (C). S1 to S6 are the transistor switches.

The rotor will immediately move to a zero torque position at 180 degrees. If we physically force the rotor in a direction opposite to the desired direction of rotation, torque will build up roughly sinusoidally and become a peak at 90 electrical degrees. Since we want the motor to be at its torque point, we back it up 30 degrees more and establish this as the point at which that particular winding condition should be switched on. If we now allow the rotor to turn, it will move back toward the 180-degree zero torque position. However, if at the 120-degree position we open S1 and close S2 we now have condition 2 in Figure 7-7 and we are operating on a new sine wave of torque versus angle. As the rotor continues to rotate, we keep sequentially switching to conditions 3, 4, 5, and so on, and pick off the top 60 degrees of the torque sine waves.

The above discussion pertains to the delta bipolar winding. The wye bipolar (Figure 7-7B) yields similar results and produces equivalent torque for a given amount of power.

There are two minor differences; the wye winding has only two windings energized at a time. This leaves one winding available for 60 degrees at a time for other use, such as for a tachometer signal. The other difference has to do with the torque wave shape. More on this later.

The other type of three-phase motor is the wye unipolar (Figure 7-7C). This simplifies the electronics since it uses only three switches instead of six. However, instead of picking off the top 60 degrees of the sine wave, it picks off the top 120 degrees. Therefore, it is 60 degrees off the maximum torque position at the switch positions, instead of 30 degrees. The torque falls off as the cosine of the angle. This amounts to 50 percent of peak torque for unipolar and only 86.6 percent for bipolar. Consequently, the peak-to-peak ripple is 50 percent for unipolar and 13.4 percent for bipolar.

Torque Waveshape

The assumption is made that the torque waveshape is sinusoidal, but this only occurs under ideal conditions. Actually, it would be better if the waveshape were trapezoidal and flat topped over at least 60 degrees (Figure 7-9).

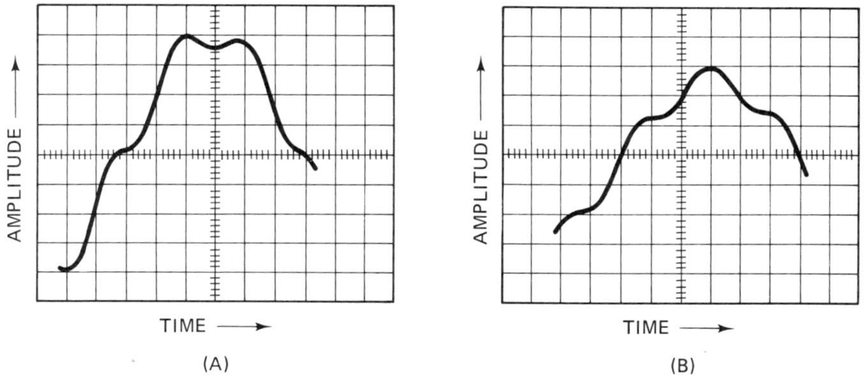

Figure 7-9 Torque waveshape is sinusoidal only under ideal conditions. It would be better if the waveshapes were trapezoidal and flat topped over at least 60°. This would minimize ripple torque. These are waveforms of the back EMF: delta (A); Y (B). The wye is a six-slot, four-pole unit.

LIMITED ROTATION BRUSHLESS DC MOTORS

By using a special single phase stator, the brushless DC motor can produce step-free, linear torque over a limited arc without electronic commutation. The angle of movement is illustrated in Figure 7-10. This type of operation finds

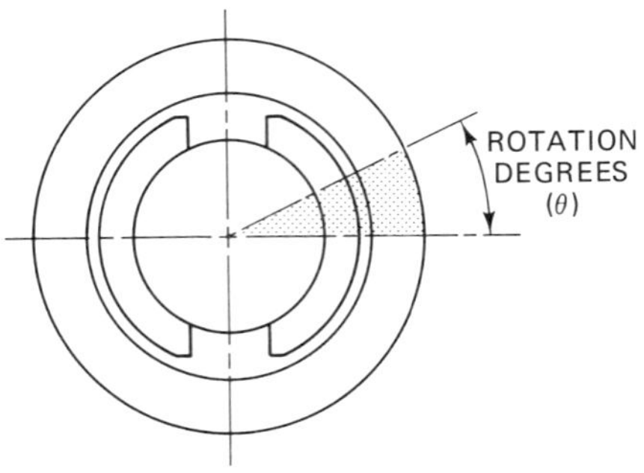

Figure 7-10 Brushless DC limited rotation. (Courtesy Motronics Corp.)

application in head positioning for computer disc memories and as valve positioners and actuators. It is also used, in general, where fast, accurate positioning is required.

Miscellaneous Data

Brushless DC motors are made in a variety of frame sizes, with some varying from 1.0 inch to 5.6 inches. The motor enclosure can be ventilated or enclosed. Output power is as high as 1/2 horsepower, but there are models available with 1/100 to 1/5 HP for continuous duty applications depending on air flow and ambient temperatures. Motor efficiencies range from 30 percent for some of the smaller unipolar designs to 85 percent. Rotation can be clockwise, counterclockwise, or reversible.

Advantages

DC motors have a performance advantage over conventional AC motors when efficiency, speed, or torque control are important factors. Maintenance and reliability problems caused by brushes and commutators have restricted DC motor usage. New advances in permanent magnets and solid-state components have made it possible to produce brushless DC motors that offer the advantages of DC motors and the ruggedness and reliability of AC motors.

8 STEPPING MOTORS

Stepping motors, also called steppers, are those whose shaft angular rotation can be controlled. In a sense all motors are stepping types since an on/off switch can initiate shaft rotation and can stop it. The difference is that with a stepping motor the rotation angle can be precisely controlled. Since the shaft turns through successive amounts of 360 degrees, the stepping motor can stop the shaft at 360 degrees or some lesser or greater amount. Further, the stepping motor can produce successive repetitions of the angular sweep originally selected. To be able to do this means the motor cannot be manually operated but requires an electronic control. Other requirements are positive braking and a minimum or absence of hunting, a condition in which the shaft seeks its precise stopping point.

Stepper motors are used whenever a precise amount of shaft rotation is essential and where the same angular rotation or any selected angular position must be repeated indefinitely. The stepper motor can be regarded as a device that is positioned between an electronic controller and some component depending on the stepping motor for the accuracy of its functioning.

DISC STEPPING MOTORS

The concept of the disc stepping motor, also called a disc stepper, is illustrated in Figure 8-1. This simplified version consists of a samarium-cobalt disc centered and mounted on a shaft free to rotate between a pair of stator poles, identified in the illustration as phase A and phase B, sometimes called phase 1

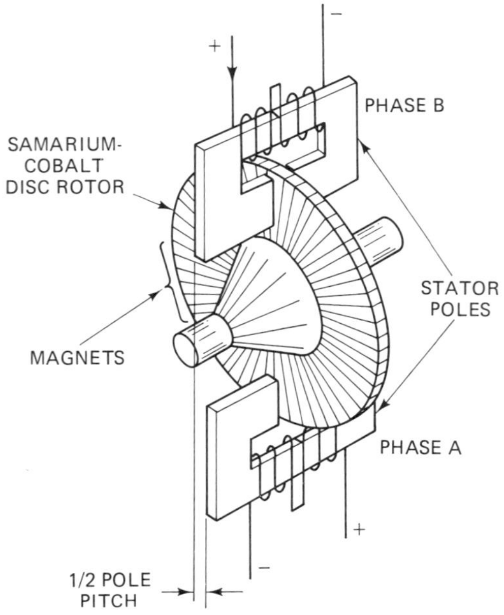

Figure 8-1 Basic stepper motors. (Courtesy Penton/PC, Inc.)

and phase 2. The unit in the illustration is a two-phase motor with each phase electrically shifted by 90 degrees. Each phase may consist of one or more windings which can be connected externally. The phase A and phase B coils can be joined in parallel, as in Figure 8-2A or in series as in Figure 8-2B. Parallel and series connections are made to provide desired performance characteristics.

The number of phases, two in a two-phase stepper, should not be confused with the number of coils, often four, as indicated in Figure 8-2. The sequence of a two-phase stepper has four states corresponding to four steps.

The excitation voltage for each of the coils in the two-phase arrangement consists of voltage pulses. Assume for the moment that a positive pulse is delivered to the coil (or coils) in phase A. If a positive pulse is supplied to phase B, the disc will rotate by one half of a pole pitch. If the input pulse polarity is transposed, that is, if the negative pulse supplied to phase A is followed by a positive pulse to phase B, the rotation of the disc will continue. To reverse the direction of rotation, all that is necessary is to reverse the pulse sequence. Half stepping the disc can be accomplished by pulsing both coils at the same time every other step.

Not all discs have the same number of magnets. The one in Figure 8-3 has 50, with each magnet only 0.7 mm long, and all of them having the same thickness as the disc. The larger the number of discrete magnets, the smaller the step angles that can be obtained, and as a result the greater the number of steps.

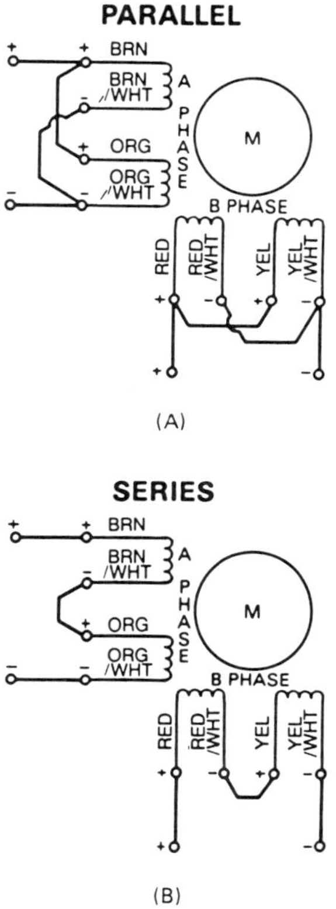

Figure 8-2 Phasing coils wired in parallel (A); in series (B). (Courtesy Portescap)

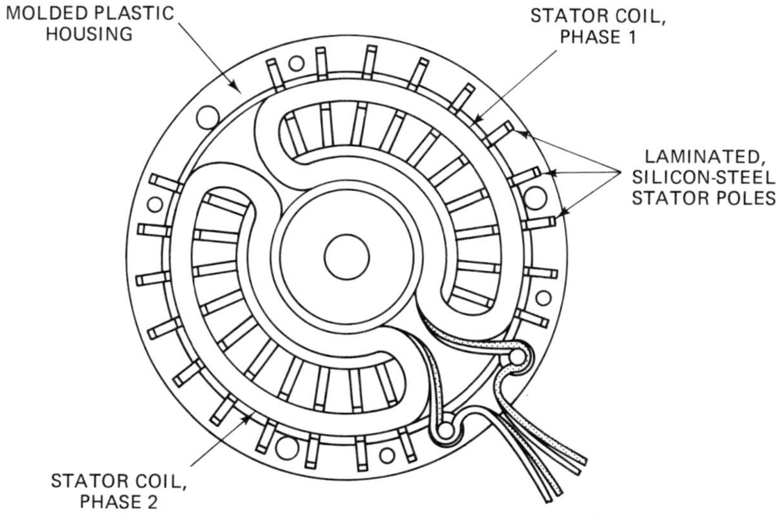

Figure 8-3 Face view of stator of disc stepper. (Courtesy Penton/PC, Inc.)

The cross-section of each magnet is approximately 1.5 × 5 mm. The diameter of the disc ranges from 30 mm to 100 mm.

Steps vs. Step Angle

As the number of steps is increased, the angle of each step becomes smaller. Thus, if a stepper can supply six steps for a single rotation of the its axis (shaft) then each step must be 60 degrees. A complete rotation of the axis is 360 degrees, and this is always equal to the product of the number of steps and the step angle: 6 × 60 = 360. The step angle can be calculated by dividing the number of steps into 360. The step angle, also called the step size, is always in degrees and is not necessarily in whole numbers, but can be a whole number plus a decimal fraction. Thus, a step size could be 3.6 degrees. Note that it is still divisible into 360 degrees.

Representative values of stepping angles are 7.5 degrees, 9, 11.25, 15 and 18 degrees. 360/7.5 = 48 steps; 360/9 = 40 steps; 360/11.25 = 32 steps; 360/15 = 24 steps; 360/18 = 20 steps. The second is the time element used in connection with steps. Steps per revolution are commonly 24, 32, and 48 steps per second.

Microstepping

It is possible to divide a full step into a number of much smaller steps (microsteps). These intermediate steps are obtained by using drive voltages or currents of variable levels.

Fundamentally, a stepper motor is a single or multiphase synchronous motor driven in an incremental mode. Incremental mode, in this instance, means that the motion of the rotor shaft is not a continuous, unbroken revolution, but proceeds in specific amounts or increments. If a rotor turns 10 degrees per input signal pulse, it will complete 36 increments in one complete rotation of its shaft.

TYPES OF STEPPING MOTORS

Just as there are various kinds of other motors, so too are there different kinds of stepping motors. The differences are not only in physical structure but in their operating characteristics, the demands imposed on them by their loads, and the precision required. Precision is a variable and in some instances can be very demanding, as, for example, the stepping motor required for computer disc drives.

Stepping motors can be classified in five major types, with each having a number of variations. All have both similar and individual characteristics. These motors include disc stepping motors, variable reluctance (VR) motors, permanent magnet motors (PM motors), hybrids, and power steppers.

Variable Reluctance (VR) Stepping Motors

Variable reluctance stepping motors do not make use of permanent magnets. These steppers (Figure 8-4) have a soft iron multiple rotor and a multi (two to five) phase stator. The flux path originates at the energized stator pole, goes through the rotor when the teeth are aligned or attracts the nearest rotor tooth and aligns it with the energized stator tooth, and then passes to the opposite energized stator pole and around the "back iron" to complete its magnetic circuit.

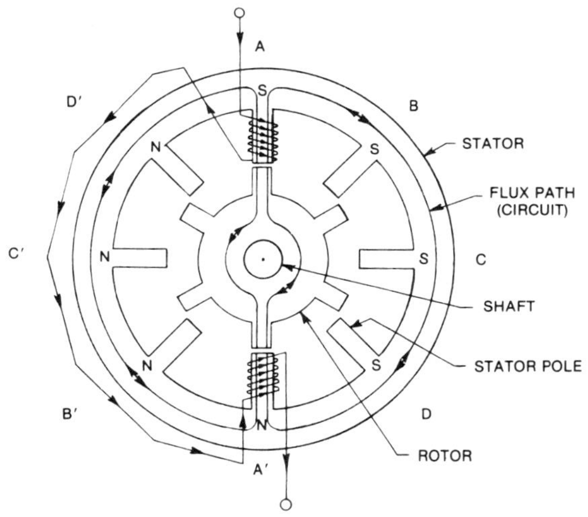

Figure 8-4 Variable reluctance stepping motor. (Courtesy Small Motor Manufacturers Association)

The number of teeth on the rotor and the number of teeth on the stator as well as the number of phases determine the step angle.

In this motor a stator tooth is wound with wire. A group of windings in specific angular relationship to each other form a phase. When a particular phase is energized a magnetic field is established in and around each winding. The strength of this magnetic field is directly related to the number of ampere-turns in each winding.

The soft iron rotor is electromagnetically attracted to the energized stator tooth, producing the electromagnetic flux field. The rotor, when directly lined up with the energized stator tooth, is in a position of minimum reluctance, that is, minimum reluctance to the existence of magnetic flux.

If torque is applied to the rotor shaft when the rotor is in a position of minimum reluctance, a counter-acting torque is developed by the existing magnetic attraction induced by the energized windings. This magnetic attraction tries to restore the rotor to its minimum reluctance position. For this reason, these motors are identified as variable-reluctance stepping motors.

A "step" takes place when one phase is de-energized and the next phase in sequence is energized. When this happens, an electromagnetic field is established. The rotor will move to a new position of minimum reluctance (motor and stator teeth lined up) because of magnetic attraction. This completes one step.

Permanent Magnet (PM) Rotor Stepping Motors

This type of stepping motor is commonly referred to as the "stamped construction" or "sheet metal" stepping motor. Another reference to this type of motor is PM (not to be confused with a synchronous inductor/hybrid PM stepping motor, a motor that uses a permanent magnet rotor). The rotor in this motor is radially magnetized with alternate poles (N-S-N-S-N-S-N).

The stator (Figure 8-5) has two halves, each of which contains a coil. The stator halves also have stator teeth to direct the magnetic flux path in and out of the permanent magnet rotor. Because of this simple design, the motor is economically competitive, but its performance is also not as good in terms of accuracy and speed.

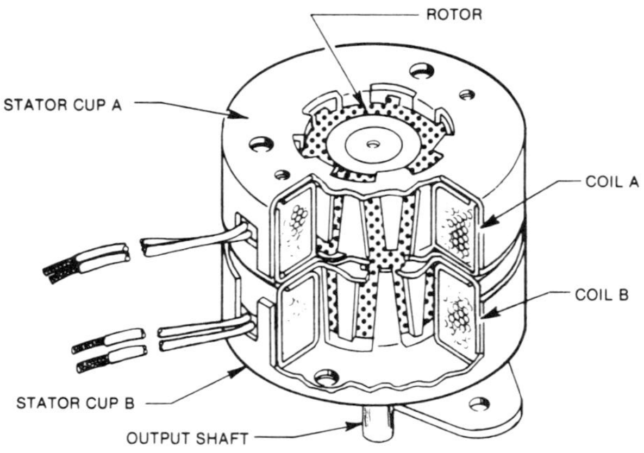

Figure 8-5 Permanent magnet stepping motor using cylindrical magnet design. (Courtesy Small Motor Manufacturers Association)

Two types of coil windings are available. The regular uses one winding, single strand, and requires a bipolar drive. A bipolar drive uses two power supplies. In this coil a reverse in current and voltage is needed to reverse the direction of flux.

The bifilar type has two windings consisting of two individual strands of wire wound on the same coil. A bifilar wound coil requires a unipolar drive using one power supply. In this coil, instead of reversing the current (and voltage) to reverse the flux path, it is just necessary to switch the drive to the other individual winding of the same coil.

The individual stator halves (cups) have uniformly spaced teeth, with 24 teeth per cup for 7.5 degrees and 12 teeth per cup for 15 degrees. The two halves are offset from each other by a half-tooth width to achieve 48 steps (7.5 degrees) with 24 steps (15 degrees) and 12 teeth per cup.

The energized coil in stator cup A sets up a magnetic circuit which starts at the down stator tooth and moves into the permanent magnet rotor, out of the permanent magnet rotor to the up stator tooth, and around the back of the coil housing to complete the circuit. When coil A is de-energized and coil B is energized, the rotor steps to the magnetic attraction (unlike poles attract) set up by coil B.

Advantages of the PM Rotor Stepping Motor

This type of motor, using a permanent magnet rotor, requires less operating power due to the presence of the permanent magnet field. This motor also has better damping (that is, settling to a dead stop) characteristics due to the presence of the permanent magnet. It is not suited for small step angles because of its construction.

Hybrid Stepping Motor

This motor is called hybrid because it uses the theories of both variable reluctance and permanent-magnet-rotor step rotors. It is also called a 1.8 degree PM stepping motor.

The hybrid uses the stator of a variable reluctance stepping motor and a permanent magnet in the rotor. This motor was originally designed as an AC two phase synchronous motor for low speed applications. It was found that the motor could step incrementally if the phase windings were excited with DC current pulses.

This motor is characterized as high efficiency. Because of the permanent magnet less power is required. The unit is capable of high step rates, available in small step angles only—1.8 degrees, 200 steps per revolution, 50 rotor teeth, four phase.

The rotor has three main parts, a permanent magnet of cylindrical form and two rotor sections on each end. The rotor sections (one north pole and one south pole set up by the permanent magnet) have uniformly spaced teeth, one tooth pitch apart (Figure 8-6). The sections are offset one from the other one half tooth pitch.

The stator has salient poles (multiple teeth on one pole) and usually bifilar wound coils. The coils are bifilar wound (or regular wound using bipolar drive) because the flux direction needs to be changed every other step to attract or repel the permanent magnet rotor. When that phase is de-energized and the next phase is energized, the rotor will rotate toward unlike and away from like stator teeth, making one step.

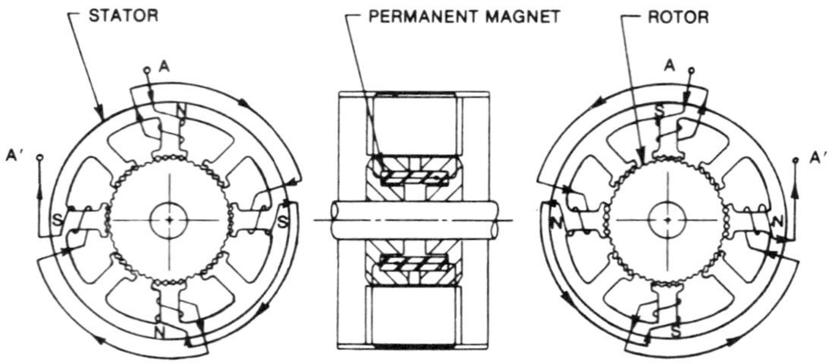

Figure 8-6 Synchronous inductor hybrid stepping motor. (Courtesy Small Motor Manufacturers Association)

The third and fourth phases are opposite wound coils (the other half of a bifilar wound coil), to produce opposite polarity (reverse flux direction) to attract or repel the permanent magnet in the third and fourth steps respectively.

The flux vector (or field) of the permanent magnet is unidirectional. In other words, the magnet does not change polarity.

STEPPING MOTOR TERMINOLOGY

As stepping motors developed they were accompanied by technical definitions. The first of these was a description of the motor itself and was characterized as a device which translates digital electrical pulses into angular positions in fixed increments. Each motor phase advances the rotor shaft one step or increment and latches it magnetically at the desired stable point to which it is stepped.

The stepping motor shaft rotation is in fixed, repeatable, and known increments, as opposed to conventional motors which have continuous free-running rotation. Position error is independent of the number of steps moved and feedback is not normally required to accomplish accuracy.

Bipolar drive. This is a drive circuit that creates an excitation that causes torque producing current reversals in the windings.

Chopper drive. This is a drive circuit with initially high voltage applied to the phase windings until the winding current reaches a predetermined set value at which time the supply is switched off. When the current has decreased to a second set value the supply is switched on.

Constant current drive. This is a drive circuit that maintains nearly constant current in the stator windings at any rotor speed.

Damping. The reduction or elimination of step overshoot is defined as damping. It is used in applications where settling down time is important.

There are a number of different types of damping, including mechanical, electrical, and viscous (fluid). The electronic method uses series resistance, capacitance, back phasing, delayed last step excitation, and others. Mechanical damping is accomplished by adding friction in various schemes including fluid dampers.

Drive or controller. This is a broad term and is used to describe the circuitry which controls the stepping motor. It usually consists of a power supply, sequencing logic, and power output switch components.

Half step. This is winding sequencing that results in steps that are one half of the specified step angle.

Holding torque. With the motor shaft at a standstill, this is the amount of torque from an external source that is required to break the shaft away from its holding position. This is measured with rated current applied to one or more phases of the motor.

Load angle. This angle is between the magnetic axis of the stator and rotor magnetic fields with the motor driving a specified load.

Pull-in rate. This rate is the constant maximum pulse rate at which the energized stepping motor will start driving a specified load without missing steps under specified drive conditions.

Pull-in torque. This is the minimum torque at which an energized stepping motor will start driving a specified load and run without losing steps on application of a fixed pulse rate.

Pull-out rate. This rate is the constant maximum pulse rate at which the energized stepping motor can run driving a specified load under specified drive conditions without losing steps.

Pull-out torque. This is the maximum torque that can be applied to the rotating shaft of a stepping motor driven at a given pulse rate under specified drive conditions without missing steps.

Ramping. This is the process of controlling the rate of change of pulse frequency to accelerate or decelerate the rotor from one speed to another. Ramping increases the capability of driving the motor and load to higher speed levels.

Residual or detent torque. Under power-off conditions with the motor shaft at a standstill, this is the amount of torque from an external source that is required to break the shaft away from its holding position. Residual torque is present only with a motor of permanent magnet design. Residual torque can be divided into friction torque and magnetic cog torque. The bearing system is the major contributor to friction torque.

Resonance. Stepping motors have certain natural frequency characteristics dependent on the motor's mechanical design and load, with the most significant dependence on the inertia of the load. This frequency can be affected by the type of drive used. In case of a severe resonant condition, the motor may lose steps and/or oscillate about a point.

Start-stop region. This is an operational envelope within which the motor can start and stop with a given rigidly attached inertial mass without incurring a step error.

Slew rate. This is an area of high speed operation about the pull-in curve where the motor can run unidirectionally without missing steps. However, it cannot instantaneously start, stop, or reverse.

Step. Step is the process involving a rotation of the motor shaft by the step angle. The symbol for step is the letter z. This unit is dimensionless.

Step accuracy. Step accuracy is defined as positional accuracy tolerance with no load applied. This figure is generally expressed as a percentage and indicates the total angular error introduced by the stepping motor in a single step movement. The error is noncumulative, that is, it does not increase as additional steps are taken.

Step angle. The motor shaft rotates a specific angular increment each time the motor receives an electrical step signal. This specific degree of rotation or increment is called the step angle. It is specified in degrees and is determined by the electromechanical construction of the motor.

Step error. The maximum positive or negative angle deviation relative to the nominal step angle which can occur when the rotor moves from one magnetic rest position to the next is known as step error.

Stepping frequency. This is the number of steps executed per second by the motor shaft at a constant control frequency.

Steps per revolution. This term describes the total number of steps required for the output shaft to rotate 360 degrees or one complete rotation. Steps per revolution can be calculated by dividing the step angle into 360 degrees.

Steps per second. Also known as the pulse rate, this is the number of angular movements accomplished by the motor in one second of time. This figure replaces the rpm figure of a standard drive motor.

Torque gradient. This is the average slope of the torque displacement curve over a specified angle through the origin.

Torque-to-inertia ratio. This ratio is determined by dividing the rated holding torque of the motor in inch-ounces by its rotor inertia in inch-ounce seconds squared. The higher the torque-to-inertia ratio, the faster the response with a given drive.

STEPPER MOTOR DRIVE CIRCUITS

The required excitation voltage for the stepper motor's stator phasing coils consists of digital pulses which are produced in a stepper motor drive circuit (Figure 8-7). The block diagram in this drawing shows the relationship of each of the components in the drive circuit.

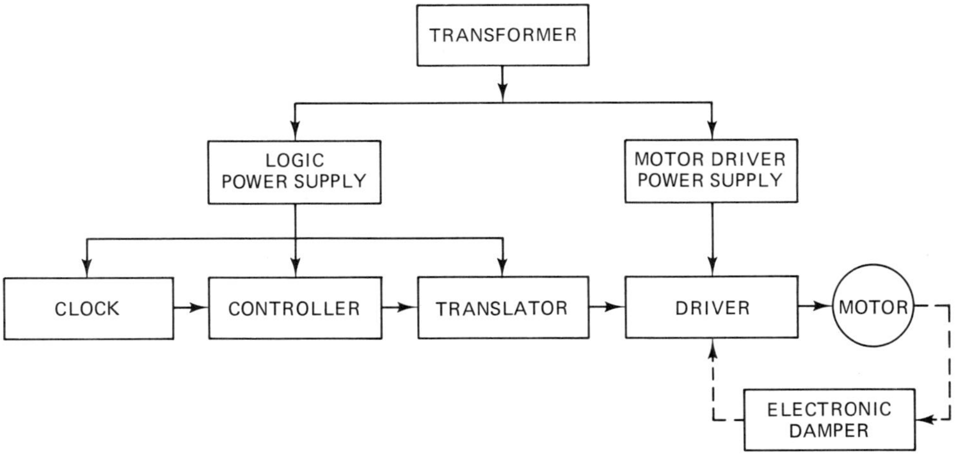

Figure 8-7 Elements of a stepper drive circuit. (Courtesy Portescap)

The unit starts with a power transformer whose primary winding is connected to the 120-volt AC line. The transformer is a step-down type since the voltage required out of the following power supplies is low-voltage DC. There are two power supplies which rectify and filter the AC voltage input. One of these is the logic power supply; the other is the motor driver power supply. The logic power supply provides the DC power suitable for the logic which consists of the clock, the controller, and the translator. The motor driver power supply furnishes the DC voltage necessary to drive the motor and the driver.

The clock indicated in the diagram has no relationship to clocks as we ordinarily know them. Instead, it is an electronic oscillator that provides a succession of pulses. Each pulse generates one increment of the controlled stepper motor. This can be a full step, a half step, or a microstep.

The controller following the clock is an electronic circuit for governing the frequency of the clock. This changes the timing of the pulses, hence the incremental frequency of the stepper motor. The timing of the clock follows the formula T = 1/f, in which T is the time in seconds and f is the pulse frequency. The time and the frequency have an inverse relationship. As the frequency is increased, the time is shortened, and vice versa. The clock frequency change is done in such a way that the motor does not lose synchronism during the frequency change.

The translator receives the clock pulses from the controller. It delivers at its output the appropriate phase control signals for correct switching of the driver. The control signals consist of a pair of pulses which are identical but which are in quadrature, that is, they are separated by a phase angle of 90 degrees. These are the signals that are supplied to the two phase coils of the stator winding.

The driver switches power in the motor phases. The driver can be either a constant voltage source (single level, bi-level, L/R) or a constant current source (linear source or chopper).

The rotating disc has inertia, and so once in motion tends to stay in motion. The electronic damper is used to damp these natural movements, acting like a brake. In order to use this circuit the motor must be equipped with velocity sensors.

Driving the disc motor with the use of what are essentially square wave pulses can result in a more or less uneven rotation, depending on the pulse rate and the load characteristics. However, the drive signal may be continuous, producing a rotating field with the motor then working as a synchronous type.

The Step Angle

By definition, one full step in a two-phase stepper corresponds to 90 electrical degrees. One full step expressed in mechanical degrees is equal to 360 electrical degrees divided by 4 times the number of pole pairs, N, magnetized on the disc magnet. A four step sequence is the sequence for two-phase steppers. One cycle corresponds to four full steps and 360 electrical degrees.

Eight-Step Sequence

This is a sequence of drive signals alternately energizing one and two phases to obtain eight stable positions in a cycle of 360 electrical degrees. A single step, for example, could be divided into eight microsteps.

Position Errors

The ideal position of the stepper rotor would be one that is angularly precise. Thus, if the rotor is to stop at exactly 7.5 degrees this would be called its theoretical position. However, there are several operating factors that militate against this. One of these is due to manufacturing tolerances. Precision can be an expensive production cost and, depending on the application of the stepper motor, may not be necessary. Various other causes of error include hysteresis error and errors due to friction (such as bearing friction).

Chopper Drive

Instead of voltage pulses, the phasing coils of the stepper rotor can receive pulses of current, supplied by a device known as a chopper. Typical values of current are two or three amperes, but smaller values are also used.

Continuous vs. Incremental Motion

The DC and AC motors described in earlier chapters can all be grouped under the general heading of continuous motors, that is, the rotor motion is controlled by mechanical means, either a manually set switch or one that is mechanically operated. With these motors, the starting or finishing angular position of the rotor is of no consequence or interest.

Incremental motors are those in which the rotor is made to turn a predetermined number of degrees, an action which can be repetitive for selected periods of time.

POWER STEPPERS

Power steppers are capable of supplying high-torque power plus an accurate stepping ability and can be made to have a positioning accuracy of plus or minus 3 percent. They are bidirectional units using permanent magnets and in some motors feature 1.8 degree steps and have high holding torques of 650 and 1150 in-oz.

As in the case of other power stepper types, the power stepper is driven by a digital control system. The construction of the stator has three phases, with each phase having the same number of poles. The polarity of the poles is such that each adjacent pole is of opposite polarity. Consequently, their magnetic paths are very short.

Power steppers are available with 54, 108, 216, and 360 poles or steps per single revolution of the shaft. The stepping rate is dependent on the drive pulse frequency. A succession of steps can be had by sequentially energizing each phase. Energizing time and the polarity of the digital pulse input is determined by the controlling drive circuitry, thus permitting stepping control, either in a forward or reverse direction.

AIR MOTORS

Not all motors are electrical. Some are known as air motors and are intended to work from compressed air, but are also capable of operating from other inert gases, such as nitrogen. These gases work with varying pressures up to 160 pounds per square inch (psi).

Air motors have rotors which revolve under the pressure of the incoming gas. They are widely used in hand-held portable tools such as those in dental, surgical, and industrial applications, in miniature torque control screwdrivers, high speed deburring tools, and in robots and remote control devices used in radioactive environments.

Air motors can have free rotational speeds up to 50,000 rpm and use right-hand rotation, but reversible models are available. Unlike electric motors, they are spark free and so can be used in explosive atmospheres. They can be operated continuously under load without heating and can be stalled repeatedly without damage. The motor is built of stainless steel and so will not corrode.

Maximum free rotational speeds are obtained at 90 psi. At this pressure air consumption is approximately 4.8 cubic feet per minute (cfm). Decreasing the air pressure from 90 psi will reduce both speed and air consumption, but it also reduces torque proportionately.

While lubrication is recommended, it is not essential, especially with low load or intermittent usage. Too much oil or the wrong oil can cause sluggish operation and loss of power.

9

COUPLINGS, BELTS, AND GEARS

DRIVES AND COUPLINGS

The torque developed by the rotating shaft of a motor can be connected directly to a load, indirectly by a coupling of some kind, or by gears and belts. When the shaft and the load are joined without the aid of some intermediary device the technique is referred to as direct drive. While there are a number of different kinds of drives, they can all be categorized under two main headings, direct drive and speed conversion drive. With direct drive, the load rotates at the same speed as the shaft. With speed conversion drive, the load may rotate at a lower or higher speed.

Direct Drive

With direct drive the load may be mounted right on the motor shaft or an extension of that shaft. A fan, such as the type used in the home, is an example. While the speed of the fan motor can be varied, the load, the fan blades, always revolve at the same speed as the armature. In this case the load speed does not depend on a speed change produced by the type of coupling used.

There are various types of direct drives, but the most commonly used include the flexible shaft, flange coupling, cushion flange coupling, flexible hose coupling. Other coupling types include belt, gear, and chain drives (Figure 9-1).

These coupling types have several advantages. They more or less isolate the load from the motor and so noise or vibration produced by the motor is not

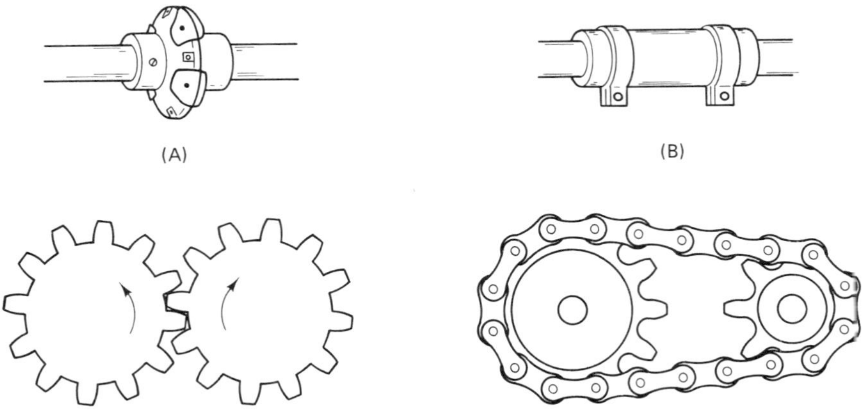

Figure 9-1 Various types of drives: flexible shaft (A); flange coupling (B); gear (C); chain (D).

transferred to the load. In some instances the load may be eccentric and so using a coupling helps minimize or eliminate wearing action on the motor bearings. In some instances the load must be located at some distance from the motor, an operating condition permitted by the use of a flexible shaft, as in the case of a dental drill.

Flexible Shaft Drive

A flexible shaft can be made from a suitable length of flexible cable free to rotate within an outer covering of natural rubber, synthetic rubber, or plastic. One end of the coupling is tightly fitted over the motor shaft or is held in place with a clamp. Similarly, the other end of the flexible drive is made to fit over a corresponding shaft extending from the load. To make sure that the flexible drive makes positive contact with the motor shaft and the load shaft, it may be equipped with a spline at both ends. The spline may have a series of projections that fit into corresponding slots on both shafts, supplying the positive contact required to permit rotation without shaft slippage. The outer covering or jacket permits the rotating cable to be held manually without discomfort or danger.

Flange Coupling

Flange coupling (Figure 9-1A) consists of a linkage between two shafts, that of the motor and that of the load. The coupling is a two piece unit with one section fastened to the motor shaft and the other to the load shaft. A flexible disc is positioned between the two coupling sections and is kept in position by a number of clips fastened to the circumference of the coupling units. The flexible disc permits a small amount of play between the two shafts, and because of its composition limits the amount of noise and vibration transfer from the motor shaft to the load shaft. With this arrangement the separation between the motor and its load is quite limited. To avoid unnecessary strain on the flange, the motor and its load are mounted so that their shafts are accurately positioned along their axes.

Cushion Flange Coupling

This type of coupling is quite similar to the flange coupling of Figure 9-1A and the flanges are the same. However, the flexible disc has now been expanded to a doughnut shape with the larger size supplying greater isolation between the motor and its load. As a result there is less noise and vibration transmission between the motor and its load.

Flexible Hose Coupling

This coupling (Figure 9-1B) consists of a short length of flexible tubing with each end fastened to the shafts of the motor and the load by screw-on clamps. The shafts should be recessed into the flexible hose so they extend beyond the clamping points. The hose should not depend solely on the clamps but should make a force fit over the two shafts. While the hose should be flexible, it should also be rigid enough to have minimum twist during operation.

Other types of drives include the gear (Figure 9-1C), chain (Figure 9-1D), and belt.

BELT DRIVES

A belt drive (Figure 9-2) uses a fabric treated with natural or synthetic rubber; rubber is vulcanized onto the fabric or cord. The belt is in the form of a completely closed loop linking the motor and its load. V-shaped wheels are mounted on the motor and load shafts with the drive belt positioned inside the V. Because the belt is tapered in the form of a V, it is wedged firmly into the sheave groove of the V-shaped wheels (pulleys). A sheave is a wheel or disk with a grooved rim, especially one used as a pulley. The transmission of energy is via the sides of the belt. The bottom of the belt, its apex, is usually not in contact with the bottom of the wheel.

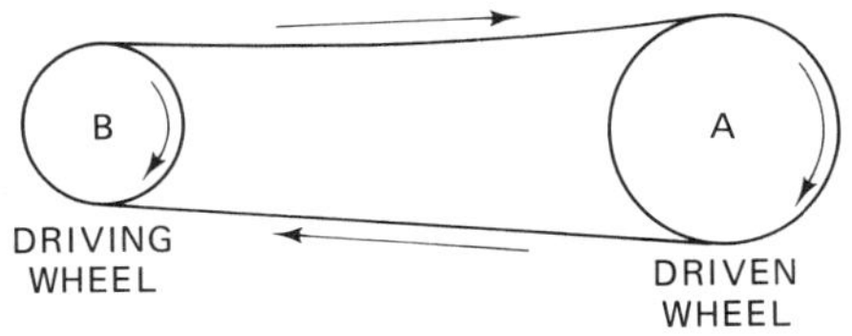

Figure 9-2 Belt drive.

This arrangement has a number of advantages. There is the possibility of greater separation between the motor and its load, much more so than in other coupling devices. The belt drive permits speed control of the load and so the rotation of the load can be the same as that of the motor or slower or faster. In some installations two different sizes of sheaves are used with the sheaves permanently joined. This is done to obtain two operating load speeds. It does require the use of two belts of different sizes.

The selection of sheaves and belts is determined by whether the motor is a light duty or fractional horsepower type, whether operation of the motor is to be intermittent or continuous, or whether service is to be medium or heavy duty, and operating speed.

V-belts are manufactured in standard sizes or the belt material can be obtained in open lengths and made to order for specific sheaves.

If the motor and load sheaves have the same diameter, the speed of the load will be the same as that of the motor. If, for example, the load sheave has twice the diameter of that used for the motor, there will be a two to one reduction in speed. The motor shaft will need to make two revolutions for each single revolution of the load shaft. Alternatively, if the reverse situation applies and the motor sheave's diameter is twice that of the load sheave, there will be a two-to-one speed increase. The load shaft will revolve twice as rapidly as that of the motor. Intermediate values of speed increase or decrease are available.

Multi-V-Belts

In many installations, particularly those involving fractional HP motors, the belt drive system consists of a pair of sheaves and a single belt. For heavier loads the system may include two or more belts. These belts can be individual units or, more often, can be a webbed type consisting of two or more. The sheaves will then consist of a double-V type for a twin webbed belt.

The advantage of using a webbed multi-V-belt is that it is less subject to slippage because of the larger contact area between the walls of the belt and those of the sheave. The disadvantage is the greater cost of the sheaves and the belt. If one of the V-belts in the webbed unit becomes damaged or develops cracks the entire belt must be discarded.

V-belt fit. The correct fit of a V-belt depends on the dimensions of the sheave. There are three possible V-belt dimensions. If it is quite evident that the belt is noticeably above the maximum dimension of the sheave, there will be a loss in the transmission of power since part of the sheave wall and the belt's V surface area will not be in contact. If the belt is such that the belt walls are below the top walls of the sheave, there will also be a loss in the transmission of power, and, in addition, there may be a certain amount of slippage. The size of the belt should be such that the flat, top surface of the belt should be slightly above the maximum vertical dimension of the sheave.

Flat Belt Drives

A flat belt drive is one in which the belt and both pulleys, the one on the motor and that on the load, are flat. This is a simpler installation than the V-belt type since the problem of fitting the belt specifically for a given pulley is eliminated. Flat belts are often used when the load is to be at some distance from the motor. To maintain good surface contact the belt is made wide. It is necessary to maintain adequate belt tension to keep it from riding or slipping. The belt should be centered on both pulley surfaces.

There are three possible flat belt conditions. The belt is too loose, or it has correct tension, or it is too tight. A loose belt can result in belt slippage and squealing. Slippage is evident if the belt "walks" to one side of the driven pulley. A loose flat belt, as in the case of the V-belt, means a loss in the transmission of power. In a severe case the load may not operate properly or at all. If the belt is too tight it will put an excessive strain on both the motor and load bearings.

V-Flat-Belt

The V-flat-belt is a combination of the flat and the V-belt. With this arrangement the pulley on the motor is a V-type while that on the load, the driven equipment, is flat. The V-flat-belt rides in the groove of the motor pulley while the inner flat surface of the belt rides on the flat pulley surface of the driven pulley. The V-belt has two horizontal, flat surfaces, one of which is wide, the other narrow. It is the narrow surface which drives the load pulley. Although this surface is referred to as narrow, it is still wide enough to make adequate contact with the flat pulley. As a result there is enough friction to prevent belt slippage. Further, the positioning of the V-belt in the motor pulley is a factor in keeping the belt running in its correct position.

DRIVE SELECTION

Using direct drive eliminates the problem of belt and pulley selection, and also means there is no need for belt installation and maintenance. The disadvantage of direct drive is that the motor and its load must both run at the same speed. Further, with direct drive both the motor shaft and that of the load revolve in the same direction. With belt drive load speed can be made to be greater or lesser than that of the motor. Also, by transposing the belt (Figure 9-3) it is possible to make the load rotate in a direction opposite that of the drive. If a multi-V-belt or a multi-V-flat-belt is used it can be in conjunction with independent pulleys mounted on the motor shaft with each of these pulleys driving independent loads. In some installations the motor pulley is belt connected to an independent axle on which are mounted a number of pulleys with each of these using belts to drive one or more loads.

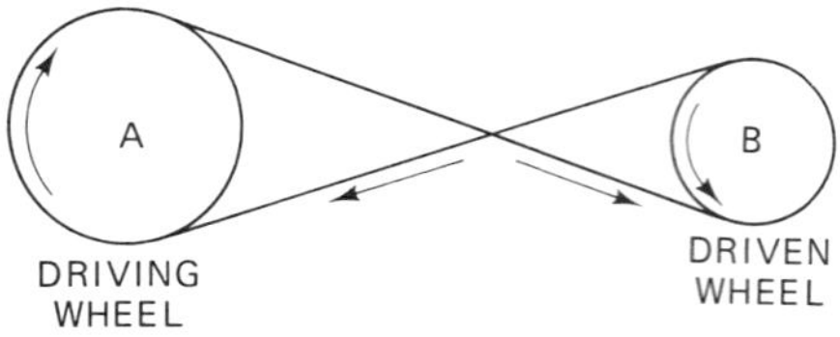

Figure 9-3 The rotation of a driven pulley can be changed by transposing the belt.

Pulley Spokes

The cross arms or spokes of pulleys can have any shape although those having an elliptical cross section (Figure 9-4) are common, a design used to reduce windage. Some pulleys, such as the crown face for a round belt, may have a solid circular face (Figure 9-4B). The angular separation of the spokes will be 30 degrees when there are a total of six spokes.

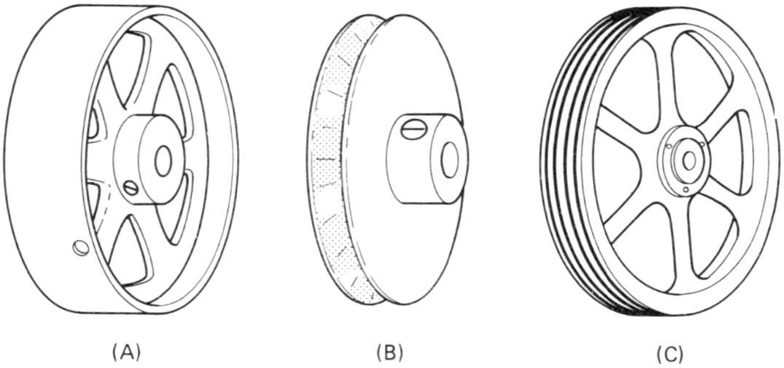

Figure 9-4 Typical pulleys: flat surface (A); V-belt sheave (B); multiple pulley (C).

The weight of the individual cross arms must be such that the wheel is balanced, thus minimizing bearing wear. With correct balance and adequate weight the pulley can supply flywheel effect.

Pulley Shapes

Pulleys are wheel-shaped but are basically differentiated by three factors: the wheel material, diameter, and design of the outer circumference. This latter characteristic can consist of one or more V-shaped sheaves, a flat surface known as a crown face, and a pulley to accommodate a round belt.

The pulley can be made to carry just a single belt or several or more. A stepped pulley (Figure 9-5) can be used to get speed control. The pulley on the motor shaft, also called the driving shaft, is belt-linked to the shaft of the load, referred to as the driven shaft. Load speed is determined by the ratio of the driven pulley to that of the driving pulley. The pulley system in this illustration is numbered to show the relationship, pulleys 1 and 2, 3 and 4, and so

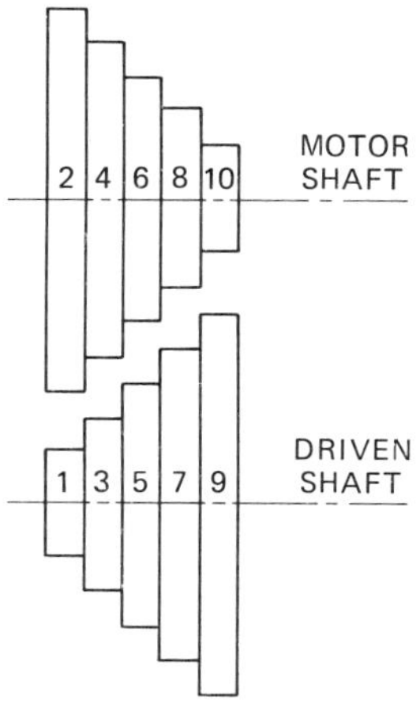

Figure 9-5 Stepped pulley.

on. With this setup, a single belt is used to operate a single load. It is also possible to use the multiple driving shaft to operate a driven shaft which, in turn, works as an intermediary, driving two or more shafts at different speeds. The drawing shown in Figure 9-5 can work with a number of belts with each of these driving different loads. Each of these loads has its own shaft.

Pulley Materials

Pulleys are generally made of cast iron or steel but a few, such as the round belt type may be brass. Iron and steel are used when substantial amounts of torque are to be transmitted. Metals are suitable when they are required to work in hostile environments, such as an acidic atmosphere.

Triple Belt System

Single and multiple pulleys can be used to drive loads at different speeds, as indicated in Figure 9-6. This particular system is intended for a clothes washer, but other drive methods are also used.

The largest pulley, called the transmission pulley, is driven indirectly via a pair of pulleys—the pulley mounted on the motor shaft plus the pump pulley. The pump pulley has two belts, one of which is connected to the motor shaft. Because of the ratio of the transmission pulley diameter to that of the pump pulley, this smaller pulley runs at a much greater speed.

The pump and motor pulleys rotate at the same speed. The motor pulley is also a double pulley type with the second belt operating a spin pulley. While there are three pulleys they all rotate in the same direction. Reversal of direction is obtainable by using a crossed belt arrangement or by using a motor that can be reversed.

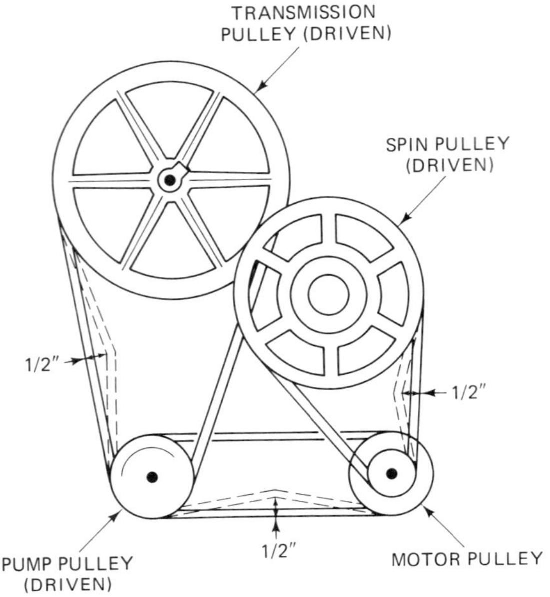

Figure 9-6 Triple belt system.

Positive Drive Belts

Slippage is one of the problems of leather belt drives, resulting in variations in speed of the driven pulley. This can be overcome by modifying the pulley and the belt, as shown in Figure 9-7, a combination of belt and gear.

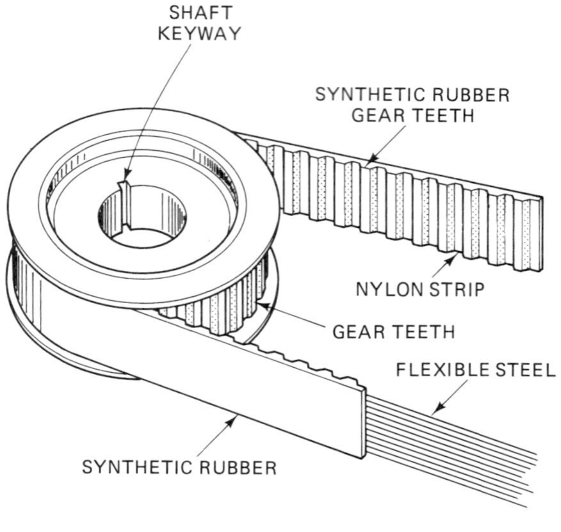

Figure 9-7 Positive drive belt.

The belt is made of a synthetic material such as Neoprene, a rubber-like substance. The inner surface of the belt is equipped with teeth which engage teeth constructed along the perimeter of the pulley. The belt is strengthened by flexible tension members made of steel and encased within the belt.

V-Belt Maintenance

V-belts should not be made exceedingly taut since this can result in excessive bearing wear, not only of the motor but the load as well. The belt should be mounted so it has no sag or slack. If the belt is slack, striking it will give a "dead" response. With correct belt tension, the belt, upon being struck, will sound "live" and will vibrate.

It is advisable to keep the belt clean and free of destructive compounds, especially oil. This latter substance reduces the area of contact between the walls of the V-belt and those of the pulley and may contribute to slippage. Further, it holds dirt which, in turn, may contain chemical substances which can attack the belt. Belts can be cleaned with a dry cloth. If this treatment appears to be inadequate, wash the belt with a sponge soaked in soapy water. Then rinse thoroughly and wipe vigorously with a clean, dry cloth. Do not depend on evaporation since that can leave a residue which may be invisible but harmful nevertheless. If, because of environmental conditions the belt becomes spattered with oil use carbon tetrachloride as a cleaning agent. Avoid inhalation of the fumes of this cleaning compound as much as possible. Following the treatment with carbon tetrachloride, wash the belt with a soapy solution of water, rinse thoroughly, and dry with a clean cloth.

With time, the belt will no longer be supple and will show a number of crack marks. The best time to make a replacement is when the motor is in normal down time. Visual inspection of the belt should be part of any maintenance schedule.

To check on belt tightness, pull on the belt with the motor not operating. The amount of stretch should be between 1/2 and 3/4 inch. If it is less than 1/2 inch it is too tight, putting an unnecessary strain on the system bearings; if more than 3/4 inch, there will be a loss in the transmission of power, with possible belt slippage.

Pulley Speed

The speed of a pulley is inversely proportional to its diameter, a concept that is illustrated in Figure 9-8. The drawing shows two pulleys, identified as A and B, with one of the pulleys smaller than the other. The diameter of each pulley is represented by the letter D. The speed relationship of the two pulleys is given by

$$\frac{N_A}{N_B} = \frac{D_B}{D_A}$$

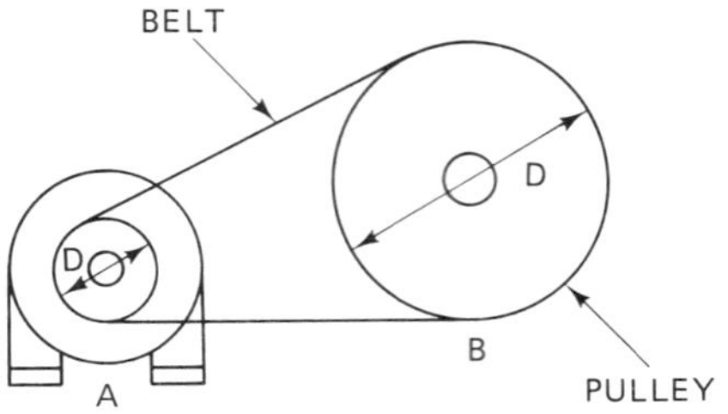

Figure 9-8 Determination of pulley speed.

where D is the diameter and N is the speed in rpm. D_A is the diameter of pulley A; D_B is the diameter of pulley B. N_A is the speed of pulley A; N_B is the speed of pulley B. The equation indicates that the speed of each pulley is inversely proportional to its diameter. Thus, if pulley A has a diameter that is one-half that of B its speed will be twice as great.

The equation is a four-part type, and so if the data of any three are known, the fourth can be easily calculated. Note that the speed of either of the pulleys is independent of the length of the belt, its type, and of the construction of the pulleys.

GEARS

Gears can be used to transmit rotational shaft energy from a motor to a load somewhat in the manner of a pulley and belt system but also similar to a direct drive.

Gears are often constructed of various kinds of metals such as alloy, carbon, and stainless steel, cast iron, aluminum, manganese, or tin bronze. Some gears are nonmetallic, fabricated from sheets of hard plastic. A gear consists of a circular blank with teeth cut, machined, stamped, or molded around its circumference. The center of the gear has a through hole to permit the gear to be mounted on a shaft.

Direction of Rotation

Unlike a pair of pulleys and a single belt, a pair of gears (Figure 9-9) result in reverse rotation. If the drive gear, the gear mounted on the motor shaft, turns clockwise, the driven gear, the one positioned on the load shaft, will rotate counterclockwise. This action is independent of the number of teeth on each gear. Rotation in the same direction can be obtained by using a third gear as indicated in Figure 9-10. Gears cannot be mounted arbitrarily without considering the effect of the gears on each other. Thus, as as indicated in Figure 9-11, the gears will lock and will not be able to rotate. This can be overcome by using four gears (Figure 9-12), and in this arrangement, all of the gears will be able to turn. Two of the gears will turn clockwise; two will turn counterclockwise, as indicated by the arrows.

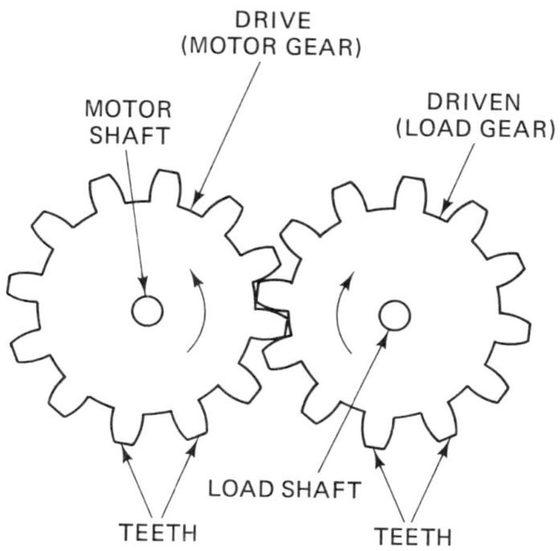

Figure 9-9 Spur gears viewed head on. These gears result in load rotation that is opposite that of the motor shaft. The driven gear may have more, fewer, or the same number of teeth as the drive gear.

Figure 9-10 Gear train with driven gear rotating in the same direction as the motor gear. The center gear is just used for obtaining a change of direction.

Figure 9-11 These gears are locked and cannot rotate.

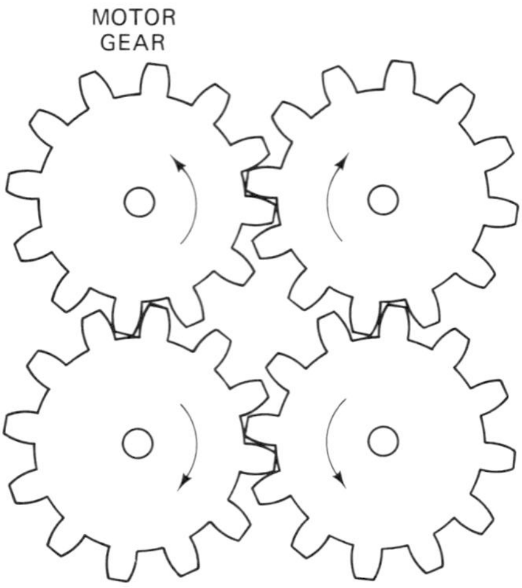

Figure 9-12 Gear train with all gears capable of rotating. It consists of a single drive gear and three driven gears. The driven gears can operate loads, or one or more can be used for obtaining a wanted direction of rotation.

Gear Train

A combination of two or more gears is sometimes referred to as a gear train. When a number of gears, usually more than two, are used, they may have an independent housing. In other cases the gears may be incorporated as part of the motor, with the combination referred to as a gear motor.

Types of Gears

There are numerous types of gears; some of the more commonly used are spur gears, helical, bevel, miter, and worm gears. In the spur gear, the teeth are parallel to each other and to the shaft on which the gear is mounted.

Gear Efficiency

The primary purpose of using gears is to transfer the horsepower output of a motor to a load such as a machine of some kind. Inevitably, in this transfer, there is a loss of some power in the gear train, mostly due to friction which shows itself in the form of heat. The ratio of these two, the horsepower available at the driven gear shaft compared to that at the input to the motor gear drive, is the gear efficiency, expressed as a decimal or as a percentage. As a percentage the maximum efficiency is 100 percent but this is never reached.

Torque Multiplication

In view of the fact that the horsepower available at the driven gear is always less than that at the motor gear, the driving gear, it may seem strange to talk about torque multiplication, also known as mechanical advantage (Figure 9-13).

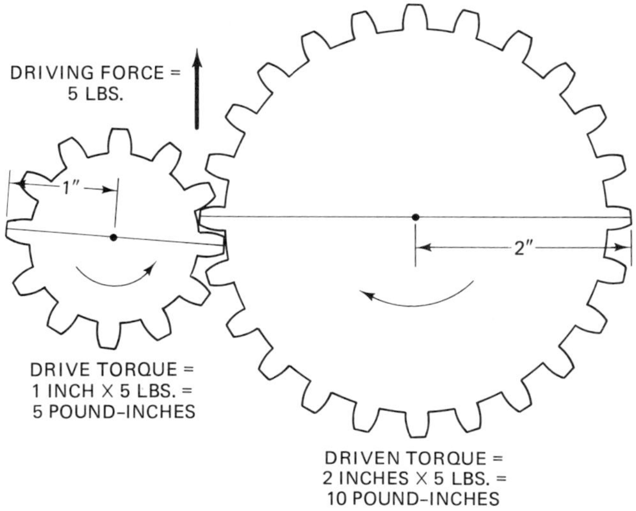

Figure 9-13 Multiplication of torque.

If both gears, drive and driven, have the same number of teeth, the drive torque and the driven torque will be exactly the same in an ideal case, that is, one having absolutely no losses. While this situation cannot exist, it is a useful concept when calculating problems involving the transfer of torque.

As an example, consider the two gears shown in Figure 9-13. The drive gear has a radius of 1 inch; that of the driven gear is 2 inches. If the smaller gear has a driving force of 5 pounds, its torque will be 1 in × 5 lbs or 5 inch-pounds. The driving force of 5 pounds is applied by the first gear to the second. The driven torque is equal to the gear radius multiplied by the driving force. The driven torque is 2 in × 5 lbs or 10 in-lb. The actual driving force is 10 inch-pounds minus any losses in the gear train.

Because the driven torque is twice that of the driving torque, the result is indicated as torque multiplication. However, the drive gear must rotate twice for every single rotation of the driven gear and for each revolution it produces 2 revolutions × 5 lbs × 1 in = 10 in-lb. This is equivalent to one revolution of the second gear.

Torque multiplication is also possible if more than two gears are involved, as illustrated in Figure 9-14. This train consists of three gears. The first gear has a radius of 1 inch; the second gear a radius of 2 inches, while the third gear has a radius of 1 inch. The driving force of the first gear is 10 pounds and so the input torque is 1 × 10 = 10 inch-pounds. The second gear has a radius of 2 inches and so its torque is 20 inch-pounds. The third gear has a radius of 1 inch and so its output is 20 inch-pounds.

Since the third gear and the first gear have the same radius, it would see that we are right back where we started. The difference is that we now have a

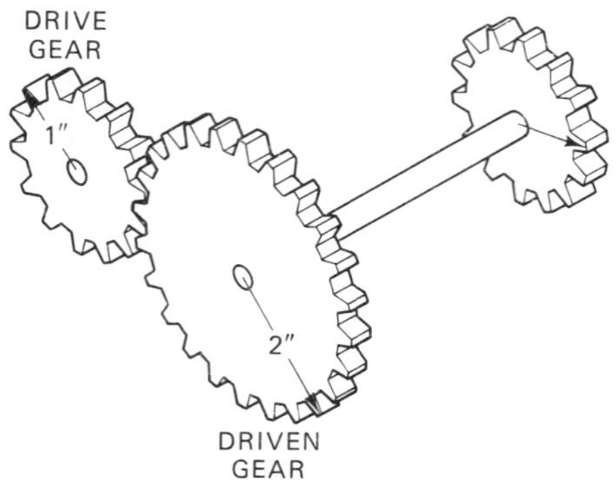

Figure 9-14 Mechanical advantage through the use of three gears.

gear with an output of 20 inch-pounds. The third gear can now drive a load having this requirement, something that could not be supplied by the first gear

Bevel Gears

The purpose of a bevel gear is to supply a change in the direction of an applied force so that the output rotation of the driven gear is at an angle of 90 degrees with that of the drive gear. The two gears of the bevel set can have the same number of teeth or the drive gear can be larger or smaller than the driven gear, and so the gears can supply a mechanical advantage. This can be calculated in the same way as that used for spur gears.

Miter Gears

Miter gears resemble bevel gears with a notable exception. The drive and driven gears always have the same number of teeth and so miter gears do not supply a mechanical advantage. Like bevel gears, miter gears supply an output force that is at right angles to the drive gear, and so this change in direction is its chief characteristic. Both bevel gears and miter gears may have straight cut teeth or the gears may be helical types. Helical cut teeth can also be used in connection with spur gears.

DRIVE SYSTEMS FOR ELECTRONICS

Drive systems are not only used in industry but in electronics as well, in tape decks, compact disc players, turntables, and so on. Various drive systems are used in these. Thus in phonograph systems, the two most commonly used drives are the belt drive and the friction drive (Figure 9-15).

In the belt drive system, a motor is connected to the turntable platter by a rubber belt. In this example, the drive wheel has a diameter of 1 inch, the driven wheel 3 inches, and so the ratio of drive wheel to driven wheel ratio is 3 to 1. Other ratios are used. The advantage of this drive system is that the belt

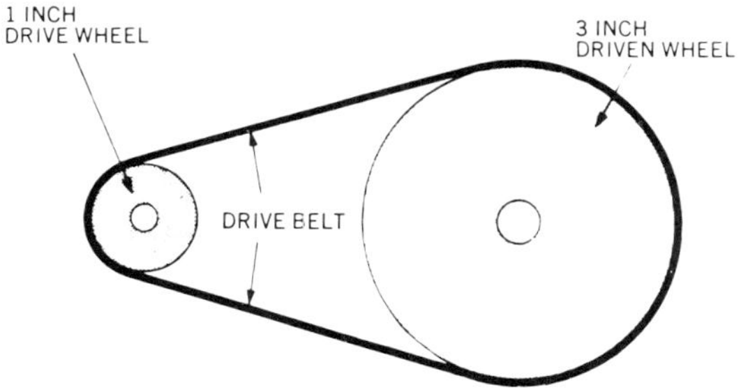

Example of belt drive.

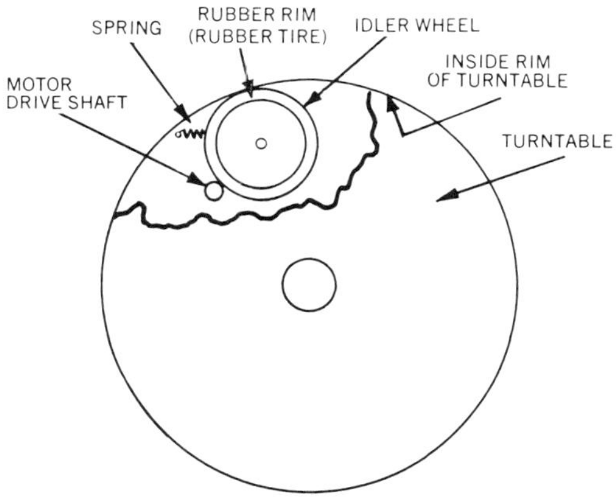

Figure 9-15 Two types of drive systems for turntables: belt drive (A); rim drive (B).

acts as a rumble filter, absorbing the vibrations from the motor. A loose belt is superior if the bearing and drive system is sufficiently friction free to permit its use without loss of speed consistency.

In the friction drive system, a motor drives a rubber wheel known as an idler wheel. The idler wheel fits against the inside rim of the turntable. Flats can develop when the idler remains pressed against the rim during nonplaying time.

INDEX

N

O